Lecture Notes in Comp
Edited by G. Goos, J. Hartmani

Springer
Berlin
Heidelberg
New York
Barcelona
Hong Kong
London
Milan
Paris
Tokyo

Andreas Wespi   Giovanni Vigna
Luca Deri (Eds.)

Recent Advances in Intrusion Detection

5th International Symposium, RAID 2002
Zurich, Switzerland, October 16-18, 2002
Proceedings

Series Editors

Volume Editors

Andreas Wespi
IBM Zurich Research Laboratory
Säumerstraße 4, 8803 Rüschlikon, Switzerland
E-mail: anw@zurich.ibm.com

Giovanni Vigna
University of California at Santa Barbara, Department of Computer Science
Santa Barbara, CA 93106-5110, USA
E-mail: vigna@cs.ucsb.edu

Luca Deri
University of Pisa, Centro Serra
Lungarno Pacinotti 43, 56100 Pisa, Italy
E-mail: deri@ntop.org

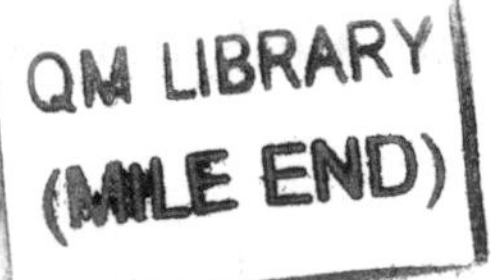

Cataloging-in-Publication Data applied for

Bibliograhpic information published by Die Deutsche Bibliothek
Die Deutsche Bibliothek lists this publication in the Deutsche Nationalbibliografie; detailed bibliographic data is available in the Internet at http://dnb.ddb.de

CR Subject Classification (1998): K.6.5, K.4, E.3, C.2, D.4.6

ISSN 0302-9743
ISBN 3-540-00020-8 Springer-Verlag Berlin Heidelberg New York

Springer-Verlag Berlin Heidelberg New York
a member of BertelsmannSpringer Science+Business Media GmbH

http://www.springer.de

Printed in Germany

Typesetting: Camera-ready by author, data conversion by PTP-Berlin, Stefan Sossna e.K.
Printed on acid-free paper SPIN: 10870986 06/3142 5 4 3 2 1 0

Preface

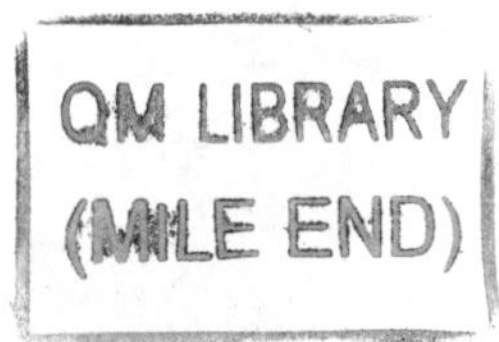

On behalf of the program committee, it is our pleasure to present to you the proceedings of the Fifth Symposium on Recent Advances in Intrusion Detection (RAID). Since its first edition in 1998, RAID has established itself as the main annual intrusion detection event, attracting researchers, practitioners, and vendors from all over the world.

The RAID 2002 program committee received 81 submissions (64 full papers and 17 extended abstracts) from 20 countries. This is about 50% more than last year. All submissions were carefully reviewed by at least three program committee members or additional intrusion-detection experts according to the criteria of scientific novelty, importance to the field, and technical quality. Final selection took place at a meeting held on May 15–16, 2002, in Oakland, USA. Sixteen full papers were selected for presentation and publication in the conference proceedings. In addition, three extended abstracts of work in progress were selected for presentation.

The program included both fundamental research and practical issues. The seven sessions were devoted to the following topics: anomaly detection, stepping-stone detection, correlation of intrusion-detection alarms, assessment of intrusion-detection systems, intrusion tolerance, legal aspects, adaptive intrusion-detection systems, and intrusion-detection analysis.

RAID 2002 also hosted a panel on "Cybercrime," a topic of major concern for both security experts and the public.

Marcus J. Ranum, the founder of Network Flight Recorder, Inc., delivered a keynote speech entitled "Challenges for the Future of Intrusion Detection".

The slides presented by the authors and the panelists are available on the RAID 2002 website, `http://www.raid-symposium.org/raid2002/`

We sincerely thank all those who submitted papers as well as the Program Committee members and the additional reviewers for their efforts. Special thanks go to the Swiss Federal Institute of Technology Zurich for hosting this year's edition of the RAID Symposium.

October 2002

Andreas Wespi
Giovanni Vigna

Preface

Organization

RAID 2002 is organized by the Swiss Federal Institute of Technology and IBM's Research Division and is held in conjunction with ESORICS 2002.

Conference Chairs

Program Chairs:	Andreas Wespi (IBM Research, Switzerland)
	Giovanni Vigna (UC Santa Barbara, USA)
General Chairs:	Günter Karjoth (IBM Research, Switzerland)
	Jörg Nievergelt (ETH Zurich, Switzerland)
Publication Chair:	Luca Deri (Centro Serra, Univ. of Pisa, Italy)
Publicity Chair:	Peter Mell (NIST, USA)
Sponsor Chair:	Diego Zamboni (IBM Research, Switzerland)

Program Committee

Matt Bishop	University of California at Davis, USA
Joachim Biskup	University of Dortmund, Germany
Frédéric Cuppens	ONERA, France
Luca Deri	Centro Serra, University of Pisa, Italy
Yves Deswarte	LAAS-CNRS, France
Tim Grance	NIST, USA
Erland Jonsson	Chalmers University of Technology, Sweden
Richard Kemmerer	UC Santa Barbara, USA
Kevin S. Killourhy	CMU, USA
Calvin Ko	NAI, USA
Jay Lala	DARPA Information Technology Office, USA
Richard Lippmann	MIT/Lincoln Lab, USA
Roy Maxion	CMU, USA
John McHugh	CMU/SEI CERT, USA
Peter Mell	NIST, USA
Vern Paxson	ICSI/LBNL, USA
Phil Porras	SRI, USA
Marty Roesch	Sourcefire, USA
Stuart Staniford	Silicon Defense, USA
Al Valdes	SRI, USA
David Wagner	UC Berkeley, USA
Diego Zamboni	IBM Research, Switzerland

Additional Reviewers

Magnus Almgren	SRI, USA
Fabien Autrel	Onera, France
Salem Benferhat	IRIT, France
Joao B. Cabrera	Scientific Systems, USA
Ramaswamy Chandramouli	NIST, USA
Nora Cuppens	France
Ulrich Flegel	University of Dortmund, Germany
Vincent Hu	NIST, USA
Klaus Julisch	IBM Research, Switzerland
Ulf Lindqvist	SRI, USA
George Mohay	Queensland University, Australia
Kymie M.C. Tan	CMU, USA
Tahlia N. Townsend	CMU, USA
Wei Zhang	Boeing, USA

Table of Contents

Stepping Stone Detection

Anomaly Detection

Correlation

Legal Aspects / Intrusion Tolerance

Assessment of Intrusion Detection Systems

Adaptive Intrusion Detection Systems

Intrusion Detection Analysis

Detecting Long Connection Chains of Interactive Terminal Sessions

Kwong H. Yung

Stanford University Statistics Department
390 Serra Mall; Stanford CA 94305-4020; USA
khyung@stat.stanford.edu

Abstract. To elude detection and capture, hackers chain many computers together to attack the victim computer from a distance. This report proposes a new strategy for detecting suspicious remote sessions, used as part of a long connection chain. Interactive terminal sessions behave differently on long chains than on direct connections. The time gap between a client request and the server delayed acknowledgment estimates the round-trip time to the nearest server. Under the same conditions, the time gap between a client request and the server reply echo provides information on how many hops downstream the final victim is located. By monitoring an outgoing connection for these two time gaps, echo-delay comparison can identify a suspicious session in isolation. Experiments confirm that echo-delay comparison applies to a range of situations and performs especially well in detecting outgoing connections with more than two hops downstream.

Keywords: Stepping stone, connection chain, intrusion detection, computer security, network security, network protocol, terminal session, delayed acknowledgment, reply echo, echo delay.

1 Introduction

Network security and intrusion detection have become important topics of active research [1,4,5,3]. As the use of the Internet becomes more common and widespread, so have network attacks and security breaches. Because the growing number of network attacks is ever more costly, network security and intrusion detection now play a crucial role in ensuring the smooth operation of computer networks.

1.1 Motivation

A skilled computer hacker launches attacks from a distance in order to hide his tracks. Before launching an actual attack with noticeable consequences, a skilled hacker will break into many computers across many political and administrative domains, to gather computer accounts. With access to multiple computer

A. Wespi, G. Vigna, and L. Deri (Eds.): RAID 2002, LNCS 2516, pp. 1–16, 2002.

$$-m \longrightarrow -m+1 \longrightarrow \ldots \longrightarrow -1 \longrightarrow 0 \longrightarrow 1 \longrightarrow \ldots \longrightarrow n-1 \longrightarrow n$$

Fig. 1. Typical connection chain. Relative to stepping stone 0, machines $-m, -m+1, \ldots, -1$ are *upstream* and machines $1, 2, \ldots, n$ are *downstream.*

accounts, the hacker can chain these computers through remote sessions, using the intermediate computers in the chain as stepping stones to his final victim.

Figure 1 shows a typical connection chain. Computer $-m$ attacks computer n, via stepping stones $-m+1, -m+2, \ldots, 0 \ldots, n-2, n-1$. Since the logs of the final victim n traces back only to the nearest stepping stone $n-1$, the originating point $-m$ of the attack cannot be determined without logs from the upstream stepping stones. Although the originating point used by the hacker may be found through repeated backtrace, this process is slow and complicated because the stepping stones belong to different political and administrative domains and often do not have logs readily available. Even when the originating attack point is found after costly investigation, the hacker will have already left and eluded capture.

This report presents a simple technique to detect interactive terminal sessions that are part of long connection chains. Once an outgoing session is found to have many hops downstream, the server machine can terminate the outgoing connection, to avoid being used as a stepping stone in the long connection chain. Moreover, the suspicious outgoing terminal session is useful as a basis for finding other sessions on the same chain.

1.2 Previous Approaches

Staniford-Chen and Heberlein [8] introduced the problem of identifying connection chains and used principal-component analysis to compare different sessions for similarities suggestive of connection chains. Because the packet contents were analyzed, the technique did not apply to encrypted sessions.

Later, Zhuang and Paxson [10] formulated the stepping stones problem and proposed a simpler approach to finding two correlated sessions, part of the same connection chain. By using only timing information of packets, the technique also applied to encrypted sessions.

Both [8] and [10] aim to match similar session logs, indicative of connection chains. Unless sessions on the *same* connection chain are in the same pool of collected sessions, however, this approach fails to identify suspicious sessions. For example, [10] analyzed logs from a large subnet at UC Berkeley. A hacker initiating an attack from a computer on the subnet would not be detected unless his chain connects back into the subnet.

The strategy of grouping similar sessions also inadvertently detects many benign, short connection chains because legitimate users often make two or three hops to their final destinations. For example, the Stanford University database research group has one well-protected computer that allows incoming connections only from trusted computers on the Stanford University network. To connect to this protected machine, an off-campus user must connect via another computer on the campus network. Because restrictions on incoming connections are quite common in heavily protected networks, short connection chains are often necessary and harmless.

1.3 New Strategy

This report presents an alternative strategy for identifying connection chains. Because sessions on a long chain behave differently than sessions on a direct connection, a suspicious session can be detected in isolation without finding other similar sessions on the same connection chain. The proposed technique makes use of delayed-acknowledgment packets, response signals found in typical protocols for interactive terminal sessions, such as telnet, rlogin, and secure shell.

Like [10], *echo-delay comparison*, the technique proposed here, relies only on timing information of packets and so applies equally well to encrypted sessions. Rather than comparing sessions in a large pool, echo-delay comparison operates on a single session and thus solves the shortcomings of [8] and [10]. A suspicious session that is part of a long connection chain can be identified even without finding another correlated session.

Clearly many factors determine the behavior of a connection, including the network, the machine, the user, and the session transcript. Therefore, isolating the distinctive properties of sessions in long connection chains is extremely difficult. Yet research in this strategy is worthwhile because looking for similar sessions from a large pool has many inherent drawbacks. Of course, these two strategies are not mutually exclusive but rather complementary.

To balance out the many uncontrolled factors in a connection, echo-delay comparison relies on the logistics of interactive terminal sessions. Because details of specific protocols involved are used, echo-delay comparison does not apply to all types of connections. Nevertheless, the technique is quite simple and can be extended to handle related protocols.

1.4 Elementary Solutions

The original Internet protocols were not designed with security as the main objective. Under the shield of anonymity offered by the Internet, malicious users can attack many computers remotely. Although the prevention and detection of attacks are now important, the widespread use of older protocols is difficult to change because the global connectivity of the Internet often requires compatibility with older software. Upgrading the vast number of computers on the Internet proves to be nearly impossible.

Perhaps the simplest way to prevent connection chains from forming is to forbid all incoming sessions from executing any outgoing terminal sessions. Some servers indeed do adopt this policy and forbid most outgoing connections. Yet implementing such a strict policy severely limits legitimate users.

Any policy blindly disabling outgoing connections is too restrictive in most settings because there are many legitimate reasons to connect via a short chain. On many networks, users are allowed external connections only through a dedicated server, which is heavily protected and closely monitored. So to connect to an outside host, users must connect via the dedicated server. In this case, the gateway server cannot blindly forbid outgoing connections.

Legitimate computer users often use short connection chains to get from one host to another. Malicious hackers generally use long connection chains to cover their tracks before executing an attack. To protect a machine from being used as a stepping stone in a long chain, a reasonable policy would be to terminate sessions that continue *multiple* hops downstream.

1.5 Outline of Report

Following this introduction, Section 2 provides a brief background of the network signals sent during an interactive terminal session between a client and a server in a direct connection. Section 3 then explains the dynamics of a connection chain and introduces two time gaps useful for detecting long connection chains. Next, Section 4 presents the mathematics for calculating and comparing the two time gaps. Afterwards, Section 5 presents two sets of experiments used to test the proposed technique. Towards the end, Section 6 discusses the advantages and limitations of the ideas proposed in this report. Finally, Section 7 summarizes this report's main conclusions.

2 Background

By comparing its incoming sessions with its outgoing sessions, a machine can determine that it is being used as a stepping stone in a connection chain. Standard protocols for remote sessions do not provide information about the length of the connection chain. So in isolation, the machine cannot in principle determine whether it is being used as part of a long chain or just a short chain.

2.1 Reply Echo

In most client implementations of interactive terminal sessions, each individual character typed by user will initiate a packet sent from the client to the server. Once the server receives the character packet, it usually echoes the character back to the client, instructing the client to display the typed character on the client screen. This reply echo is the typical response of the server to a non-special character from the client.

When the user types a carriage return, the carriage return received at the server usually triggers the server to execute a special command. After executing the command, the server then sends back command output. Figure 2 illustrates a typical exchange between a client and a server on a direct connection.

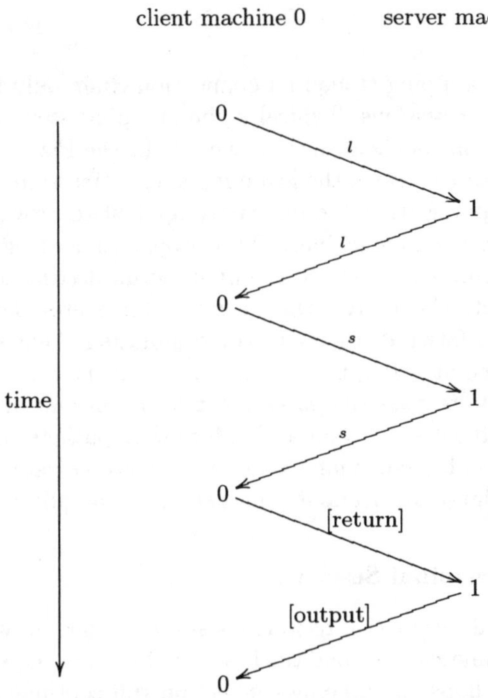

client machine 0 server machine 1

time

Fig. 2. Interactive session on direct connection. Client 0 sends the `ls` command to server 1 and receives back the directory listing. The vertical time axis is not drawn to scale.

2.2 Delayed Acknowledgment

Most often, the server responds soon enough with a nonempty packet, which also functions to acknowledge the client request. If the requested command requires a long execution time, then the server times out and sends a so-called *delayed acknowledgment.* This delayed-acknowledgment packet contains no content but signals to the client that the server indeed received the client request. Thus, the delayed acknowledgment functions to keep the conversation between the client and the server alive, at the expense of sending an empty packet. The server sends a delayed acknowledgment only when it cannot send a nonempty response in time. The server implementation determines the actual delay tolerance before a delayed acknowledgment is sent from the server to client.

Similarly, when the server sends a nonempty packet to the client, the client must acknowledge the server. Usually this acknowledgment to the server is sent by the client along with the next character packet. If the user is slow to type the next character, however, the client times out and sends a delayed acknowledgment to the server. The client implementation determines the delay tolerance before a delayed acknowledgment is send from the client to the server.

3 Theory

A machine used as a stepping stone in a connection chain only knows about its incoming and outgoing sessions. Typically the stepping stone only passes the packet information from its client onto its server. In the Figure 1, the stepping stone 0 acts as a server to receive the incoming connection from upstream client machine -1. The stepping stone 0 then acts as a client to forward the outgoing connection onto downstream machine 1. This propagation of signals starts from the very first originating client $-m$ to the final victim destination n.

After the final victim n receives the packet and executes the command, the output from n is then forwarded back to the originating client $-m$ in a similar manner. The intermediate stepping stones $-m+1, -m+2, \ldots, n-2, n-1$ act mainly as conduits to pass the packets between $-m$ and n. Along the way, packets may be fragmented or coalesced. Moreover, packets may be lost and retransmitted. Because human typing generates character packets separated by relatively large time intervals, character packets are generally not coalesced.

3.1 Interactive Terminal Sessions

Protocols for standard interactive terminal sessions do not provide information about an outgoing connection beyond the first hop. So there is no certainty about how many additional hops an outgoing connection will continue downstream. In this scenario, the client machine only communicates with the nearest server one hop downstream and does not know about additional servers several hops downstream.

After sending out a character packet to the server, the client waits for a reply echo from the server. If the final victim machine is many hops downstream from the client, then the nearest server must forward the character packet downstream. Upon receiving the character packet, the final victim then sends the reply echo to the client via the client's nearest server. To the client, the reply echo appears to come from the nearest server.

3.2 Dynamics of Connection Chains

In a connection chain with many hops downstream, there is a long time gap between the client request and the server reply echo because the nearest server must forward the client request to the final victim and then pass the reply echo back to the client. A long delay between the client request and the server reply echo also results if the client and the nearest server are separated by a noisy connection or if the server is just slow. *Consequently, a long echo delay alone is not sufficient to suggest that there are many hops downstream.*

As soon as the nearest server receives the client request, the nearest server forwards the request. If the nearest server cannot pass the reply echo from the final victim back to the client in time, the nearest server sends a delayed acknowledgment to the client in the meantime. So if there are many hops downstream of the nearest server, the client will first receive the delayed acknowledgment

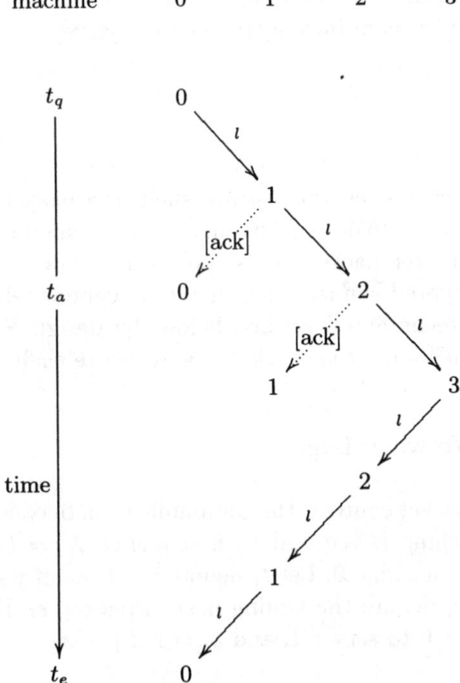

Fig. 3. Interactive session on connection chain. Client 0 issues a character packet containing letter l. Downstream servers 1 and 2 forward the packet to the final victim 3. After executing the packet, the final victim 3 sends the reply echo back to the client 0, via the stepping stones 1 and 2. In the above scenario, client 0 logs three packets, at times $t_q, t_a,$ and t_e.

from the nearest server before receiving the reply echo. Figure 3 illustrates the dynamics of a connection chain in the simplest scenario.

3.3 Two Time Gaps

In the simplest scenario, a client sending out a character packet can record three different signals at three points in time,

$$t_q < t_a < t_e. \tag{1}$$

Here t_q is the time of character request sent from the client, t_a is the time of delayed acknowledgment received from the server, and t_e is the time of reply echo received from the server. All three times refer to when a packet leaves or arrives at the client. Figure 3 illustrates these three time points.

The delayed-acknowledgment gap $t_a - t_q$ provides an overestimate of the travel time between the client and the nearest server. The reply-echo gap $t_e - t_q$ provides an estimate of the travel time between the client and the final victim.

The difference between these two gaps provide an estimate of the number of additional hops downstream beyond the nearest server.

4 Analysis

In encrypted connections, as with secure shell, the packet contents are obfuscated and cannot be compared. Thus, delay times cannot be calculated easily by matching a character packet to its reply echo. Any technique that applies equally well to encrypted connections, therefore, cannot rely on the packet content. The analysis technique described below, by design, applies to interactive terminal sessions, including telnet, rlogin, and secure shell.

4.1 Reducing Network Logs

After ignoring the packet content, the communication between the client machine and the server machine is reduced to a sequence $X = (x_1, x_2, \ldots, x_i, \ldots)$ of packets recorded at machine 0. Let t_i denote the time of packet x_i, as recorded at machine 0. Let o_i denote the issuing host of packet x_i. Here $o_i = 0$ if packet x_i is sent from client 0 to server 1, and $o_i = 1$ if packet x_i is sent from server 1 to client 0.

The calculations described below provide estimates of the time gaps instead of the exact values. The packet content is not used in these simplified calculations. Instead, packet header provided the necessary information: the logging time, the issuing host, the receiving host, the port numbers, and the packet flags.

4.2 Estimating Gap of Reply Echo

The gap $t_e - t_q$ between the client request and the server reply echo is estimated by considering only *nonempty* packets, essentially those packets whose flags are not delayed-acknowledgment flags. Define $(x_{i_k}) = (x_j : x_j \text{ nonempty})$ as the subsequence of packets with nonempty content. Then the set

$$E(X) = \{t_{i_{k+1}} - t_{i_k} : (o_{i_k}, o_{i_{k+1}}) = (0,1)\} \tag{2}$$

captures all the gaps for a (0,1)-transition in the packet sequence X.

Certain (0,1)-transitions correspond to the server execution of a client command. Other (0,1)-transitions correspond to the reply echoes to the client's single-character packets. Because a user can type a short burst of characters before the first character is echoed back, a (0,1)-gap generally measures the time between a character, followed by the server reply echo to an earlier character in the burst of characters typed. The set E of (0,1)-gaps include both long command-execution gaps and also include short gaps for reply-echo lags. Thus, the distribution (0,1)-gaps in E has wide variance and skew.

4.3 Gap of Delayed Acknowledgment

The gap $t_a - t_q$ between the client request and the server delayed acknowledgment is calculated by considering the sequence of all packets. Delayed-acknowledgment packets from the server to the client is easy to identify from the packet header. Each such delayed acknowledgment can be matched to the most recent nonempty packet from the client. More precisely, let a_i be the gap between packet i and its most recent nonempty packet from the client, defined as

$$a_i = \min\{t_i - t_l : l < i, o_l = 0, x_l \text{ nonempty}\} \quad (3)$$
$$= t_i - t_{[i]}, \quad (4)$$

where $[i] = \max\{l < i : o_l = 0, x_l \text{ nonempty}\}$. Then the calculated set A of delayed-acknowledgment gap is

$$A(X) = \{a_i : o_i = 1, x_i \text{ delayed acknowledgment}\}. \quad (5)$$

This distribution of these delayed-acknowledgment gaps in A tends to be sharply focused, compared the distribution of (0,1)-gaps in E. In any case, $\alpha(X) = \max A(X)$, the maximum of the delayed-acknowledgment gaps in A, provides an over-estimate the round-trip time between the client and the server.

4.4 Comparison of Delay Gaps

Because the set E of (0,1)-gaps contains many different types of gaps, the distribution of these (0,1)-gaps depend on many factors. In general though, if there are many slow connections downstream, then the (0,1)-gaps tend to be large. Comparing the maximum delayed-acknowledgment gap α to the distribution of (0,1)-gaps in E provides insight into the connections downstream.

A simple comparison of α to E is the quantile $\text{quan}(E, \alpha)$ of α with respect to E. This quantile is robust estimate of how large E is compared to α. If the downstream delay is large, then $\text{quan}(E, \alpha)$ would be small.

5 Results

Experiments under a variety of settings were conducted to test the technique proposed in this report. The author had root access to one machine at Stanford University and user access to many remote machines. Connection chains were tested on these accounts, with different numbers and orderings of computers. All the results presented here use the secure shell protocol for the connections.

5.1 Experimental Setting

The author's machine ST ran the Red Hat 7.2 distribution of Linux. This logging machine used Snort [7] 1.8.3 to collect network packet dumps for incoming and outgoing connections.

Most of the remote machines were located throughout the US, and several were located in Europe. They ran a variety of operating systems, including Linux, FreeBSD, Solaris, VMS, and S390. All of them supported incoming and outgoing terminal sessions, with support for secure shell, either version 1 or version 2.

Each logged session lasted between one to ten minutes. In these experiments, the network dumps were not analyzed online but after the experiments were completed. The simple analysis on the network packet dumps, however, could have been performed in an online setting.

In all experimental setups, the logging machine ST acted as a client 0 to the nearest server 1. Simple one-pass online algorithms were used to calculate the two delay gaps from connection between the logging client and the nearest server.

5.2 Experimental Results

Echo-delay comparison proved useful under many situations, as confirmed by many experiments. Two different sets of experimental results are reported here. Both sets used the machine ST to record network logs. In the first set, the recording machine was a *stepping stone* in the connection chain. In the second set the recording machine was the *originating point* of the chain.

5.3 Recording on Stepping Stone

In this set of experiments, each connection chain had a topology of the form $-1 \longrightarrow 0 \longrightarrow 1 \longrightarrow \ldots \longrightarrow n$, where machine 0 refers to the recording machine ST. The length of the chain and the identity of the machines varied. The same sequence of commands were executed in all the sessions. To control for variability in network congestion, all the experiments were conducted in the same time frame, within one hour.

Table 1 shows three groups, each of three chains. There are nine network logs, corresponding to the nine chains presented. Because the outgoing connection from client 0 is ST-rs, Essentially same delayed-acknowledgment gap α is used for all nine chains. The distributions E of (0,1)-gaps differ.

The second group of three chains replicates the first group of three chains. Yet the comparison quantiles differ slightly because there are still uncontrolled variability between the two replications.

In the third group of three chains, the quantile $\text{quan}(E, \alpha)$ is not monotonically increasing as the chain length decreases. On the other hand, the quantile $\text{quan}(E, 2\alpha)$ is better-behaved. The three groups also seem more consistent on the quantile $\text{quan}(E, 2\alpha)$. Apparently, the $\text{quan}(E, 2\alpha)$ feature discriminates more sharply the three chain lengths.

By rejecting chains whose $\text{quan}(E, 2\alpha)$ is less than 0.95, the two long chains of each group would be identified as suspicious. Decreasing this fraction would be less restrictive but also allow more suspicious chains. In any case, the rejection

Table 1. Quantile behavior under varying topology but identical session transcript. In each chain, the leftmost initial denotes the originating point, and the rightmost initial denotes the final victim. All network logs were recorded on machine ST. All sessions executed the same sequence of commands, using only the simple command line under character mode.

quan(E, α)	quan($E, 2\alpha$)	connection chain
0.46	0.73	e5 → ST → rs → e13 → zi → e14 → vm → e15
0.63	0.91	e5 → ST → rs → e13 → zi → e14
0.77	1.00	e5 → ST → rs → e13
0.48	0.61	e5 → ST → rs → e13 → zi → e14 → vm → e15
0.54	0.80	e5 → ST → rs → e13 → zi → e14
0.69	1.00	e5 → ST → rs → e13
0.34	0.57	e5 → ST → rs → e6 → zi → e7 → vm → e8
0.63	0.88	e5 → ST → rs → e6 → zi → e7
0.57	1.00	e5 → ST → rs → e6

region S of suspicious chain, based on the quan($E, 2\alpha$) value alone, would have the form

$$S(c) = \{\text{packet sequence } X : \text{quan}(E(X), 2\alpha(X)) \leq c\}, \tag{6}$$

where $0 \leq c \leq 1$ is an adjustable parameter. A larger value of c enlarges the rejection region S and generally results in more false alarms.

5.4 Recording on Originating Point

In this second set of experiments shown in Table 2, the recording machine ST acted as the originating point in all the chains. The length of the chain varied, but the ordering of machines remained fixed. The session transcript also varied substantially. To control for variability in network congestion, all the experiments were conducted in the same time frame, within two hours.

The bar graph in Figure 4 plots the quan($E, 2\alpha$) values versus the chain length for the 14 sessions in Table 2. On the whole, the quantile quan($E, 2\alpha$) decreases as the chain length increases. Deviations from the general trend are expected because these sessions do not have a common transcript.

In this set of experiments, not only does the chain length vary, but so does the session transcript. Using rejection region based on the quan($E, 2\alpha$) alone, as in Equation 6, gives mixed results. Figure 5 shows the ROC curves for three different tests. To test for more than two hops downstream, the rejection region of the form in Equation 6 provides a perfect test. To test for more than five hops downstream or to test for more than nine hops, however, the results are uniformly worse.

In test samples with variability in session transcript, the simple rejection region based on quan($E, 2\alpha$) alone still works reasonably well, especially in testing

Table 2. Quantile behavior under varying session transcripts but fixed topology. The 14 logs were all recorded from the original point ST. The session transcripts involved shell commands and more complicated commands within the EMACS editor. Each log had a different session transcript.

quan($E, 2\alpha$)	connection chain
0.40	ST→e2→zi→e3→cp→e4→ls→sp→e6→cs→e7→xb→df→bs→e8
0.36	ST→e2→zi→e3→cp→e4→ls→sp→e6→cs→e7→xb→df→bs
0.28	ST→e2→zi→e3→cp→e4→ls→sp→e6→cs→e7→xb→df
0.39	ST→e2→zi→e3→cp→e4→ls→sp→e6→cs→e7→xb
0.42	ST→e2→zi→e3→cp→e4→ls→sp→e6→cs→e7
0.44	ST→e2→zi→e3→cp→e4→ls→sp→e6→cs
0.21	ST→e2→zi→e3→cp→e4→ls→sp→e6
0.68	ST→e2→zi→e3→cp→e4→ls→sp
0.57	ST→e2→zi→e3→cp→e4→ls
0.45	ST→e2→zi→e3→cp→e4
0.70	ST→e2→zi→e3→cp
0.62	ST→e2→zi→e3
0.92	ST→e2→zi
0.99	ST→e2

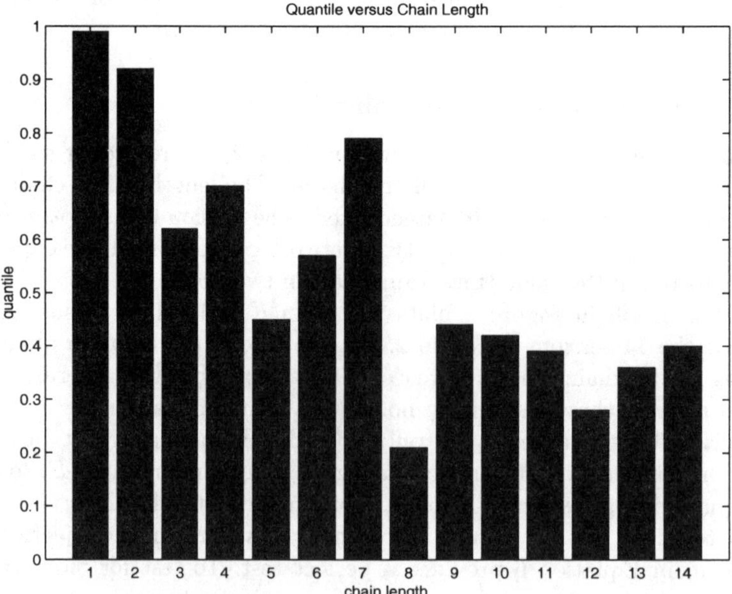

Fig. 4. Plot of quan($E, 2\alpha$) versus chain length.

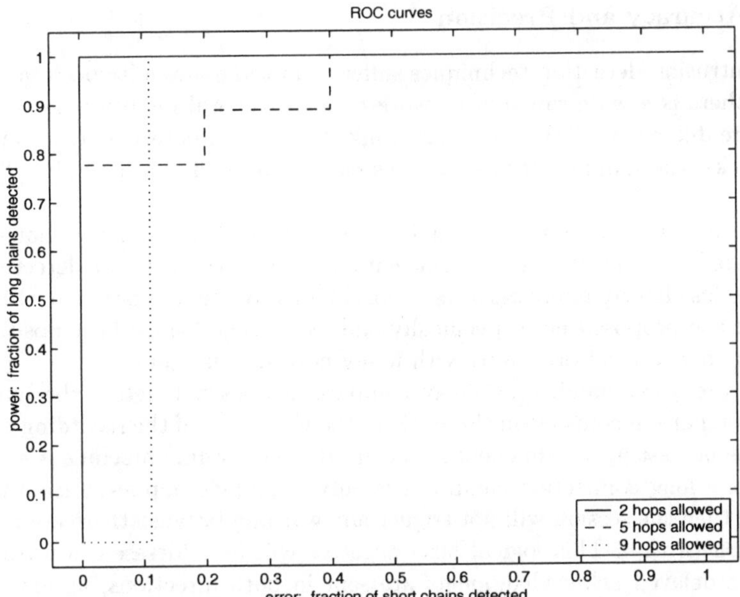

Fig. 5. ROC curves for rejection region $S(c)$.

for more than two hops downstream. As a test for more than five hops downstream or as a test for more than nine hops downstream, the performance of the simple rejection region based on quan$(E, 2\alpha)$ alone deteriorates.

When there are many hops downstream, the variability introduced by the machines downstream adds considerable complication. As expected, tolerance for a big number of downstream hops is more difficult to implement than tolerance for a small of downstream hops. In practice, the low tolerance policy that rejects sessions with more than two hops downstream is more useful and realistic.

6 Discussion

The rejection region in Equation 6 classifies an outgoing connection as having too many hops downstream if the outgoing connection has a packet sequence X whose $2\alpha(X)$ value is too small compared to the gaps in set $E(X)$. To test for an outgoing connection with more than two hops, experiments indicate that the using cut-off parameter $c = 0.9$ would give reasonably accurate results without high false alarms.

The experiments used a machine on the Stanford University network as the logging machine. Although a wide range of experiments under different settings showed that the cut-off parameter $c = 0.9$ performed well, machine and network properties can vary considerably. For best results, each deployment can train on its own machines and networks to determine the optimal cut-off to meet prescribed requirements for accuracy and precision.

6.1 Accuracy and Precision

Most intrusion-detection techniques suffer from too many false positives [4]. Because there is a wide range of network attacks, general intrusion-detection systems are difficult to design. By narrowing the scope of detection to special types of attacks, the number of false attacks can be lowered considerably, [2] for example.

The problem of connection chains is well-defined and narrow enough to be considered as a specific attack. Yet, earlier techniques [8,10] for detecting stepping stones identify many harmless, short chains common in practice. Echo-delay comparison proposed here specifically addresses these logical false positives and detects only connection chains with many hops downstream.

On the other hand, echo-delay comparison does not detect the number of *upstream* hops in connection chain. From the viewpoint of the recording machine, only the nearest upstream client is known. If the recording machine is a stepping stone in a long connection chain but is only one single hop away from the final victim, then the session will not trigger any warning because there are not many hops *downstream*. This logical false negative will be addressed in future work. Because delayed acknowledgments are sent in both directions, an extension of the current work may prove useful in detecting many hops upstream.

6.2 Hacker Intervention

In the ever-raging battle between hackers and intrusion-detection systems, intelligent hackers always search for new ways to elude detection. Echo-delay comparison and previous approaches [8,10] are all susceptible to hacker intervention. In fact, in a theoretical sense [6], any intrusion detector relying solely on network logs can be circumvented by carefully manipulating network signals.

The time gap between the client request and the delayed acknowledgment of the nearest server provides an overestimate of the travel time for one hop downstream. Since not all downstream hops are equally noisy, two problems may arise. A fast connection between the client and the nearest server may over-amplify the slow hops downstream. This configuration does not benefit the hacker trying to avoid detection though.

Likewise, a slow connection between the client and the nearest server may mask the fast hops downstream. If the detector is used at its optimal settings, to detect more than two hops downstream, then there is minimal leeway for hiding many hops behind the slow first connection. Hiding many quick connections on machines within close proximity would defeat the purpose of using a connection chain.

A knowledgeable hacker can manipulate the network signals. To elude detection, the hacker may delay and suppress the delayed-acknowledgment signal and the reply-echo signal. Because the analysis in this paper uses aggregate statistics, targeting a few signals will not thwart the detector. Manipulating many signals simultaneously without adversely affecting the dynamics of the connection chain would be difficult even for the skilled hacker.

7 Summary

Echo-delay comparison monitors an outgoing connection to estimate two important time gaps. First, the gap between the client request and the server delayed acknowledgment estimates the round-trip travel time between the client and the server. Second, the gap between the client request and the server reply echo estimates the how far downstream the final victim is away. Together these two time gaps provide a simple test to identify a session whose final victim is many hops downstream.

Unlike previous approaches for detecting stepping stones, echo-delay comparison works in isolation, without matching for similar sessions on the same connection chain. Moreover, this new strategy will allow benign, short connection chains common in practice. Echo-delay comparison makes use network signals found in interactive terminal sessions, such as telnet, rlogin, and secure shell. Experiments demonstrate that the technique is effective under a wide range of conditions and performs especially well in identifying sessions with more than two hops downstream.

Acknowledgments. This research project was funded in part by the US Department of Justice grant 2000-DT-CX-K001. Jeffrey D. Ullman of the Stanford University Computer Science Department introduced the author to the field of intrusion detection and offered invaluable advice throughout the past year. Jerome H. Friedman of the Stanford University Statistics Department provided important feedback in several discussions. The author is grateful for their help and extends his delayed acknowledgment, long overdue.

References

1. Stefan Axelsson. "Intrusion Detection Systems: A Survey and Taxonomy." Technical Report 99-15, Department of Computer Engineering, Chalmers University, March 2000.
2. Robert K. Cunningham, et al. "Detecting and Deploying Novel Computer Attacks with Macroscope." *Proceeding of the 2000 IEEE Workshop on Information Assurance and Security.* US Military Academy, West Point, NY, 6–7 June, 2001.
3. Harold S. Javitz and Alfonso Valdes. "The NIDES Statistical Component: Description and Justification." Technical report, Computer Science Laboratory, SRI International. Menlo Park, California, March 1993.
4. Richard P. Lippmann, et al. "Evaluating Intrusion Detection Systems: The 1998 ARPA Off-Line Intrusion Detection Evaluation." *Proceedings of DARPA Information Survivability Conference and Exposition.* DISCEX '00, Jan 25–27, Hilton Head, SC, 2000. `http://www.ll.mit.edu/IST/ideval/index.html`
5. Peter G. Neumann and Phillip A. Porras. "Experience with EMERALD to Date." *1st USENIX Workshop on Intrusion Detection and Network Monitoring*, pages 73-80. Santa Clara, California, USA, April 1999.
6. Thomas H. Ptacek and Timothy H. Newsham. "Insertion, Evasion, and Denial of Service: Eluding Network Intrusion Detection." Secure Networks, Inc., January 1998. `http://www.aciri.org/vern/PtacekNewsham-Evasion-98.ps`

7. Martin Roesch. "Snort: Lightweight intrusion detection for networks." 13th Systems Administration Conference (LISA'99), pages 229–238. USENIX Associations, 1999.
8. Stuart Staniford-Chen and L. Todd Heberlein. "Holding Intruders Accountable on the Internet." *Proceedings of the 1995 IEEE Symposium on Security and Privacy*, pages 39–49. Oakland, CA, May 1995.
9. W. Richard Stevens. *TCP/IP Illustrated Volume 1: The Protocols.* Addison-Wesley: Reading, Massachusetts, 1994.
10. Yin Zhang and Vern Paxson. "Detecting stepping stones." *Proceedings of 9th USENIX Security Symposium.* August 2000.

Multiscale Stepping-Stone Detection: Detecting Pairs of Jittered Interactive Streams by Exploiting Maximum Tolerable Delay

David L. Donoho[1], Ana Georgina Flesia[1], Umesh Shankar[2], Vern Paxson[3], Jason Coit[4], and Stuart Staniford[4]

[1] Department of Statistics, Stanford University
Sequoia Hall, 390 Serra Mall, Stanford, CA 94305-4065 USA
{donoho,flesia}@stanford.edu

[2] Department of Computer Science, University of California at Berkeley
567 Soda Hall, Berkeley, CA 94704
ushankar@cs.berkeley.edu

[3] International Computer Science Institute
1947 Center St. suite 600, Berkeley, CA 94704-1198
vern@icir.org

[4] Silicon Defense
203 F Street, suit E, Davis, CA95616, USA
{stuart,jasonc}@silicondefense.com

Abstract. Computer attackers frequently relay their attacks through a compromised host at an innocent site, thereby obscuring the true origin of the attack. There is a growing literature on ways to detect that an interactive connection into a site and another outbound from the site give evidence of such a "stepping stone." This has been done based on monitoring the access link connecting the site to the Internet (Eg. [7,11, 8]). The earliest work was based on connection content comparisons but more recent work has relied on timing information in order to compare encrypted connections.

Past work on this problem has not yet attempted to cope with the ways in which intruders might attempt to modify their traffic to defeat stepping stone detection. In this paper we give the first consideration to constraining such intruder *evasion*. We present some unexpected results that show there are theoretical limits on the ability of attackers to disguise their traffic in this way for sufficiently long connections.

We consider evasions that consist of local jittering of packet arrival times (without addition and subtraction of packets), and also the addition of superfluous packets which will be removed later in the connection chain (chaff).

To counter such evasion, we assume that the intruder has a "maximum delay tolerance." By using wavelets and similar multiscale methods, we show that we can separate the short-term behavior of the streams – where the jittering or chaff indeed masks the correlation – from the long-term behavior of the streams – where the correlation remains.

It therefore appears, at least in principle, that there is an effective countermeasure to this particular evasion tactic, at least for sufficiently long-lived interactive connections.

A. Wespi, G. Vigna, and L. Deri (Eds.): RAID 2002, LNCS 2516, pp. 17–35, 2002.

Keywords: Network intrusion detection. Evasion. Stepping Stone. Interactive Session. Multiscale Methods. Wavelets. Universal Keystroke Interarrival Distribution.

1 Introduction

Perpetrators launching network intrusions over the Internet of course wish to evade surveillance. Of the many methods they use, one of the most common and effective is the construction of *stepping stones*. In this technique, the attacker uses a series of compromised hosts as relay machines and constructs a chain of interactive connections running on these hosts using protocols such as Telnet or SSH. The commands typed by the attacker on their own host are then passed along, unmodified, through the various hosts in the chain. The ultimate victim of the attack sees traffic coming from the final host in the chain, and because this is not the actual origin of the attack, little is revealed about the real location of the attacker.

An investigator seeking to locate the perpetrator would appear to be stymied by the need to execute a lengthy and administratively complex 'traceback' procedure, working back host by host, figuring out each predecessor in the chain step-by-step (based on whatever log records may be available at each stepping-stone site). For discussion of the use of stepping-stone attacks in high profile cases – and the difficulty of unraveling them – see for example [6] or [3].

An alternate paradigm for stepping-stone detection entails the installation of a *stepping-stone monitor* at the network access point of an organization (such as a university or other substantial local network). The monitor analyzes properties of both incoming and outgoing traffic looking for correlations between flows that would suggest the existence of a stepping stone [7,11,8]. See Figure 1.

This tradition of work has all been concerned with traceback of interactive connections: traceback of short non-interactive connections is harder and is presently unaddressed in the literature. Nor do we address it here. However, the interactive traceback problem is of interest, since there are many tasks that attackers must perform interactively. If the hacker has a goal beyond just compromising machines for zombies, if he or she really wishes to exploit a particular site for criminal ends, then the creative exploration and understanding of the newly compromised site requires a significant amount of human time, and for this one or more interactive sessions are highly desirable.

Attackers who are aware of the risk of monitors looking for stepping stones can attempt to evade detection of their stepping stones by modifying the streams crossing the network access point so that they appear unrelated. Since the stepping-stone hosts are under their control, we must assume that attackers can arbitrarily modify their traffic in such evasion attempts. A wide spectrum of possible evasions might be considered; in the worst case, the information the attacker truly wishes to transmit could be embedded steganographically in connections that appear to be quite unrelated, both as regards content and traffic properties. On the other hand, such evasions might be very inconvenient to design, implement and use. It is of considerable interest to understand how well

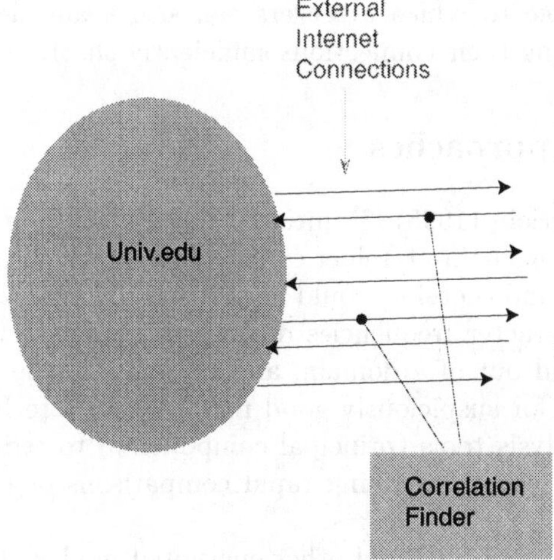

Fig. 1. Stepping-Stone Monitor

the various evasion techniques can work, and under what circumstances they can be defeated by monitoring techniques – particularly the ones that are easiest to deploy.

In this article we consider evasions based on keeping within the Telnet/SSH connection paradigm – which is obviously the most convenient for the attacker – and simply modifying the traffic being handled through the Telnet or SSH connections. We discuss a class of evasions to the monitoring techniques based on local timing perturbations, in which the stepping stone simply adds delay in more or less random amounts to traffic crossing the stepping stone. However, we assume that there is a maximum tolerable delay that the attacker is willing to introduce (since humans are not able to work effectively over interactive connections with very long latencies). We give a theoretical result that such *packet conserving* evasions are ineffective against appropriate correlation based on multiscale analysis using wavelets, at least in the limit of long connections.

We then consider the case of attackers who add extra packets into their connection, but still wish to have it look interactive. Again, we show that for long enough connections, it will be possible to correlate the two connections despite the added packets.

This suggests that the attacker wishing to evade stepping stone detection would be ill-advised to rely solely on local timing jitter or addition of chaff packets to connections. Based on our analysis, it appears that the most likely approach will require abandoning connection chains that use only standard interactive protocols, such as Telnet or SSH, for more sophisticated schemes, that can steganographically add traffic to connections that look like something else. These tools will be correspondingly harder to install and use. However, we also note that our results are primarily asymptotic, and require further analysis to determine the degree to which attackers can still evade detection within the framework by keeping their connections sufficiently short.

2 Previous Approaches

Staniford and Heberlein (1995) [7] initiated the literature of stepping-stone detection by considering chain-of-Telnet connections, in which the content is transmitted in the clear, and therefore could be statistically analyzed. Their approach was to tabulate character frequencies during set time intervals over all Telnet connections into and out of a domain, and to compare the tables of character frequencies looking for suspiciously good matches. As a technical feature, they used statistical analysis tools (principal components) to reduce the dimensionality of the feature vector, enabling rapid comparisons of features of different connections.

The increasing use of SSH and other encrypted modes of communication in recent years makes it important to develop a tool which does not require access to the actual transmitted content. Zhang and Paxson (2000) [8] developed an *activity*-based rather than *content*-based approach, which in particular could be used with chain-of-SSH connections. Their approach was based on the observation that interactive sessions have a strong "on–off" structure. Given this, one can then monitor the "off" periods of connections looking for suspicious coincidences of connections making nearly simultaneous transitions to an "on" period. They developed an on-line algorithm for looking for stepping stones and showed they could do this in practice at a large site (though it turned out that most of the stepping stones found were innocent).

Yoda and Etoh (2000) [11] also considered the problem of comparing interactive network connections that might be encrypted and they too relied on timing properties of the connection. However, they based their approach on looking at the average time lag between one connection and another, minimized over possible time offsets. They developed a quadratic time off-line algorithm for doing this comparison, and showed at least a preliminary ability to discriminate stepping stones pairs from unrelated connection pairs based on a threshold of about three seconds for the average delay between connections.

The mechanisms developed by Zhang/Paxson and Yoda/Etoh are both vulnerable to attackers perturbing the timing structure of their connections in order to defeat the stepping stone detection. In this paper we demonstrate that, for some types of perturbations, such vulnerabilities do not appear fundamental.

3 Next Generation Evasions

Existing approaches to detecting stepping-stones are subject to evasions, and the purpose of this paper is to analyze the capabilities of some of these evasions. The central issue is that attackers will have available the ability to effect *stream transformations* on the hosts in a stepping-stone chain, altering the relays from performing pure 'passthru' to instead modifying the stream in some way.

For example, in a Unix context, one could introduce filters for "chaff embedding" and "chaff stripping." Imagine a filter ⟨`enchaff`⟩ that merges the standard input with meaningless 'chaff' input from another source, so that the standard input content comprises only a sub-sequence of the output stream; ⟨`dechaff`⟩ extracts the embedded meaningful sub-sequence; and ⟨`passthru`⟩ simply copies standard input to standard output. By chaining together such filters, one can conceptually arrange that the content transmitted over the connection incoming to a site will not obey the sequencing and volume relationships of the outgoing connection, even though the semantic content is identical.

In fact, writing such filters in Unix is not quite trivial, due to buffering and pseudo-tty issues. But clearly they are realizable without a great deal of effort. Accordingly, we need to consider the possible impact of stream transformations used to evade stepping-stone detectors.

In short, the challenge for the next generation of stepping-stone monitors is to detect correlated activity between two streams when

- One or both streams may be transformed
- It is not possible to examine content for correlations

4 The Constraint Hypothesis

Our research began with the hypothesis that, while arbitrary stream transformations might conceivably be very effective at evading detections, certain constraints on interactive sessions might prevent the use of effective transformations.

For interactive connections, we argue for the following two constraints:

- *Latency constraints.* Ultimately, the chain of interactive connections is tied to a human user, for whom it will be annoying/tiring/error-prone to have to wait a long time for the results of their typing to be echoed or processed. Hence, we posit a *maximum tolerable delay* limiting what a stream transformation can impose; anything longer will be just too painful for the user.
- *Representative traffic.* Typing (and "think time" pauses) by humans manifests certain statistical regularities in the corresponding interpacket spacings, sharply distinct from machine-driven network communication. In particular, interpacket spacings above 200 msec (the majority) are well-described as reflecting a Pareto distribution with shape parameter $\alpha \approx 1.0$ [9]. A stream transformation which upsets this regularity can in principle call attention to itself and become itself a source of evident correlation between ingress and egress connections.

We can summarize these constraints as: *(i)* the original stream and its transformation must be synchronized to within a certain specific maximum tolerable delay, and *(ii)* the stream interarrival times must have the same distribution as the universal Pareto distribution described above.

This second constraint is particularly powerful. It seems difficult to add chaff to a stream without destroying invariant distributional properties, so in most of the remainder of this paper we consider schemes which do not add chaff. That is, we consider transforms that *conserve character counts*: each character in one stream corresponds to one character in the other stream. Such conservative transforms can only alter the interarrival times between items in the input stream and the output stream, and can be thought of as simply *jittering* the times to mask the similarity of the two streams.

We must, however, note an important caveat regarding constraint *(ii)*, which is that the Pareto distribution emerges as a general property when we analyze the statistics of many interactive interpacket times aggregated together. However, the *variation* in the distribution seen across different individual interactive sessions has not yet been characterized in the literature. It is possible that sufficient variation exists such that an attacker could inject chaff that significantly alters the distribution of the interpacket timings without an anomaly detector being able to flag the altered distribution as anomalous. With that possible limitation in mind, we now investigate what we can do to thwart evasion if in fact that the attacker cannot pursue such alterations. Later, we will return to the issue of detecting correlations despite the addition of chaff.

Assuming that the attacker is confined to conservative transforms, can they actually be used to hide the common source of two streams? To answer this question, we now examine possible evasion transforms that conform with the above assumptions.

One aproach is to use a transform which re-randomizes interarrival times. In words, we take a stream and 'strip it' of its identity by changing all the inter-keystroke times to a stochastically independent set of inter-keystroke times. Formally,

- Stream 1 contains Characters $c_1, ..., c_n$ at Times $t_1, ..., t_n$.
- Stream 2 contains the same Characters $c_1, ..., c_n$, at Times $u_1, ..., u_n$.
- The interarrival times $t_i - t_{i-1}$ are known to be independent and identically distributed (i.i.d.) according to a known distribution function F.
- Stream 2 is defined by interarrival times $(u_i - u_{i-1})$ which are also i.i.d. F, independently of (t_i).

This approach certainly removes all correlations between the two streams, but has two major flaws. First, it is not *causal*: it is possible that $u_i < t_i$ for certain characters i and $u_{i'} > t_{i'}$ for other characters i', while properly speaking, one of the streams must occur strictly after the other. (Which one occurs after the other depends on the monitor's location with respect to the position of the transformation element.)

It might appear that the difficulty with causality can be addressed by conceptually shifting the transformed stream in time by a fixed amount, preserving its distribution but ensuring that we always have $u_i \geq t_i$. But a second problem remains: the two streams become unboundedly out-of-sync. Indeed, the difference between the two cumulative counting functions behaves as a random walk, and so we know from simple probability calculations that for this scheme, $|t_n - u_n|$ fluctuates unboundedly as $n \to \infty$, and that $Var\,(t_n - u_n) \geq constant \cdot n$ (essentially because $t_n - u_n$ is a sum of n i.i.d. random variables). It follows that for any given buffer length, eventually the delay between the two streams will surpass that length; and that for any tolerable perceptual delay, eventually the delay caused by the transcoding will exceed that length. In summary, transcoding to remove all correlations leads to complete desynchronization over time, violating the maximum tolerable delay constraint.

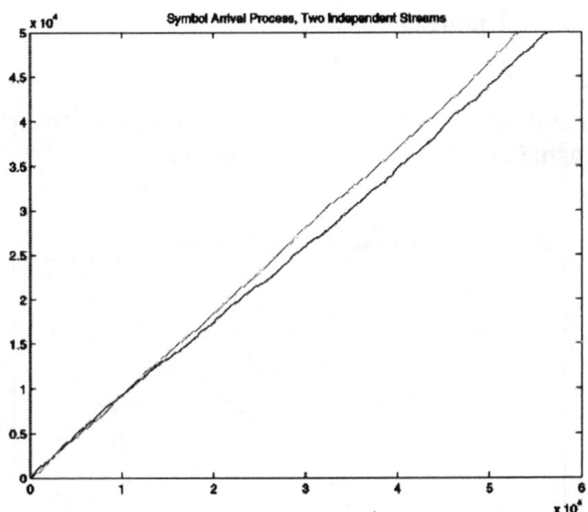

Fig. 2. Divergence of Independent Streams. The two cumulative counting functions diverge arbitrarily as time progresses.

The point is illustrated in the figure 2, which shows a simulation of the transformation discussed above, where keystroke arrivals are drawn from the empirical distribution given in [9]. We can see that after about 5,000 symbols, the two streams are about 500 characters out of sync.

How about partial randomization of keystroke arrival times? Consider the following local jittering algorithm, which we call *dyadic block reshuffling.* Given Stream 1 with arrival times t_i, this approach creates Stream 2 with arrival times u_i that never differ from those in Stream 1 by more than a certain specific guaranteed amount, but which are completely independent at fine levels. The approach has the following general structure:

- For dyadic intervals $[k2^j, (k+1)2^j)$, $N^1_{j,k}$ in Stream 1, compute arrival counts $N^1_{j,k}$.
- For a given 'scale' j_0, create Stream 2 so that $N^2_{j,k} = N^1_{j,k}$, for all $j \geq j_0$, all k.
- Method: identify all arrivals in $I_{j,k}$ and select random uniform arrivals in same interval.

The approach is illustrated in figures 3 and 4. The first one depicts the algorithm operating at a specific medium scale j_0: This sort of local shuffling does not suffer

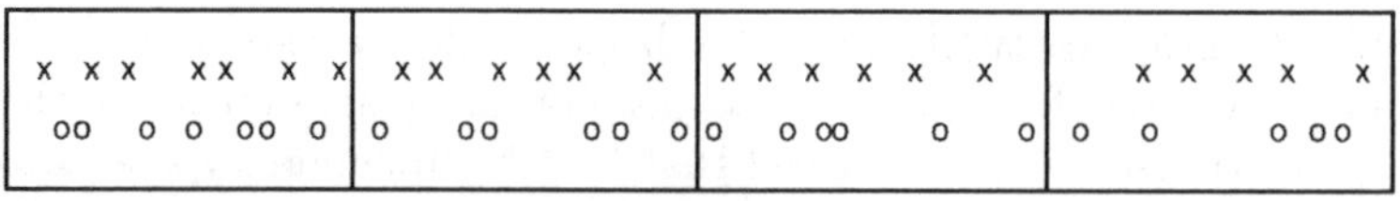

x x x xx x x	x x x x x x	x x x x x x	x x x x x
oo o o oo o	o oo oo o	o o oo o o	o o o oo

Fig. 3. Dyadic Block Reshuffling. Row of 'x': arrival times in original stream. Row of 'o': arrival times in transformed stream. Black Boxes: equi-spaced blocks of time. There are just as many times in each block for each stream. Times in transformed stream are chosen uniformly at random within block

from de-synchronization; as figure 4 shows, the two cumulative character counts functions cross regularly, at least once for each box.

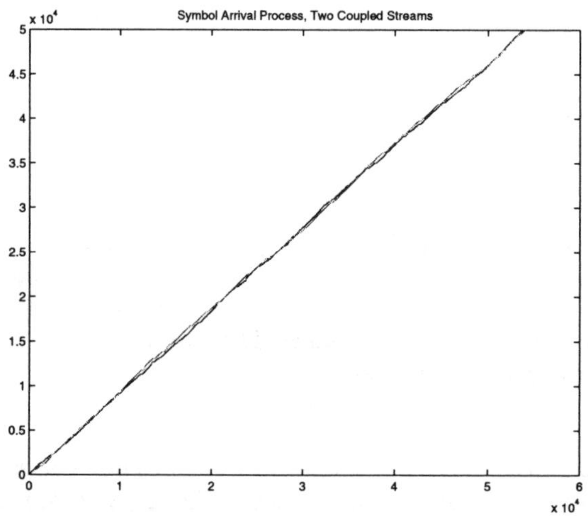

Fig. 4. Non-divergence of streams under dyadic block reshuffling

However, the constraint which is responsible for this crossing mechanism – $N^1_{j,k} = N^2_{j,k}$ at scales $j \geq j_0$ – says also that on sufficiently coarse scales the two counting functions agree, so there are measurable correlations between the two streams. We are confronted by a tradeoff:

- Pick j_0 at fine scale – we get tolerable delay but high correlation
- Pick j_0 at coarser scale – we get worse delay but reduced correlation.

Is this tradeoff inevitable? For example, are there local jitterings which are more cleverly constructed and which avoid such correlations?

We can formulate our central question in the following terms. Let $N_1(t)$ be the *cumulative character counting function* on the untransformed stream:

$$N_1(t) = \# \text{ of symbols in Stream 1 on } [0, t)$$

and similarly let $N_2(t)$ be the character counting function on the transformed stream. Our earlier discussion imposes specific constraints on these functions:

1. *Causality.* Characters cannot emerge from the transformed stream before they have emerged from the original stream. I.e., we must always have:

$$N_2(t) \leq N_1(t).$$

(The ordering of the inequality here is a convention; we could as well impose the reverse inequality, since, as discussed above, the direction of the causality between the two streams depends on the location of the monitor with respect to the transformation element.)

2. *Maximum Tolerable Delay.* Per the previous discussion, owing to human factors characters must emerge from the second stream within a time interval Δ after they emerged from the first stream:

$$N_2(t + \Delta) \geq N_1(t).$$

We then ask:

1. Do Causality & Maximum Tolerable Delay combine to imply noticeable correlations between properties of stream 1 and stream 2?

2. If so, what properties should we measure in order to observe such correlations?

5 Main Result

Our principal result is a theoretical one, showing that *multiscale analysis of stream functions* N_i *will reveal, at sufficiently long time scales, substantial correlations.* To make this precise, we introduce a systematic multiscale machinery. Good references on multiscale analysis and wavelets abound, but we are particularly fond of [4,5].

To begin, we fix a wavelet $\psi(t)$ which is either a 'bump' (like a bell curve) taking only positive values or a 'wiggle' taking both positive and negative values. See Figure 5 for some examples of each. We form a multiscale family of translates and dilates of ψ

$$\psi_{a,b} = \psi((t - b)/a)/a^p$$

Here the parameter p controls the kind of analysis we are doing. If ψ is a 'bump', we use $p = 1$; if ψ is a wiggle, we use $p = 1/2$. (The rationale for the different choices of p is given in the appendix.)

For computational reasons, we limit ourselves to a special collection of times and scales: the dyadic family $a = 2^j$, $b = k \cdot 2^j$. We can then use the fast wavelet transform to rapidly compute the *wavelet coefficients* of each stream function N_i, defined by

$$\alpha^i_{j,k} = \langle \psi_{a,b}, N_i \rangle$$

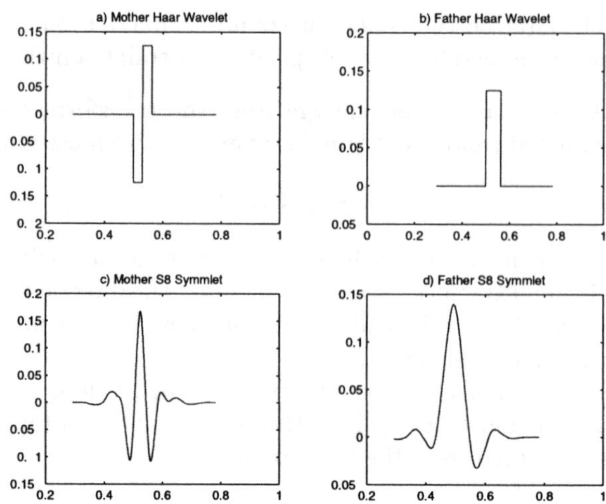

Fig. 5. Some wavelet waveforms: a) 'wiggle' (Mother Haar Wavelet); b) 'bump' (Father Haar wavelet); c) Mother S8 Symmlet; d) Father S8 Symmlet

where $\langle f, g \rangle$ denotes the inner product

$$\langle f, g \rangle = \sum_t f(t)g(t).$$

When the wavelet is a 'bump', these are also called 'scaling coefficients'; when the wavelet is a 'wiggle' these are commonly called 'wavelet coefficients'. See Figure 6.

With this machinery, our central question becomes: if N_1 and N_2 obey the causality/maximum tolerable delay constraints, how similar are $\alpha^1_{j,k}$ and $\alpha^2_{j,k}$? In essence, we are analyzing the character counting functions of both streams across different time scales, looking for how similarly the character arrivals cluster at each time scale.

Our analysis follows two specific branches, depending on the choice of ψ.

- *Analysis by Multiscale Block Averages.* Here we choose ψ to be a very simple 'bump' – actually the "boxcar" function $\psi(t) = 1_{[0,1]}$ depicted in Figure 5, panel (b). As indicated above, we choose $p = 1$, and it then turns out that the coefficients amount to simple averages of the data over blocks of various lengths and locations. Accordingly, we call this choice of ψ as corresponding to the analysis of "multiscale block averages."

  We analyze the stream functions $N_i(t)$ via the dyadic boxcar family

  $$\psi_{j,k}(t) = \psi((t - k2^j)/2^j)/2^j.$$

  How similar are $\alpha^1_{j,k}$ and $\alpha^2_{j,k}$?

  Our strategy for analysis is to estimate two quantities at each scale level j:

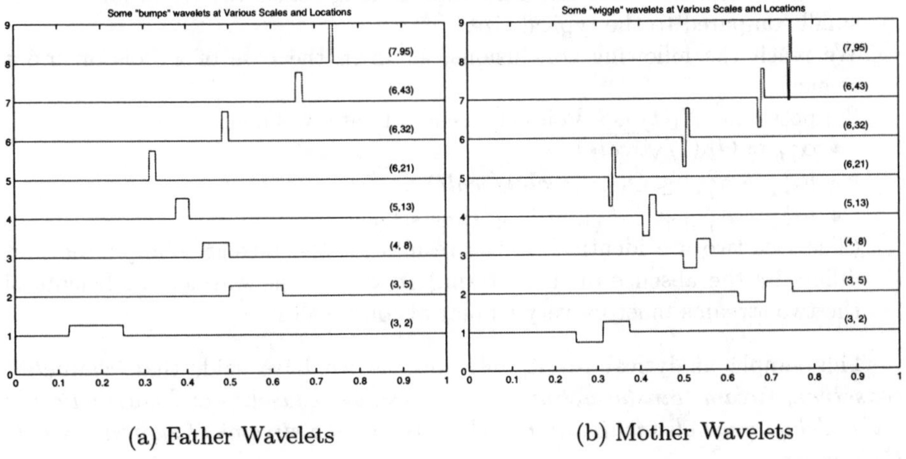

(a) Father Wavelets

(b) Mother Wavelets

Fig. 6. Scale/Location Families: 'bumps' and 'wiggles' at various scales and locations

- The typical size of $\alpha^1_{j,k}$ at scale j; and
- The maximal deviation of $\alpha^1_{j,k} - \alpha^2_{j,k}$ at scale j.

We then compare these, and it will turn out that at long scales the deviation term is small compared to the typical size. Using analysis developed in the next section, we can then reach the following conclusions, explained here for the case of a Poisson input stream, where the analysis is simpler.
Suppose that $N_1(t)$ is a Poisson stream at rate λ. Then

- $\alpha^1_{j,k} \approx \lambda \pm O_P(1/\sqrt{scale})$, where $O_P()$ denotes the asymptotic order in probability.
- $|\alpha^1_{j,k} - \alpha^2_{j,k}| \leq O_P(\log(scale)/scale)$
- $|\alpha^1_{j,k} - \alpha^2_{j,k}| \ll |\alpha^1_{j,k}|$, at long time scales.

In words, the scaling coefficients of the two streams must be very similar at long time scales.

– *Multiscale Block Differences.* Here we choose ψ to be a very simple 'wiggle' – actually the Haar wavelet $\psi(t) = 1_{[1/2,1)} - 1_{[0,1/2)}$ depicted in Figure 5, panel (a); this is a simple difference of boxcars. As indicated above, we therefore choose $p = 1/2$, and it then turns out that the coefficients amount to simple scaled differences of averages of the data over blocks of various lengths and locations. Accordingly, we call this choice of ψ as corresponding to the analysis of "multiscale block differences."
We analyze the stream functions $N_i(t)$ via the dyadic Haar family

$$\psi_{j,k}(t) = \psi((t - k2^j)/2^j)/2^{j/2}.$$

How similar are $\alpha^1_{j,k}$ and $\alpha^2_{j,k}$?
Our strategy for analysis is again to estimate two quantities:

- The typical size of $\alpha^1_{j,k}$ at level j; and
- The maximal deviation of $\alpha^1_{j,k} - \alpha^2_{j,k}$ at level j.

We then compare these and find that at long scales the deviation term is small compared to the typical size.
We reach the following conclusions, again in the case of a Poisson input stream.
Suppose that $N_1(t)$ is a Poisson stream at rate λ. Then
- $\alpha^1_{j,k} \approx O_P(1/\sqrt{scale})$.
- $|\alpha^1_{j,k} - \alpha^2_{j,k}| \leq O(\log(scale)/scale)$
- $|\alpha^1_{j,k} - \alpha^2_{j,k}| \ll |\alpha^1_{j,k}|$, at long time scales.

(The last two are identical to the results for the boxcar 'bump'; the first differs by the absence of the λ term.) In words: the wavelet coefficients of the two streams must be very similar at long scales.

This simple analytical result indicates, as we have said, that *character-conserving stream transformations which maintain causality and maximum tolerable delay, must also maintain correlations between streams at sufficiently long time scales.*

As stated so far, the result applies just to Poisson input streams. In the appendix we discuss extending the result to Pareto streams. This extension is of significant practical import, because the Pareto distribution, which as a model of network keystroke interarrivals is well supported by empirical data [9], is radically different in variability from the exponential distribution.

6 Analysis

In this section we develop some of the machinery used to support the results outlined in the previous section. Our first analytical tool for developing this result is a simple application of integration by parts. Let $\Psi = \Psi(t)$ be a function that is piecewise differentiable and which vanishes outside a finite interval. (This condition holds for both the Boxcar and the Haar wavelets). Then from integration by parts

$$\int \Psi \, \mathrm{d}N_1 - \int \Psi \, \mathrm{d}N_2 = \int \Psi \, \mathrm{d}(N_1 - N_2) = -\int (N_1 - N_2)(t) d\Psi(t)$$

so

$$\left| \int \Psi \, \mathrm{d}N_1 - \int \Psi \, \mathrm{d}N_2 \right| \leq TV(\Psi) \cdot \max\{|(N_1 - N_2)(t)| : t \in \mathrm{supp}(\Psi)\}$$

where $supp(\psi)$ – the support of ψ – the part of the t-axis where ψ is nonzero; and $TV(\Psi)$ – the Total Variation – is informally the sum of all the ups and downs in the graph of Ψ; formally, for a smooth function $\Psi(t)$, $TV(\Psi) = \int |\Psi'(t)| dt$, while for piecewise smooth functions, the total variation includes also the sum of the jumps across discontinuities.

Our second analytical tool has to do with properties of extreme values of stochastic processes. Causality and Maximum Tolerable Delay imply that

$$N_1(t) \geq N_2(t) \geq N_1(t - \Delta)$$

hence,

$$|N_1(t) - N_2(t)| \leq N_1(t) - N_1(t - \Delta)$$

and so

$$|N_1(t) - N_2(t)| \leq \max\{N_1(t + \Delta) - N_1(t) : t, t + \Delta \in \text{supp}(\Psi)\}.$$

In words, the difference between N_1 and N_2 is controlled by the volume in N_1. We now use results about extremes of Poisson processes. If N_1 is the set of cumulative arrivals of a Poisson counting process, then

$$\max\{N_1(t + \Delta) - N_1(t) : t, t + \Delta \in [a, b]\} \leq O_P(\log(b - a)) \cdot E\{N_1(t + \Delta) - N_1(t)\}$$

For more details see [2] and [1].

Based on these two analytical tools, we can easily obtain the results in the previous section:

- *Calculation for Multiscale Block Averages.* This is based on the following ingredients. First, symbols emerge at Poisson arrival times $t_1, \ldots, t_N$, with rate λ. Second, the 'bump' has mean 1 and so $E[\alpha^1_{j,k}] = \lambda$ (as one might guess). Third, $Var[\alpha^1_{j,k}] = Const \cdot \lambda/scale$, which is the usual $1/n$-law for variances of means. Consequently, the random fluctuations of the scaling coefficients obey

$$\alpha^1_{j,k} \approx \lambda \pm c/\sqrt{scale}$$

  To calculate the maximum fluctuation of $\alpha^1_{j,k} - \alpha^2_{j,k}$, we observe that $TV(\psi_{j,k}) \leq 4/scale$ and $\sup\{N_1(t + \Delta) - N_1(t) : t \in [a, a + scale]\} = O_P(\log(scale))$, giving the key conclusion $|\alpha^1_{j,k} - \alpha^2_{j,k}| \leq O(\log(scale)/scale)$.
- *Calculation for Multiscale Block Differences.* Again we assume that symbols emerge at Poisson arrival times $t_1, \ldots, t_N$, with rate λ. Second, the 'wiggle' has mean 0 and so $E[\alpha^1_{j,k}] = 0$ (again as one might guess). Third, $Var[\alpha^1_{j,k}] = Const \cdot \lambda/scale$, which is the usual $1/n$-law for variances of means. Consequently, the random fluctuations of the wavelet coefficients obey

$$\alpha^1_{j,k} \approx \pm c/\sqrt{scale}$$

  The calculation of the maximum fluctuation of $\alpha^1_{j,k} - \alpha^2_{j,k}$ is as for multiscale block averages above.

We again note that this analysis, as it stands, applies only to Poisson streams. Further analysis, given in the appendix, indicates that the same type of analysis can apply to Pareto streams and many others as well.

7 Simulation

To illustrate the above results, consider a simple transcoder: *local inter-keystroke shuffling* (LIS). This transcoder works as follows. We buffer symbols for M consecutive symbols or Δ milliseconds, whichever comes first. Suppose that the times of the symbol arrivals into the incoming stream buffer are t_1,..., t_m, so that necessarily $m \leq M$ and $t_m - t_1 < \Delta$. We then compute the interarrival times of the symbols in the buffer, $\delta_1 = t_2 - t_1$, $\delta_2 = t_3 - t_2$, ... , $\delta_{m-1} = t_m - t_{m-1}$.

Given the number $\delta_1, \ldots, \delta_{m-1}$, we perform a random shuffling of the times, obtaining $\varepsilon_1, \ldots, \varepsilon_{m-1}$. Then we define a second set of times by

$$u_1 = t_m,\ u_2 = u_1 + \varepsilon_1,\ \ldots\ u_i = u_{i-1} + \varepsilon_{i-1},\ \ldots,$$

and we output symbols in the second stream at times u_i (we ignore here the processing time required for calculating the u_i, which is surely trivial in this case). Figure 7 illustrates the type of transformation obtained by LIS.

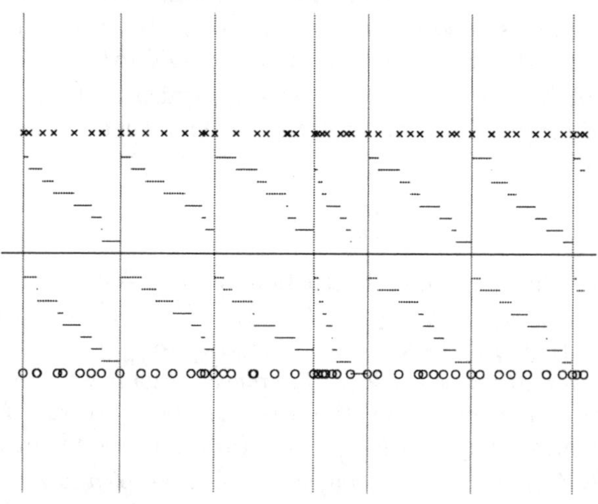

Fig. 7. LIS Transform: Row of 'x' – arrival times in original stream; Row of 'o' – arrival times in transformed stream; vertical lines demarcate zones of 8 characters; top group of horizontal lines – lengths depict inter-keystroke times; bottom group – lengths depict same times, but shuffled within one zone. *Note, the bottom set of boxes should be shifted in time over to the right by Δ; this will be fixed in the final version of the figure.*

A new stream synthesized by LIS has these properties:

- *Identical Distribution.* Whatever the underlying distribution of inter-keystroke times δ_i associated with (t_i), the new stream (u_i) has inter-keystroke times with the same distribution, because the actual inter-keystroke times are just the same numbers arriving in a different order.

- *Causality.* Characters arrive in Stream 2 later than Stream 1

$$t_i \leq u_i$$

- *Controlled delay.* Characters do not arrive much later in Stream 2:

$$t_i < u_i < t_i + 2\Delta.$$

Thus, there is no possibility that a statistical traffic anomaly detector can have cause for flagging a stream 2 produced by LIS as aberrant traffic. Also, by controlling the parameter Δ, we control the maximum tolerable delay.

To study the properties of multiscale detectors in the LIS setting, we use Monte-Carlo simulation. In our simulation experiments, we used for (t_i) samples from the empirical distribution of inter-keystroke times described in [9]. We created a stream several minutes in length and we transcoded this stream three times at three different delay parameters: 100 msec, 200 msec, and 300 msec.

From the streams, we created time series. We selected 256 second stretches of each stream, divided the time axis into 1/64th second intervals, and we counted the number of character arrivals (always 0 or 1, of course) within each interval. This gave us equally spaced series of $16384 = 2^{15}$ 0's and 1's. We then used the freely available wavelet transform routines in WaveLab [10] to perform a wavelet analysis using the Haar wavelets.

The table below reports correlations between wavelet coefficients of the original stream and the three transformed streams. It contains the empirical correlations between the wavelet coefficients at various scales, defining the correlation coefficient at scale j by

$$Corr(j) = \sum_k \alpha^1_{j,k}\alpha^2_{j,k} / \left(\sum_k (\alpha^1_{j,k})^2 \cdot \sum_k (\alpha^2_{j,k})^2 \right)^{1/2}$$

where each sum is over all coefficients at a given scale j.

Scale	Δ =100 ms	200 ms	300 ms
32 sec	0.9964	0.9599	0.9382
64 sec	0.9965	0.9654	0.9371
128 sec	0.9966	0.9695	0.9458

These results are typical and repeatable. Correlations, of course, cannot exceed 1.0. So these correlations, which approach 1 at sufficiently long scales, are rather large. Evidently, given about 1 minute's worth of data on two jittered streams, we can obtain a substantial signal by correlation of wavelet coefficients. Note particularly the very high correlations when jittering is less than .1 sec.

8 Detecting Evasions in the Presence of Chaff

In the analysis since Section 4, we have assumed that any stream transformation being used to disguise correlations was conservative – that is, it exactly preserves the number and content of keystrokes, but perhaps alters their timing.

We now discuss the more general situation when this is not the case. A reasonable way to begin modeling the general situation is to say that we have two cumulative character counting functions N_1 and N_2, and that now

$$N_2(t) = N_1'(t) + M(t),$$

where $N_1'(t)$ is the cumulative counting function of the input character arrival times, perhaps after a conservative stream transformation, and M is the cumulative counting function of chaff arrival times. In short, as we watch characters coming in, some are from the input stream – only jittered – and others are chaff.

Suppose that we once again compute the statistic $Corr(j)$ at each scale. What happens? The results of course change. Suppose for simplicity that the chaff arrives according to the universal keyclick interarrival process, and that the chaff arrival process is stochastically independent of the N_1 process. Then one can show that instead of

$$Corr(j) \to 1, \quad j \to \infty,$$

we actually have

$$Corr(j) \to \rho, \quad j \to \infty,$$

where $0 < \rho < 1$. Here ρ may be interpreted as a 'signal/(signal+noise)' ratio, meaning that the limiting value ρ can be very small if the relative amount of chaff is very large, whereas it will be close to 1 if the fraction of chaff is negligible.

Nevertheless, no matter how small ρ might be, *any* nonzero value for ρ will be detectable, at least for sufficiently long-lived connections. Indeed, for large enough n, it will be clear that the empirical fluctuations in correlations due to statistical sampling effects are simply too small to cause observed values of $Corr(j)$ which are substantially nonzero. Given more space, we would provide a detailed analysis of this effect, showing that the mathematics predicts a substantial correlation between wavelet coefficients of N_1 and of N_2. The analysis is entirely parallel to the analysis given in earlier sections.

In short, although certainly the presence of chaff causes a more complex problem, the statistical tools and diagnostic approaches suggested for the no-chaff case seem to be equally applicable to the chaff case.

9 Discussion

This paper has considered basic 'proof of concept' issues. In particular, we have not discussed here the systems-level issues of working in a setting where there may be many hundreds of active Telnet or SSH connections into and out of a large site (say, a university or corporate network), and it is required to monitor and detect stepping stones in real time. A typical issue would be to consider as a monitoring interval a specific length of time (e.g. 4 minutes), and calculate the proper height of the threshold for the 'stepping stone alarm' in order that in looking at thousands of pairs of connections which are truly uncorrelated, we control the false alarm rate to a tolerable number of false alarms per day. Obviously, it would be very important to study such issues carefully.

We have discussed here the fact that real interactive sessions seem to have inter-keystroke times whose distribution is Pareto in the upper tail. In the analysis section of this paper we have considered Poisson streams, which are easy to analyze. In the appendix, we show that the analysis can generalize to other streams. This points out that an accurate theoretical model for inter-keystroke timing is in order, so that we can focus attention and develop mathematical analysis associated with that model. Such a correct model would be extremely useful in practical terms, for example in systems-level work where it could used for false alarm calibration purposes.

Two particular components of the inter-keystroke timing model which should be considered more closely: (a) the correlation structure of adjacent/nearby inter-keystroke times; and (b) the chance of seeing many characters in a very short interval. The significance of these can be gleaned from the discussion in the appendix. Knowing more about either or both components would help mathematical analysis and simulation accuracy.

There are also other sources of information that we haven't discussed – the key one being the two-way nature of interactive sessions. There is far more information than just the keystrokes on the forward path through the stepping stones, there are also the echoes and command output on the reverse path, and it should be possible to use information about these to substantially improve detection.

Acknowledgments. DLD and AGF would like to thank NSF ANI-008584 (ITR). AGF would like to thank the Statistics Department of UC Berkeley for its hospitality. JC and SS would like to thank DARPA contract N66001-00-C-8045.

References

1. Aldous, D.L.: Probability Approximations Via the Poisson Clumping Heuristic. Springer-Verlag, New York. January 1989
2. Lindgren, G., Leadbetter, M.R., and Rootzen, H.: Extremes and related properties of stationary sequences and processes. Springer, New York (1983). Russian translation; Nauka: Moscow (1988).
3. Shimomura, T. and Markoff, J.: Takedown. The pursuit and capture of Kevin Mitnick, America's most wanted computer outlaw–by the man who did it. Hyperion. December 1995.
4. Mallat, S.: A Wavelet Tour of Signal Processing. Academic Press. Second Edition, 2000.
5. Meyer, Y.:Wavelets: Algorithms and Applications. SIAM.May 1993
6. Stoll, C.: The Cuckoo's Egg: Tracking a Spy through the Maze of Computer Espionage. Pocket Books. October 2000
7. Staniford-Chen, S. and Heberlein, L.: Holding Intruders Accountable on the Internet. Proceedings of the 1995 IEEE Symposium on Security and Privacy, Oakland, CA (1995)

8. Zhang, Y. and Paxson, V.: Detecting stepping stones. Proceedings of the 9th USENIX Security Symposium, Denver, Colorado, August 2000. *http://www.aciri.org/vern/papers/stepping-sec00.ps.gz*
9. Paxson, V. and Floyd, S.: Wide-Area Traffic: The Failure of Poisson Modeling. IEEE/ACM Transactions on Networking, Vol. 3(3),June 1995, 226–244
10. Wavelab Toolbox for Wavelet Analysis. Requires Matlab. *http://www-stat.stanford.edu/ wavelab*
11. Yoda, K. and Etoh, H.: Finding a Connection Chain for Tracing Intruders, In: Guppens, F.,Deswarte, Y., Gollamann, D. and Waidner, M. (eds): 6th European Symposium on Research in Computer Security - ESORICS 2000 LNCS -1985, Toulouse, France, Oct 2000

Appendix

Explanation of the Bumps/Wiggles Dichotomy

Analysis by "multiscale bumps" provides, as explained above, a collection of multiscale block averages. In other words, the coefficients measure the *rate* of typing of the given stream.

Analysis by "multiscale wiggles" provides, as explained above, a collection of multiscale differences of block averages. In other words, the coefficients measure the *changes* in the rate of typing of the given stream.

It is our opinion that measuring changes in rate, and noticing the times those occur, provides more reliable evidence for the identity of two streams; so we believe that analysis by multiscale wiggles (i.e. what is ordinarily called simply wavelet analysis) will give more reliable information indicating the identity of the two streams.

(There is one other advantage of wavelet analysis: the possibility of developing detectors for non-keystroke conserving schemes which work by the multiplexing of constant-rate chaff together with the original stream. Suppose that two streams differ in that stream 2 contains stream 1 along with characters from an independent chaff source of constant rate (e.g. Poisson with Rate 20 char/sec). It can be shown, by elaborating the point of view here, that the wavelet coefficients at sufficiently long scales will have a dependable correlation < 1, but which is stable and nonzero, and determined by a kind of statistical signal/chaff ratio. So we might notice that two streams which should be completely uncorrelated actually exhibit correlations which are definitely nonzero).

The different normalization of the wavelet coefficients in the two cases has to do with the appropriate means of interpretation of each type of coefficient. Averages are directly interpretable in the units of the phenomenon being measured, no matter what the scale of the average. Differences are not so universally interpretable; the convention $p = 1/2$ ensures that they are normalized according to the square-root of interval size rather than interval size. The rationale is that for typical point processes, the coefficients at different scales will then be of comparable size.

Generalization to Non-poisson Streams

The reader will note that only in two places in the argument of Section 6 did we use the Poisson process assumption

The first was

$$\max\{N_1(t+\Delta)-N_1(t):t,t+\Delta\in[a,b]\}\le O_P(\log(b-a))\cdot E\{N_1(t+\Delta)-N_1(t)\}.$$

This condition says that *within any maximum tolerable delay interval, we are very unlikely to see character counts dramatically greater than the average character counts.* This inequality is extremely easy to satisfy and many point processes will obey it. It is also the case that real data will obey it. We might for example stipulate that no actual human person is ever going to exceed an absolute maximum of K characters in Δ no matter how long we wait. If we do, the above inequality will automatically be true, because $\log(b-a)$ grows unboundedly with observation period, while K is an absolute constant.

Incidentally, the Pareto nature of the upper half of the inter-keystroke timing distribution described in [9] is entirely compatible with this inequality. Indeed, the Pareto upper tail is responsible for occasional long dead spots in a stream, where no characters emerge. It is the lower tail – near zero inter-keystroke spacing – that determines whether the needed condition holds. The Poisson assumption makes the inter-keystroke distribution have a density $e^{-t/\lambda}/\lambda$ which is of course bounded near $t=0$; this boundedness implies that there will not typically be large numbers of events in a short time. It seems safe to say that this aspect of the Poisson distribution accurately models real streams of keystrokes.

The second fact used was (in, say, the multiscale block averages case)

$$Var[N_1(0,T)]\asymp Const\cdot T$$

which says that the fluctuation in the number of events per unit time within an interval grows like the square root of the interval size. This will be true for many stationary point processes.

Now, the Pareto nature of the upper half of the inter-keystroke timing distribution described in [9], and the possibility of a non-i.i.d. behavior of inter-keystroke times can modify this inequality, even making the variability grow like a power T^β with $\beta\neq 1$. A more detailed analysis shows that even though the variance scaling exponents could be different, the fundamental behavior of the corresponding terms in the analysis would be the same.

Since our simulations indicate that the multiscale diagnostics work very well in the Pareto case, we omit further discussion of the mathematical details of the extension.

Detecting Malicious Software by Monitoring Anomalous Windows Registry Accesses

Frank Apap, Andrew Honig, Shlomo Hershkop, Eleazar Eskin, and Sal Stolfo

Department of Computer Science
Columbia University, New York NY 10027, USA
{fapap, arh, shlomo, eeskin, sal}@cs.columbia.edu

Abstract. We present a host-based intrusion detection system (IDS) for Microsoft Windows. The core of the system is an algorithm that detects attacks on a host machine by looking for anomalous accesses to the Windows Registry. The key idea is to first train a model of normal registry behavior on a windows host, and use this model to detect abnormal registry accesses at run-time. The normal model is trained using clean (attack-free) data. At run-time the model is used to check each access to the registry in real time to determine whether or not the behavior is abnormal and (possibly) corresponds to an attack. The system is effective in detecting the actions of malicious software while maintaining a low rate of false alarms

1 Introduction

Microsoft Windows is one of the most popular operating systems today, and also one of the most often attacked. Malicious software running on the host is often used to perpetrate these attacks. There are two widely deployed first lines of defense against malicious software, virus scanners and security patches. Virus scanners attempt to detect malicious software on the host, and security patches are operating systems updates to fix the security holes that malicious software exploits. Both of these methods suffer from the same drawback. They are e.ective against known attacks but are unable to detect and prevent new types of attacks.

Most virus scanners are signature based meaning they use byte sequences or embedded strings in software to identify certain programs as malicious [10, 24]. If a virus scanner's signature database does not contain a signature for a specific malicious program, the virus scanner can not detect or protect against that program. In general, virus scanners require frequent updating of signature databases, otherwise the scanners become useless [29]. Similarly, security patches protect systems only when they have been written, distributed and applied to host systems. Until then, systems remain vulnerable and attacks can and do spread widely.

In many environments, frequent updates of virus signatures and security patches are unlikely to occur on a timely basis, causing many systems to remain vulnerable. This leads to the potential of widespread destructive attacks caused by malicious software. Even in environments where updates are more frequent,

A. Wespi, G. Vigna, and L. Deri (Eds.): RAID 2002, LNCS 2516, pp. 36–53, 2002.

the systems are vulnerable between the time new malicious software is created and the time that it takes for the software to be discovered, new signatures and patches created by experts, and the ultimate distribution to the vulnerable systems. Since malicious software may propagate through email, often the malicious software can reach the vulnerable systems long before the updates are available.

A second line of defense is through IDS systems. Host-based IDS systems monitor a host system and attempt to detect an intrusion. In the ideal case, an IDS can detect the effects or behavior of malicious software rather then distinct signatures of that software. Unfortunately, the commercial IDS systems that are widely in use are based on signature algorithms. These algorithms match host activity to a database of signatures which correspond to known attacks. This approach, like virus detection algorithms, require previous knowledge of an attack and is rarely effective on new attacks. Recently however, there has been growing interest in the use of data mining techniques such as anomaly detection, in IDS systems [23,25]. Anomaly detection algorithms build models of normal behavior in order to detect behavior that deviates from normal behavior and which may correspond to an attack [9,12]. The main advantage of anomaly detection is that it can detect new attacks and can be an effective defense against new malicious software. Anomaly detection algorithms have been applied to network intrusion detection [12,20,22] and also to the analysis of system calls for host based intrusion detection [13,15,17,21,28]. There are two problems to the system call approach to host based IDS which inhibits their use in actual deployment. The first is that the computational overhead of monitoring all system calls is very high, which degrades the performance of a system. The second is that system calls themselves are irregular by nature, which makes it difficult to differentiate between normal and malicious behaviors, which may cause a high false positive rate.

In this paper, we examine a new approach to host IDS that monitors a program's use of the Windows Registry. We present a system called RAD (Registry Anomaly Detection), which monitors the accesses to the registry in real time and detects the actions of malicious software.

The Windows Registry is an important part of the Windows operating system and is very heavily used, making it a good source of audit data. By building a sensor on the registry and applying the information gathered to an anomaly detector, we can detect activity that corresponds to malicious software. The main advantages of monitoring the Windows Registry is that the activity is regular by nature, can be monitored with low computational overhead, and almost all system activities interact with the registry.

Our anomaly detection algorithm is a registry-specific version of PHAD (Packet Header Anomaly Detection), an anomaly detection algorithm originally presented to detect anomalies in packet headers [25]. We show that the data generated by a registry sensor is useful in detecting malicious behavior. We shall describe how various malicious programs use the registry, and what data can be gathered from the registry to detect these malicious activities. We then apply an anomaly detection algorithm to this data to detect abnormal registry behavior which corresponds to the actions of malicious software. By showing the results of an experiment and detailing how various malicious activities use the registry, we show that the registry is a good source of data for intrusion detection. The paper will also discuss the modifications of the PHAD algorithm as it is applied in the RAD system.

We present results of experiments evaluating the RAD system and demonstrate that it is effective in detecting attacks while maintaining a low rate of false alarms.

2 Modeling Registry Accesses

2.1 The Windows Registry

In Microsoft Windows, the registry file is a database of information about a computer's configuration. The registry contains information that is continually referenced by many different programs. Information stored in the registry includes the hardware installed on the system, which ports are being used, profiles for each user, configuration settings for programs, and many other parameters of the system. It is the main storage location for all configuration information for many Window programs. The Windows Registry is also the source for all security information: policies, user names, and passwords. The registry also stores much of the important run-time configuration information that programs need to run.

The registry is organized hierarchically as a tree. Each entry in the registry is called a key and has an associated value. One example of a registry key is

```
HKCU\Software\America Online\AOL Instant Messenger (TM)
\CurrentVersion\Users\aimuser\Login\Password
```

This is a key used by the AOL instant messenger program. This key stores an encrypted version of the password for the user name `aimuser`. Upon start up the AOL instant messenger program queries this key in the registry in order to retrieve the stored password for the local user. Information is accessed from the registry by individual registry accesses or queries. The information associated with a registry query is the key, the type of query, the result, the process that generated the query and whether the query was successful. One example of a query is a read for the key shown above. For example, the record of the query is:

Process: `aim.exe`
Query: `QueryValue`
Key: `HKCU\Software\America Online\AOL Instant Messenger (TM)\CurrentVersion\Users\aimuser\Login\Password`
Response: `SUCCESS`
ResultValue: `" BCOFHIHBBAHF"`

The Windows Registry is an effective data source to monitor attacks because many attacks show up as anomalous registry behavior. Many attacks take advantage of Windows' reliance on the registry. Indeed, many attacks themselves rely on the Windows Registry in order to function properly.

Many programs store important information in the Registry, notwithstanding the fact that other programs can arbitrarily access the information. Although some versions of Windows include security permissions and Registry logging, both features are rarely used (because of the computational overhead and the complexity of the configuration options).

2.2 Analysis of Malicious Registry Accesses

Most Windows programs access a certain set of Registry keys during normal execution. Furthermore, most users use a certain set of programs routinely while running their machines. This may be a set of all programs installed on the machine or more typically a small subset of these programs. Another important characteristic of Registry activity is that it tends to be regular over time. Most programs either only access the registry on start-up and shutdown, or access the registry at specific intervals. This regularity makes the registry an excellent place to look for irregular, anomalous activity, since a malicious program may substantially deviate from normal activity and can be detected.

Many attacks involve launching programs that have never been launched before and changing keys that have not been changed since the operating system had first been installed by the manufacturer. If a model of the normal registry behavior is computed over clean data, then these kinds of registry operations will not appear in the model. Furthermore malicious programs may need to query parts of the registry to get information about vulnerabilities. A malicious program can also introduce new keys that will help create vulnerabilities in the machine.

Some examples of malicious programs and how they produce anomalous registry activity are described below.

- **Setup Trojan:** This program when launched adds full read/write sharing access on the file system of the host machine. It makes use of the registry by creating a registry structure in the networking section of the Windows keys. The structure stems from `HKLM\Software\Microsoft\Windows\CurrentVersion\Network\LanMan`. It then creates typically eight new keys for its own use. It also accesses `HKLM\Security\Provider` in order to find information about the security of the machine to help determine vulnerabilities. This key is not accessed by any normal programs during training or testing in our experiments and its use is clearly suspicious in nature.
- **Back Orifice 2000:** This program opens a vulnerability on a host machine, which grants anyone with the back orifice client program complete control over the host machine. This program does make extensive use of the registry, however, it uses a key that is very rarely accessed on the Windows system.

HKLM\Software\Microsoft\VBA\Monitors was not accessed by any normal programs in either the training or test data, which allowed our algorithm to determine it as anomalous. This program also launches many other programs (LoadWC.exe, Patch.exe, runonce.exe, bo2k_1_o_intl.e) as part of the attack all of which made anomalous accesses to the Windows Registry.
- **Aimrecover:** This is a program that steals passwords from AOL users. It's actually a very simple program that simply reads the keys from the registry where the AOL Instant Messenger program stores the user names and passwords. The reason that these accesses are anomalous is because Aimrecover is accessing a key that usually is accessed by a different program that created that key.
- **Disable Norton:** This is a very simple exploitation of the registry that disables Norton Antivirus. This attack toggles one record in the registry, the key HKLM\SOFTWARE\INTEL\LANDesk\VirusProtect6\CurrentVersion\Storages\Files\System\RealTimeScan\OnOff. If this value is set to 0 then Norton Antivirus real time system monitoring is turned off. Again this is anomalous because of its access to a key that was created by a different program.
- **L0phtCrack:** This program is probably the most popular password cracking program for Windows machines. It retrieves the hashed SAM file containing the passwords for all users and then uses either a dictionary or brute force approach to find the passwords. This program also uses flaws in the Windows encryption scheme which allows the program to discover some of the characters in the password. This program uses the registry by creating its own section in the registry. This will consist of many create key and set value queries, all of which will be on keys that did not exist previously on the host machine and therefore have not been seen before.

Another important piece of information that can be used in detecting attacks, all programs observed in our data set, and presumably all programs in general, cause Windows Explorer to access a specific key. The key

HKLM\Software\Microsoft\Windows NT\CurrentVersion\Image File Execution Options\processName

where processName is the name of the process being executed, is a key that is accessed by Explorer each time an application is run. Therefore we have a reference point for each specific application being launched to determine malicious activity. In addition many programs add themselves in the auto-run section of the Windows Registry under

HKLM\Software\Microsoft\Windows \CurrentVersion\Run .

While this is not malicious in nature, this is a rare event that can definitely be used as a hint that a system is being attacked. Trojan programs such as Back Orifice utilize this part of the registry to auto load themselves on each boot.

Anomaly detectors do not look for malicious activity directly. They look for deviations from normal activity. It is for this reason that any deviation from normal activity will be declared an attack by the system. The installation of a new program on a system will be viewed as anomalous activity. Programs often create new sections of the registry and many new keys on installation. This will cause a false alarm, much like adding a new machine to a network may cause an alarm on an anomaly detector that analyzes network traffic. There are a few possible solutions to this problem. Malicious programs are often stealthy and install quietly so that the user does not know the program is being installed. This is not the case with most user initiated (legitimate) application installations that make themselves (loudly) known. The algorithm could be modified to ignore alarms while the install shield program was running because that would mean that the user is aware that a new program is being installed. Another option is to simply prompt the user when a detection occurs so that the user can let the anomaly detection system know that a legitimate installed program is under way and that therefore the anomaly detection model needs to be updated with a newly available training set gathered in real time. This is a typical user interaction in many application installations where user feedback is requested for configuration information.

3 Registry Anomaly Detection

The RAD system has three basic components: an audit sensor, a model generator, and an anomaly detector. The sensor logs each registry activity to either a database where it is stored for training, or to the detector to be used for analysis. The model generator reads data from the database and creates a model of normal behavior. The model is then used by the anomaly detector to decide whether each new registry access should be considered anomalous.

In order to detect anomalous registry accesses, RAD generates a model of normal registry activity. A set of five features are extracted from each registry access. Using these feature values over normal data, a model of normal registry behavior is generated. This model of normalcy consists of a set of consistency checks applied to the features. When detecting anomalies, the model of normalcy determines whether the values in features of the current registry access are consistent with the normal data or not. If new activity is not consistent, the algorithm labels the access as anomalous.

3.1 RAD Data Model

The RAD data model consists of five features directly gathered from the registry sensor. The five raw features used by the RAD system are as follows.

- **Process:** This is the name of process accessing the registry. This is useful because it allows the tracking of new processes that did not appear in the training data.
- **Query:** This is the type of query being sent to the registry, for example, `QueryValue`, `CreateKey`, and `SetValue` are valid query types. This allows the identification of query types that have not been seen before. There are many query types but only a few are used under normal circumstances.

- **Key:** This is the actual key being accessed. This allows our algorithm to locate keys that are never accessed in the training data. Many keys are used only once for special situations like system installation. Some of these keys can be used to create vulnerabilities.
- **Response:** This describes the outcome of the query, for example `success`, `not found`, `no more`, `buffer overflow`, and `access denied`.
- **Result Value:** This is the value of the key being accessed. This will allow the algorithm to detect abnormal values being used to create abnormal behavior in the system.

Table 1. Registry Access Records. Two registry accesses are shown. The first is a normal access by AOL Instance Messenger to the key where passwords are stored. The second is a malicious access by AIMrecover to the same key. The final column shows which fields register as anomalous. Note that the pairs of features must be used to detect the anomalous behavior of `AIMrecover.exe`. This is because under normal circumstances only `AIM.exe` accesses the key that stores the AIM password. Another process accessing this key generates an anomaly.

Feature	aim.exe	aimrecover.exe
Process	aim.exe	aimrecover.exe
Query	QueryValue	QueryValue
Key	HKCU\Software\America Online \AOL Instant Messenger (TM) \CurrentVersion\Users \aimuser\Login\Password	HKCU\Software\America Online \AOL Instant Messenger (TM) \CurrentVersion\Users \aimuser\Login\Password
Response	SUCCESS	SUCCESS
Result Value	" BCOFHIHBBAHF"	" BCOFHIHBBAHF"

3.2 RAD Anomaly Detection Algorithm

Using the features that we monitor from each registry access, we train a model over features extracted from normal data. That model allows us to classify registry accesses as either normal or malicious.

Any anomaly detection algorithm can be used to perform this modeling. Since we aim to monitor a significant amount of data in real time, the algorithm must be very efficient. We apply a probabilistic algorithm described in Eskin, 2002 [14] and here we provide a short summary of the algorithm. The algorithm is similar to the heuristic algorithm that was proposed by Chan and Mahoney in the PHAD system [25], but is more robust.

In general, a principled probabilistic approach to anomaly detection can be reduced to density estimation. If we can estimate a density function p(x) over the normal data, we can define anomalies as data elements that occur with low probability. In practice, estimating densities is a very hard problem (see the discussion in Schölkopf et al., 1999 [26] and the references therein.) In our setting, part of the problem is that each of the features have many possible values. For example, the Key feature has over 30, 000 values in our training set. Since there are so many possible feature values relatively rarely does the same exact record occur in the data. Data sets with this characterization are referred to as sparse.

Since probability density estimation is a very hard problem over sparse data, we propose a different method for determining which records from a sparse data set are anomalous. We define a set of consistency checks over the normal data. Each consistency check is applied to an observed record. If the record fails any consistency check, we label the record as anomalous.

We apply two kinds of consistency checks. The first consistency check evaluates whether or not a feature value is consistent with observed values of that feature in the normal data set. We refer to this type of consistency check as a first order consistency check. More formally, each registry record can be viewed as the outcome of 5 random variables, one for each feature, X_1, X_2, X_3, X_4, X_5. Our consistency checks compute the likelihood of an observation of a given feature which we denote $P(X_i)$.

The second kind of consistency check handles pairs of features as motivated by the example in Table 1. For each pair of features, we consider the conditional probability of a feature value given another feature value. These consistency checks are referred to as second order consistency checks. We denote these likelihoods $P(X_i|X_j)$. Note that for each value of X_j, there is a different probability distribution over X_i.

In our case, since we have 5 feature values, for each record, we have 5 first order consistency checks and 20 second order consistency checks. If the likelihood of any of the consistency checks is below a threshold, we label the record as anomalous.

What remains to be shown is how we compute the likelihoods for the first order $(P(X_i))$ and second order $(P(X_i|X_j))$ consistency checks. Note that from the normal data, we have a set of observed counts from a discrete alphabet for each of the consistency checks. Computing these likelihoods reduces to simply estimating a multinomial. In principal we can use the maximum likelihood estimate which just computes the ratio of the counts of a particular element to the total counts. However, the maximum likelihood estimate is biased when there is relatively small amounts of data. When estimating sparse data, this is the case. We can smooth this distribution by adding a virtual count to each possible element. This is equivalent to using a Dirichlet estimator [11]. For anomaly detection, as pointed out in Mahoney and Chan, 2001 [25], it is critical to take into account how likely we are to observe an unobserved element. Intuitively, if we have seen many different elements, we are more likely to see unobserved elements as opposed to the case where we have seen very few elements.

To estimate our likelihoods we use the estimator presented in Friedman and Singer, 1999 [16] which explicitly estimates likelihood of observing a previously unobserved element. The estimator gives the following prediction for element i

$$P(X = i) = \frac{\alpha + N_i}{k^0\alpha + N} C \tag{1}$$

if element i was observed and

$$P(X = i) = \frac{1}{L - k^0}(1 - C) \tag{2}$$

if element i was not previously observed. α is a prior count for each element, N_i is the number of times i was observed, N is the total number of observations, k^0 is the number of different elements observed, and L is the total number of possible elements or the alphabet size. The scaling factor C takes into account how likely it is to observe a previously observed element versus an unobserved element. C is computed by

$$C = \left(\sum_{k=k^0}^{L} \frac{k^0\alpha + N}{k\alpha + N} m_k \right) \left(\sum_{k \geq k^0} m_k \right)^{-1} \tag{3}$$

where $m_k = P(S = k)\frac{k!}{(k-k^0)!}\frac{\Gamma(k\alpha)}{\Gamma(k\alpha+N)}$ and $P(S = k)$ is a prior probability associated with the size of the subset of elements in the alphabet that have nonzero probability. Although the computation of C is expensive, it only needs to be done once for each consistency check at the end of training.

The prediction of the probability estimator is derived using a mixture of Dirichlet estimators each of which represent a different subset of elements that have non-zero probability. Details of the probability estimator and its derivation are given in [16] and complete details of the anomaly detection algorithm are given in [14].

Note that this algorithm labels every registry access as either normal or anomalous. Programs can have anywhere from just a few registry accesses to several thousand. This means that many attacks will be represented by large numbers of records where many of those records will be considered anomalous.

Some records are anomalous because they have a value for a feature that is inconsistent with the normal data. However, some records are anomalous because they have an inconsistent combination of features although each feature itself may be normal. Because of this, we examine pairs of features. For example, let us consider the registry access displayed in Table 1. The basic features for the normal program `aim.exe` versus the malicious program `aimrecover.exe` do not appear anomalous. However, the fact that the program `aimrecover.exe` is accessing a key that is usually associated with `aim.exe` is in fact an anomaly. Only by examining the combination of the two raw features can we detect this anomaly.

4 Architecture

The basic architecture of the RAD system consists of three components, the registry auditing module (RegBAM), the model generator, and the real-time anomaly detector. An overview of the RAD architecture is shown in Figure 1.

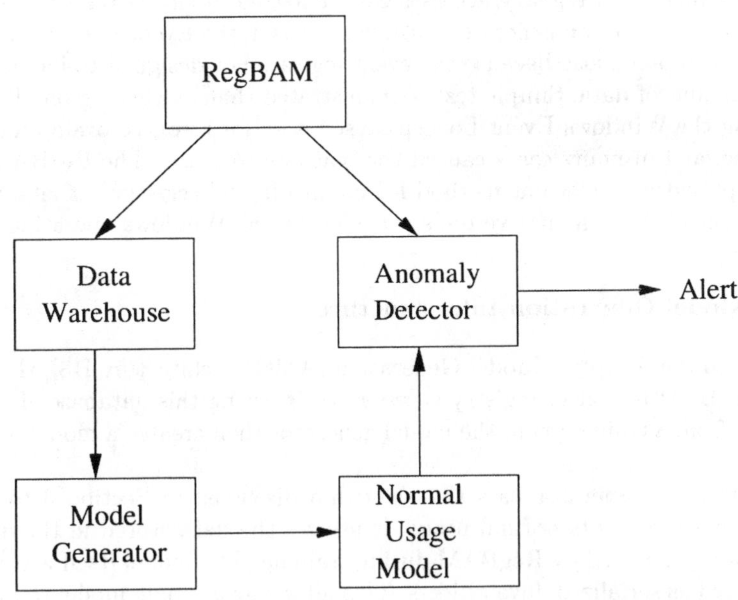

Fig. 1. The RAD System Architecture. RegBAM outputs to the data warehouse during training model and to the anomaly detector during detection mode.

4.1 Registry Basic Auditing Module

The RAD sensor is composed of a Basic Auditing Module (BAM) for the RAD system which monitors accesses to the registry. BAMs implement an architecture and interface for sensors across the system. They include a hook into the audit stream (in this case the registry) and various communication and data-buffering components. BAMs use an XML data representation similar to the IDMEF standard (of the IETF) for IDS systems [19]. BAMs are described in more detail in [18].

The Registry BAM (RegBAM) runs in the background on a Windows machine as it gathers information on registry reads and writes. RegBAM uses Win32 hooks to tap into the registry and log all reads and writes to the registry. RegBAM is akin to a wrapper and uses a similar architecture to that of SysInternal's Regmon [27]. After gathering the registry data, RegBAM can be configured for two distinct purposes. One use is as the audit data source for model generation.

When RegBAM is used as the data source, the output data is sent to a database where it is stored and later used by the model generator described in Section 4.2 [18]. The second use of RegBAM, is as the data source for the real-time anomaly detector described in Section 4.3. While in this mode, the output of RegBAM is sent directly to the anomaly detector where it is processed in real time. An alternative method to collect the registry accesses is to use the Windows auditing mechanism. All registry accesses can be logged in the Windows Event Log. Each read or write can generate multiple records in the Event Log. However, this method is problematic because the event logs are not designed to handle such a large amount of data. Simple tests demonstrated that by turning on all registry auditing the Windows Event Logger caused a major resource drain on the host machine, and in many cases caused the machine to crash. The RegBAM application provides an efficient method for monitoring all registry activity, with far less overhead than the native tools provided by the Windows operating system.

4.2 Model Generation Infrastructure

Similar to the Adaptive Model Generation (AMG) architecture [18], the system uses RegBAM to collect registry access records. Using this database of collected records from a training run, the model generator then creates a model of normal usage.

The model generator uses the algorithm discussed in Section 3 to build a model that represents normal usage. It utilizes the data stored in the database which was generated by RegBAM during training. The model itself is comprised and stored as serialized Java objects. This allows for a single model to be generated and to be easily distributed to additional machines. Having the model easily deployed to new machines is a desirable feature, since in a typical network, many Windows machines have similar usage patterns. This allows the same model to be used for multiple host machines.

4.3 Real-Time Anomaly Detector

For real time detection, RegBAM feeds live data for analysis by an anomaly detector. The anomaly detector will load the normal usage model created by the model generator and begin reading each record from the output data stream of RegBAM. The algorithm discussed in Section 3 is then applied against each record of registry activity. The score generated by the anomaly detection algorithm is compared by a user configurable threshold to determine if the record should be considered anomalous. A list of anomalous registry accesses are stored and displayed as part of the detector. A user configured threshold allows the user to customize the alarm rate for the particular environment. Lowering the threshold, will result in more alarms being issued. Although this can raise the false positive rate, it can also increase the chance of detecting new attacks.

4.4 Efficiency Considerations

In order for a system to detect anomalies in a real time environment it can not consume excessive system resources. This is especially important in registry attack detection because of the heavy amount of traffic that generated by applications interacting with the registry. While the amount of traffic can vary greatly from system to system, in our experimental setting (described below) the traffic load was about 50,000 records per hour. Our distributed architecture is designed to minimize the resources used by the host machine. It is possible to spread the work load on to several separate machines, so that the only application running on the host machine is the lightweight RegBAM. However this will increase network load due to the communication between components. These two concerns can be used to configure the system to create the proper proportion between host system load and network load. The RegBAM module is a far more efficient way of gathering data about registry activity than full auditing with the Windows Event Log.

5 Evaluation and Results

The system was evaluated by measuring the detection performance over a set of collected data which contains some attacks. Since there are no other existing publicly available detection systems that operate on Windows registry data we were unable to compare our performance to other systems directly.

5.1 Data Generation

In order to evaluate the RAD system, we gathered data by running a registry sensor on a host machine. Since there are no publicly available data sets containing registry accesses, we collected our own data. Beyond the normal execution of standard programs, such as Microsoft Word, Internet Explorer, and Winzip, the training also included performing housekeeping tasks such as emptying the Recycling Bin and using the Control Panel. All simulations were done by hand to simulate a real user. All data used for this experiment is publicly available online in text format at `http://www.cs.columbia.edu/ids/rad`. The data includes a time stamp and frequency of the launched programs in relation to each other.

The training data collected for our experiment was collected on Windows NT 4.0 over two days of normal usage (in our lab). We informally de.ne "normal" usage to mean what we believe to be typical use of a Windows platform in a home setting. For example, we assume all users would log in, check some internet sites, read some mail, use word processing, then log off. This type of session is assumed to be relatively "typical" of many computer users. Normal programs are those which are bundled with the operating systems, or are in use by most Windows users. Creating realistic testing environments is a very hard task and testing the system under a variety of environments is a direction for future work.

The simulated home use of Windows generated a clean (attack-free) dataset of approximately 500,000 records. The system was then tested on a full day of test data with embedded attacks executed. This data was comprised of approximately 300,000 records most of which were normal program executions interspersed with attacks. The normal programs run between attacks were intended to simulate an ordinary Windows session. The programs used were Microsoft Word, Outlook Express, Internet Explorer, Netscape, AOL Instant Messenger, and others.

The attacks run include publicly available attacks such as aimrecover, browslist, bok2ss (back orifice), `install.exe` `xtxp.exe` both for backdoor.XTCP, l0phtcrack, runattack, whackmole, and setuptrojan. Attacks were only run during the one day of testing throughout the day. Among the twelve attacks that were run, four instances were repetitions of the same attack. Since some attacks generated multiple processes there are a total of seventeen distinct processes for each attack. All of the processes (either attack or normal) as well as the number of registry access records in the test data is shown in Table 3.

The reason for running some of the attacks twice, was to test the effectiveness of our system. Many programs act differently when executed a second time within a windows session. In the experiments reported below our system was less likely to detect a previously successful attack on the second execution of that attack. The reason is that a successful attack creates permanent changes to the registry and hence on subsequent queries the attack no longer appears irregular. Thus the next time the same attack is launched it is more difficult to detect since it interacts less with the registry.

We observed that this is common for both malicious and regular applications since many applications will do a much larger amount of registry writing during installation or when first executed.

5.2 Experiments

The training and testing environments were set up to replicate a simple yet realistic model of usage of Windows systems. The system load and the applications that were run were meant to resemble what one may deem typical in normal private settings.

We trained the anomaly detection algorithm presented in Section 3 over the normal data and evaluated each record in the testing set. We evaluate our system by computing two statistics. We compute the *detection rate and the false positive rate.*

The normal way to evaluate the performance of RAD would be to measure detection performance over processes labeled as either normal or malicious. However, with only seventeen malicious processes at our disposal in our test set, it is difficult to obtain a robust evaluation for the system. We do discuss the performance of the system in terms of correctly classified processes, but also measure the performance in terms of the numbers of records correctly and incorrectly classified. Future work on RAD will focus on testing over long periods of time to measure significantly more data and process classifications as well as alternative means of alarming on processes. (For example, a process may be declared an attack on the basis of one anomalous record it generates, or perhaps on some number of anomalous records.) There is also an interesting issue to be investigated regarding the decay of the anomaly models that may be exhibited over time, perhaps requiring regenerating a new model.

The detection rate reported below is the percentage of records generated by the malicious programs which are labeled correctly as anomalous by the model. The false positive rate is the percentage of normal records which are mislabeled anomalous. Each attack or normal process has many records associated with it. Therefore, it is possible that some records generated by a malicious program will be mislabeled even when some of the records generated by the attack are accurately detected. This will occur in the event that some of the records associated with one attack are labeled normal. Each record is given an anomaly score, S, that is compared to a user defined threshold. If the score is greater than the threshold, then that particular record is considered malicious. Fig. 2 shows how varying the threshold affects the output of the detector. The actual recorded scores plotted in the Figure are displayed in Table 2.

Table 2. Varying the threshold score and its effect on False Positive Rate and Detection Rate.

Threshold Score	False Positive Rate	Detection Rate
6.847393	0.001192	0.005870
6.165698	0.002826	0.027215
5.971925	0.003159	0.030416
5.432488	0.004294	0.064034
4.828566	0.005613	0.099253
4.565011	0.006506	0.177161
3.812506	0.009343	0.288687
3.774119	0.009738	0.314301
3.502904	0.011392	0.533084
3.231236	0.012790	0.535219
3.158004	0.014740	0.577908
2.915094	0.019998	0.578442
2.899837	0.020087	0.627001
2.753176	0.033658	0.629136
2.584921	0.034744	0.808431
2.531572	0.038042	0.869797
2.384402	0.050454	1.000000

Table 3 is sorted in order to show the results for classifying processes. From the table we can see if the threshold is set at 8.497072, we would label the processes `LOADWC.EXE` and `ipccrack.exe` as malicious and would detect the Back Orifice and IPCrack attacks. Since none of the normal processes have scores that high, we would have no false positives. If we lower the threshold to 6.444089, we would have detected several more processes from Back Orifice and the BrowseList, BackDoor.xtcp, SetupTrojan and AimRecover attacks. However, at this level of threshold, the following processes would be labeled as false positives: `systray.exe`, `CSRSS.EXE`, `SPOOLSS.EXE`, `ttssh.exe`, and `winmine.exe`. As we have mentioned, our future work on RAD will model and measure a Windows system for a far longer period of time over many more processes in order to generate a meaningful ROC curve in terms of processes. The measurements reported next are cast in terms of registry query records.

By varying the threshold for the inconsistency scores on records, we were able to demonstrate the variability of the the detection rate and false positive rate. We plot the false positive rate versus the detection rate in an ROC (Receiver Operator Characteristic) curve shown in Figure 2 and Table 2.

Many of the false positives were from processes that were simply not run as a part of the training data but were otherwise normal Windows programs. A thorough analysis of what kinds of processes generate false positives is a direction for future work.

Part of the reason why the system is successfully able to discriminate between malicious and normal records is that accesses to the Windows Registry are very regular, which makes normal registry activity relatively easy to model.

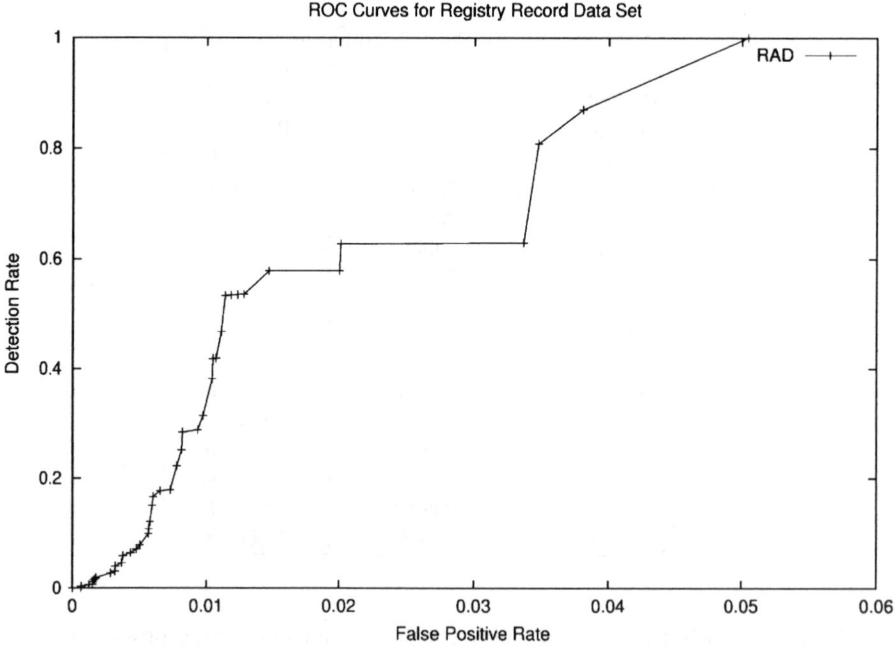

Fig. 2. Figure showing varying the threshold on the data set.

Table 3. Information about all processes in testing data including the number of registry accesses and the maximum and minimum score for each record as well as the classification. The top part of the table shows this information for all of the attack processes and the bottom part of the table shows this information for the normal processes. The reference number (by the attack processes) give the source for the attack. Processes that have the same reference number are part of the same attack. [1] AIMCrack. [2] Back Orifice. [3] Backdoor.xtcp. [4] Browse List. [5] Happy 99. [6] IPCrack. [7] L0pht Crack. [8] Setup Trojan.

Program Name	Number of Records	Maximum Record Value	Minimum Record Value	Classification
LOADWC.EXE[2]	1	8.497072	8.497072	ATTACK
ipccrack.exe[6]	1	8.497072	8.497072	ATTACK
mstinit.exe[2]	11	7.253687	6.705313	ATTACK
bo2kss.exe[2]	12	7.253687	6.62527	ATTACK
runonce.exe[2]	8	7.253384	6.992995	ATTACK
browselist.exe[4]	32	6.807137	5.693712	ATTACK
install.exe[3]	18	6.519455	6.24578	ATTACK
SetupTrojan.exe[8]	30	6.444089	5.756232	ATTACK
AimRecover.exe[1]	61	6.444089	5.063085	ATTACK
happy99.exe[5]	29	5.918383	5.789022	ATTACK
bo2k_1_0_intl.e[2]	78	5.432488	4.820771	ATTACK
_INS0432._MP[2]	443	5.284697	3.094395	ATTACK
xtcp.exe[3]	240	5.265434	3.705422	ATTACK
bo2kcfg.exe[2]	289	4.879232	3.520338	ATTACK
l0phtcrack.exe[7]	100	4.688737	4.575099	ATTACK
Patch.exe[2]	174	4.661701	4.025433	ATTACK
bo2k.exe[2]	883	4.386504	2.405762	ATTACK
systray.exe	17	7.253687	6.299848	NORMAL
CSRSS.EXE	63	7.253687	5.031336	NORMAL
SPOOLSS.EXE	72	7.070537	5.133161	NORMAL
ttssh.exe	12	6.62527	6.62527	NORMAL
winmine.exe	21	6.56054	6.099177	NORMAL
em_exec.exe	29	6.337396	5.789022	NORMAL
winampa.exe	547	6.11399	2.883944	NORMAL
PINBALL.EXE	240	5.898464	3.705422	NORMAL
LSASS.EXE	2299	5.432488	1.449555	NORMAL
PING.EXE	50	5.345477	5.258394	NORMAL
EXCEL.EXE	1782	5.284697	1.704167	NORMAL
WINLOGON.EXE	399	5.191326	3.198755	NORMAL
rundll32.exe	142	5.057795	4.227375	NORMAL
explore.exe	108	4.960194	4.498871	NORMAL
netscape.exe	11252	4.828566	-0.138171	NORMAL
java.exe	42	4.828566	3.774119	NORMAL
aim.exe	1702	4.828566	1.750073	NORMAL
findfast.exe	176	4.679733	4.01407	NORMAL
TASKMGR.EXE	99	4.650997	4.585049	NORMAL
MSACCESS.EXE	2825	4.629494	1.243602	NORMAL
IEXPLORE.EXE	194274	4.628190	-3.419214	NORMAL
NTVDM.EXE	271	4.59155	3.584417	NORMAL
CMD.EXE	116	4.579538	4.428045	NORMAL
WINWORD.EXE	1541	4.457119	1.7081	NORMAL
EXPLORER.EXE	53894	4.31774	-1.704574	NORMAL
msmsgs.exe	7016	4.177509	0.334128	NORMAL
OSA9.EXE	705	4.163361	2.584921	NORMAL
MYCOME 1.EXE	1193	4.035649	2.105155	NORMAL
wscript.exe	527	3.883216	2.921123	NORMAL
WINZIP32.EXE	3043	3.883216	0.593845	NORMAL
notepad.exe	2673	3.883216	1.264339	NORMAL
POWERPNT.EXE	617	3.501078	-0.145078	NORMAL
AcroRd32.exe	1598	3.412895	0.393729	NORMAL
MDM.EXE	1825	3.231236	1.680336	NORMAL
ttermpro.exe	1639	2.899837	1.787768	NORMAL
SERVICES.EXE	1070	2.576196	2.213871	NORMAL
REGMON.EXE	259	2.556836	1.205416	NORMAL
RPCSS.EXE	4349	2.250997	0.812288	NORMAL

6 Conclusions

By using registry activity on a Windows system, we were able to label all processes as either attacks or normal, with relatively high accuracy and low false positive rate, for the experiments performed in this study. We have shown that registry activity is regular, and described ways in which attacks would generate anomalies in the registry. Thus, an anomaly detector for registry data may be an effective intrusion detection system augmenting other host-based detection systems. It would also improve protection of systems in cases of new attacks that would otherwise pass by scanners that have not been updated on a timely basis.

We plan on testing the system under a variety of environments and conditions to better understand its performance. Future plans include combining the RAD system with another detector that evaluates Windows Event Log data. This will allow for various data correlation algorithms to be used to make more accurate system behavior models which we believe will provide a more accurate anomaly detection system with better coverage of attack detection. Part of our future plans for the RAD system include adding data clustering and aggregation capabilities. Aggregating alarms will allow for subsets of registry activity records to be considered malicious as a group initiated from one attack rather than individual attacks. We also plan to store the system registry behavior model as part of the registry itself. The motivation behind this, is to use the anomaly detector to protect the system behavior model from being maliciously altered, hence making the model itself secured against attack. These additions to the RAD system will make the system a more complete and effective tool for detecting malicious behavior on the Windows platform.

References

1. Aim Recovery. `http://www.dark-e.com/des/software/aim/index.shtml`.
2. Back Orifice. `http://www.cultdeadcow.com/tools/bo.html`.
3. BackDoor.XTCP. `http://www.ntsecurity.new/Panda/Index.cfm?FuseAction=Virus&VirusID=659`.
4. BrowseList. `http://e4gle.org/files/nttools/`, `http://binaries.faq.net.pl/security tools`.
5. Happy 99. `http://www.symantex.com/qvcenter/venc/data/happy99.worm.html`.
6. IPCrack. `http://www.geocities.com/SiliconValley/Garage/3755/toolicq.html`, `http://home.swipenet.se/~w-65048/hacks.htm`.
7. L0pht Crack. `http://www.atstack.com/research/lc`.
8. Setup Trojan. `http://www.nwinternet.com/~pchelp/bo/setuptrojan.txt`.
9. V. Barnett and T. Lewis. *Outliers in Statistical Data*. John Wiley and Sons, 1994.
10. Fred Cohen. *A Short Course on Computer Viruses*. ASP Press, Pittsburgh, PA, 1990.
11. M. H. DeGroot. *Optimal Statistical Decisions*. McGraw-Hill, New York, 1970.
12. D. E. Denning. An intrusion detection model. *IEEE Transactions on Software Engineering*, SE-13:222–232, 1987.
13. Eleazar Eskin. Anomaly detection over noisy data using learned probability distributions. In *Proceedings of the Seventeenth International Conference on Machine Learning (ICML-2000)*, 2000.

14. Eleazar Eskin. Probabilistic anomaly detection over discrete records using inconsistency checks. Technical report, Columbia University Computer Science Technical Report, 2002.
15. Stephanie Forrest, S. A. Hofmeyr, A. Somayaji, and T. A. Longstaff. A sense of self for unix processes. pages 120–128. IEEE Computer Society, 1996.
16. N. Friedman and Y. Singer. E.cient bayesian parameter estimation in large discrete domains, 1999.
17. S. A. Hofmeyr, Stephanie Forrest, and A. Somayaji. Intrusion detect using sequences of system calls. *Journal of Computer Security*, 6:151–180, 1998.
18. Andrew Honig, Andrew Howard, Eleazar Eskin, and Salvatore Stolfo. Adaptive model generation: An architecture for the deployment of data minig-based intrusion detection systems. In *Data Mining for Security Applications.* Kluwer, 2002.
19. Internet Engineering Task Force. Intrusion detection exchange format. In `http://www.ietf.org/html.charters/idwg-charter.html`, 2000.
20. H. S. Javitz and A. Valdes. The nides statistical component: Description and justification. Technical report, SRI International, 1993.
21. W. Lee, S. J. Stolfo, and P. K. Chan. Learning patterns from unix processes execution traces for intrusion detection. pages 50–56. AAAI Press, 1997.
22. W. Lee, S. J. Stolfo, and K. Mok. Data mining in work flow environments: Experiences in intrusion detection. In *Proceedings of the 1999 Conference on Knowledge Discovery and Data Mining (KDD-99)*, 1999.
23. Wenke Lee, Sal Stolfo, and Kui Mok. A data mining framework for building intrusion detection models. 1999.
24. MacAfee. Homepage: macafee.com. *Online publication*, 2000. `http://www.mcafee.com`.
25. M. Mahoney and P. Chan. Detecting novel attacks by identifying anomalous network packet headers. Technical Report CS-2001-2, Florida Institute of Technology, Melbourne, FL, 2001.
26. B. Schölkopf, J. Platt, J. Shawe-Taylor, A. J. Smola, and R. C. Williamson. Estimating the support of a high-dimensional distribution. Technical Report 99–87, Microsoft Research, 1999. To appear in *Neural Computation*, 2001.
27. SysInternals. Regmon for Windows NT/9x. *Online publication*, 2000. `http://www.sysinternals.com/ntw2k/source/regmon.shtml`.
28. Christina Warrender, Stephanie Forrest, and Barak Pearlmutter. Detecting intrusions using system calls: alternative data models. pages 133–145. IEEE Computer Society, 1999.
29. Steve R. White. Open problems in computer virus research. In *Virus Bulletin Conference*, 1998.

Undermining an Anomaly-Based Intrusion Detection System Using Common Exploits

Kymie M.C. Tan, Kevin S. Killourhy, and Roy A. Maxion

Dependable Systems Laboratory
Computer Science Department
Carnegie-Mellon University
Pittsburgh, Pennsylvania 15213 USA

Abstract. Over the past decade many anomaly-detection techniques have been proposed and/or deployed to provide early warnings of cyber-attacks, particularly of those attacks involving masqueraders and novel methods. To date, however, there appears to be no study which has identified a systematic method that could be used by an attacker to undermine an anomaly-based intrusion detection system. This paper shows how an adversary can craft an offensive mechanism that renders an anomaly-based intrusion detector blind to the presence of on-going, common attacks. It presents a method that identifies the weaknesses of an anomaly-based intrusion detector, and shows how an attacker can manipulate common attacks to exploit those weaknesses. The paper explores the implications of this threat, and suggests possible improvements for existing and future anomaly-based intrusion detection systems.

1 Introduction

In recent years, a vast arsenal of tools and techniques has been accumulated to address the problem of ensuring the availability, integrity and confidentiality of electronic information systems. Such arsenals, however, are frequently accompanied by equally vast "shadow" arsenals of tools and techniques aimed specifically at subverting the schemes that were designed to provide system security. Although a shadow arsenal can be viewed negatively as a formidable threat to the security of computer systems, it can also be viewed positively as a source of knowledge for identifying the weaknesses of current security tools and techniques in order to facilitate their improvement.

A small part of the security arsenal, and the focus of this work, is the anomaly-based intrusion-detection system. Anomaly-based intrusion-detection systems have sought to protect electronic information systems from intrusions or attacks by attempting to detect deviations from the normal behavior of the monitored system. The underlying assumption is that such deviations may indicate that an intrusion or attack has occurred (or may still be occurring) on the system. Anomaly detection – detecting deviations from normal – is one of two fundamental approaches used in systems that seek to automate the detection of attacks or intrusions; the other approach is signature-based detection. Anomaly

A. Wespi, G. Vigna, and L. Deri (Eds.): RAID 2002, LNCS 2516, pp. 54–73, 2002.

detection is typically credited with a greater potential for addressing security problems such as the detection of attempts to exploit new or unforeseen vulnerabilities (novel attacks), and the detection of abuse-of-privilege attacks, e.g., masquerading and insider misuse [1].

The promise of the anomaly-detection approach and its incorporation into a number of current automated intrusion-detection strategies (e.g., AT&T's ComputerWatch, SRI's Emerald, SecureNet, etc. [1]) underscores the importance of studying how attackers may fashion counter-responses aimed at undermining the effectiveness of anomaly-based intrusion-detection systems. Such studies are important for two reasons:

- to understand how to strengthen the anomaly-based intrusion-detection system by identifying its weaknesses; and
- to provide the necessary knowledge for guiding the design and implementation of a new generation of anomaly-based intrusion detectors that are not vulnerable to the weaknesses of their forebears.

This paper lays out a method for undermining a well-known anomaly-based intrusion-detection system called stide [2], by first identifying the weaknesses of its anomaly-detection algorithm, and then by showing how an attacker can manipulate common attacks to exploit those weaknesses, effectively hiding the presence of those attacks from the detector's purview. Stide was chosen primarily because it is freely available to other researchers via the Internet. Its accessibility not only encourages independent verification and replication of the work performed here, but it also builds on, and contributes to, a large body of previously published work that uses the stide detection mechanism.

To undermine an anomaly-based intrusion detector, an attacker needs to know the three elements described in Table 1. These elements set the framework for the paper.

Table 1. Elements of methodology for undermining.

1.	Detection coverage (specifically, blind spots) of an anomaly detector.
2.	Where and how an attack manifests in sensor data.
3.	How to shift the manifestation from a covered spot to a blind one.

2 Approaches to Undermining Anomaly Detectors

There are two approaches that would most obviously cause an anomaly detector to miss detecting the anomalous manifestation of an attack. The first of these two items describes the approach commonly found in the literature; the second describes the approach adopted by this study.

– modify the normal to look like the attack, i.e., incorporate the attack manifestations into the model of normal behavior; or
– modify the attack to make it appear as normal behavior.

In the intrusion detection literature, the most cited way to undermine an anomaly-based intrusion detection system is to incorporate undesired, intrusive behavior into the training data, thereby falsely representing "normal" behavior [1,9,10]. By including intrusive behavior explicitly into the training data, the anomaly detector is forced to incorporate the intrusive behavior into its internal model of normal and consequently lose the ability to flag future instances of that intrusive behavior as anomalous. Note that the anomaly detector is viewed as that component of an anomaly-based intrusion detection system solely responsible for detecting deviations from normal; it performs no diagnostic activities.

For example, one way to incorporate intrusive behavior into an anomaly detector's model of normal behavior is to exploit the fact that behavior can change over time. Changing behavior requires the anomaly detector to undergo periodic on-line retraining. Should the information system undergo attacks during the retraining process, then the anomaly detector could inadvertently incorporate undesired attack behavior into its model of normal behavior [1]. The failure of an anomaly-based intrusion detector to detect intrusions or attacks can typically be attributed to contaminated training data, or to updating schemes that incorporate new normal behavior too quickly.

Undermining an anomaly-based intrusion detection system by simply incorporating intrusive behavior into its training data is too imprecise and abstract a method as to be practically useful to an attacker. Identifying and accessing the segment, feature or attribute of the data that will be used to train an anomaly detector, and then surreptitiously and slowly introducing the intrusive behavior into the training dataset, may require time, patience and system privileges that may not be available to an attacker. Moreover, such a scheme does not provide the attacker with any guarantees as to whether or not the act of subversion has been, or will be, successful. The incorporation of intrusive behavior into the training data is no guarantee that the anomaly detector will be completely blind to the attack when the attack is actually deployed. The attacker has no knowledge of how the anomaly detector perceives the attack, i.e., how the attack truly manifests in the data, and no knowledge concerning the conditions that may impede or boost the anomaly detector's ability to detect the manifestation of the attack. For example, even if intrusive behavior were to be incorporated into an anomaly detector's model of normal, it is possible that when the attack is actually deployed, it will interact with other conditions in the data environment (conditions that may not have been present during the training phase), causing anomalous manifestations that *are* detectable by the anomaly detector. It is also possible for the anomalous manifestation of an attack to be detectable *only* when the detector uses particular parameter values. These points illustrate why it is necessary to determine precisely what kinds of anomalous events an anomaly detector may or may not be able to detect, as well as the conditions that enable it to do so.

These issues are addressed by determining the coverage of the anomaly detector in terms of anomalies (not in terms of attacks); this forms the basis of the approach. Only by knowing the kinds of anomalies that are or are not detectable by a given anomaly detector is it possible to modify attacks to manifest in ways that are not considered abnormal by a given anomaly detector. The coverage of an anomaly detector serves as a guide for an attacker to know precisely *how* to modify an attack so that it becomes invisible to the detector.

3 Detection Coverage of an Anomaly Detector

Current evaluation techniques attempt to establish the detection coverage of an anomaly-based intrusion detection system with respect to its ability to detect attacks [21,4,5] , but without establishing whether or not the anomalies detected by the system are attributable to the attack. Typically, claims that an anomaly-based intrusion detector is able to detect an attack are based on *assumptions* that the attack must have manifested in a given stream of data, that the manifestation was anomalous, and that the anomaly detector was able to detect that specific kind of anomaly.

The anomaly-based evaluation technique described in this section establishes the detection coverage of stide [2,21] with respect to the types of anomalous manifestations that the detector is able to detect. The underlying assumption of this evaluation strategy is that no anomaly detection algorithm is perfect. Before it can be determined whether an anomaly-based intrusion detector is capable of detecting an attack, it must first be ascertained that the detector is able to detect the anomalous manifestation of the attack.

This section shows how detection coverage (in terms of a coverage map) can be established for stide. For the sake of completeness, and to facilitate a better understanding of the anomaly-based evaluation strategy, the stide anomaly-detection algorithm is described, followed by a description of the anomaly-based evaluation strategy used to establish stide's detection coverage. A description and explanation of the results of the anomaly-based evaluation is given.

3.1 Brief Description of the Stide Anomaly Detector

Stide operates on fixed-length sequences of categorical data. It acquires a model of normal behavior by sliding a detector window of size DW over the training data, storing each DW-sized sequence in a "normal database" of sequences of size DW. The degree of similarity between test data and the model of normal behavior is based on observing how many DW-sized sequences from the test data are identical matches to any sequences from the normal database. The number of mismatches between sequences from the test data and the normal database is noted. The anomaly signal, which is the detector's response to the test data, involves a user-defined parameter known as the "locality frame" which determines the size of a temporally local region over which the number of mismatches is summed up. The number of mismatches occurring within a locality frame is referred to as the locality frame count, and is used to determine the extent to which the test data are anomalous. A detailed description of stide and its origins, can be found in [2,21].

3.2 Evaluation Strategy for Anomaly Detectors

It is not difficult to see that stide will only detect "unusual" or foreign sequences – sequences that do not exist in the normal database. Its similarity metric establishes whether or not a particular sequence exists in a normal database of sequences of the same size. Such a scheme means that any sequence that is foreign to the normal database would immediately be marked as an anomaly. However, this observation alone is not sufficient to explain the anomaly detector's performance in the real world. There are two other significant issues that must be considered before the performance of the anomaly detector can be understood fully. Specifically:

- how foreign sequences actually manifest in categorical data; and
- how the interaction between the foreign sequences and the anomaly detection algorithm affects the overall performance of the anomaly detector.

In order to obtain a clear perspective of these two issues, a framework was established in [12,19] that focused on the architecture and characteristics of anomalous sequences, e.g., foreign sequences. The framework defined the anomalous sequences that a sliding-window anomaly detector like stide would likely encounter, and provided a means to describe the structure of those anomalous sequences in terms of how they may be composed from other kinds of subsequences. The framework also provided a means to describe the interaction between the anomalous sequences and the sliding window of anomaly-detection algorithms like stide. Because the framework established how each anomalous sequence was constructed and composed, it was possible to evaluate the detection efficacy of anomaly detectors like stide on synthetic data with respect to examples of clearly defined anomalous sequences. The results of the evaluation showed the detection capabilities of stide with respect to the various foreign sequences that may manifest in categorical data, and how the interaction between the foreign sequences in categorical data and the anomaly-detection algorithm affected the overall performance of the anomaly detector.

3.3 Stide's Performance Results

The most significant result provided by the anomaly-based evaluation of stide was that there were conditions that caused the detector to be completely blind to a particular kind of foreign sequence that was found to exist (in abundance) in real-world data [19]: a minimal foreign sequence. A minimal foreign sequence is foreign sequence whose proper subsequences all exist in the normal data. Put simply, a minimal foreign sequence is a foreign sequence that contains within it no smaller foreign sequences.

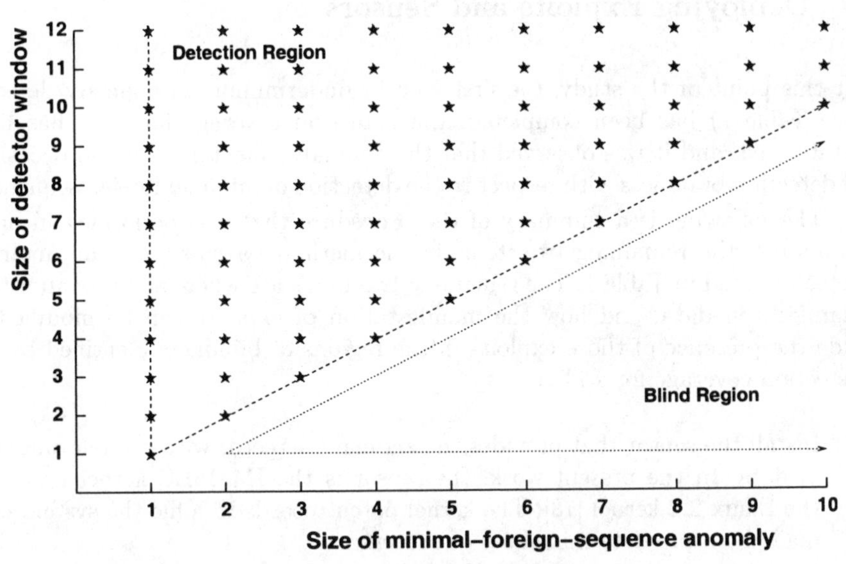

Fig. 1. The detector coverage (detection map) for stide; A comparison of the size of the detector window (rows) with the ability to detect different sizes of minimal foreign sequence (columns). A star indicates detection.

For stide to detect a minimal foreign sequence, it is imperative that the size of the detector window is set to be equal to or larger than the size of the minimal foreign sequence. The consequence of this observation can be seen in Figure 1 which shows stide's detection coverage with respect to the minimal foreign sequence. This coverage map for stide, although previously presented in [19], is shown again here as an aid to the reader's intuition for the coverage map's essential role in the subversion scheme.

The graph in the figure plots the size of the minimal foreign sequence on the x-axis and the size of the detector window on the y-axis. Each star marks the size of the detector window that successfully detected a minimal foreign sequence whose corresponding size is marked on the x-axis. The term *detect* for stide means that the minimal foreign sequence must have caused as at least one sequence mismatch. The diagonal line shows the relationship between the detector window size and the size of the minimal foreign sequence, a relationship that can be described by the function, $y = x$. The figure also shows a region of blindness in the detection capabilities of stide with respect to the minimal foreign sequence. This means that it is possible for a foreign sequence to exist in the data in such a way as to be completely invisible to stide. This weakness will presently be shown to be exploitable by an attacker.

4 Deploying Exploits and Sensors

At this point of the study, the first step in undermining an anomaly detector (see Table 1) has been completed; the detection coverage for stide has been established, and it was observed that the anomaly detector exhibited occasions of detection blindness with respect to the detection of minimal foreign sequences.

The following is a summary of the procedure that was performed in order to address the remaining two items in the method for subverting an anomaly detector listed in Table 1. The remaining two items are where and how an attack manifests in data, and how the manifestation of exploits can be modified to hide the presence of those exploits in the regions of blindness identified by the detection coverage for stide.

1. Install the sensor that provides the anomaly detector with the relevant type of data. In the present work, the sensor is the IMMSEC kernel patch for the Linux 2.2 kernel [18]. The kernel patch records to a file the system calls made by a pre-determined set of processes.
2. Download the `passwd` and `traceroute` exploits and determine the corresponding system programs that these exploits misuse.
3. Execute the system program under normal conditions to obtain a record of normal usage, to obtain normal data. An account of what is considered normal conditions and normal usage of the system programs that correspond to both exploits is described in section 5.2.
4. Deploy the exploits against the host system to obtain the data recording the occurrence of the attacks.
5. Identify the precise manifestation of the attacks in the sensor data.
6. Using the normal data obtained from step 3, and the intrusive data obtained from step 4, deploy stide to determine if the anomaly detector is capable of detecting the unmodified exploits that were simply downloaded, compiled and executed. This is performed in order to establish the effectiveness of the subversion process. If stide is able to detect the unmodified exploits but not the modified exploits, then the subversion procedure has been effective.
7. Using information concerning the kind of events that stide is blind to, modify the attacks and show that it is possible to make attacks that were once detectable by stide, undetectable for detector window sizes one through six.

5 Where and How an Attack Manifests in the Data

This section addresses the second item in the list of requirements for undermining an anomaly detector – establishing where and how an attack manifests in sensor data (see Table 1) – by selecting two common exploits, deploying them, and establishing how and where they manifest in the sensor data. Steps 2 to 5 of the method laid out above are covered by this section.

5.1 Description and Rationale for the Exploits Chosen

The attacks selected for this study are examples of those that stide is designed to detect, i.e., attacks that exploit privileged UNIX system programs. UNIX system programs typically run with elevated privileges in order to perform tasks that require the authority of the system administrator – privileges that ordinary users are not typically afforded. The authors of stide have predominantly applied the detector towards the detection of abnormal behavior in such privileged system programs, because exploiting vulnerabilities to misuse privileged system programs can potentially bestow those extra privileges on an attacker [6].

Two attacks were chosen arbitrarily out of several that fulfill the requirement of exploiting UNIX system programs. The two attacks chosen will be referred to as the `passwd` and `traceroute` exploits. The `passwd` exploit takes advantage of a race condition between the Linux kernel and the `passwd` system program; the `traceroute` exploit takes advantage of a vulnerability in the `traceroute` system program.

`passwd` is a system program used to change a user's password [3]. The program allows an ordinary user to provide his or her current password, along with a new password. It then updates a system-wide database of the user's information so that the database contains the new password. The system-wide database is commonly referred to as the `/etc/passwd` or the `/etc/shadow` file. A user does not normally have permission to edit this file, so `passwd` must run with root privileges in order to modify that file. The exploit that misuses the `passwd` system program does so by employing a race condition that is present in the Linux kernel to debug privileged processes.

Normally, the `passwd` system process performs only a restricted set of actions that consists of editing the `/etc/passwd` and/or the `/etc/shadow` file. However, the `passwd` system process can be made to do more, because of a race condition in the Linux kernel which allows an unprivileged process to debug a system process. Using an unprivileged process, an attacker can alter or "debug" the `passwd` system process and force it to execute a command shell, granting the attacker elevated privileges.[1] Details of race conditions in the Linux kernel are given in [13] and [17]. The `passwd` exploit was obtained from [15].

The `traceroute` network diagnostic utility is a system program that is usually employed by normal users to gather information about the availability and latency of the network between two hosts [7]. To accomplish this task, the `traceroute` system program must have unrestricted access to the network interface, a resource provided only to privileged system programs. However, a logic error in the `traceroute` system program allows an attacker to corrupt the memory of the process by specifying multiple network gateways on the command line [16]. The `traceroute` exploit uses this memory corruption to redirect the process to instructions that execute a command shell with the elevated privileges of the `traceroute` system program [8]. More detail on this memory corruption vulnerability is provided in [16]. The `traceroute` exploit was obtained from [8].

[1] In industry parlance, the instructions injected by the exploit are termed "shellcode", and the shell in which an intruder gains elevated privileges is a "rootshell."

Several key features make certain attacks or exploits likely candidates for subverting sequence-based anomaly detectors such as stide. The subversion technique presented in this paper is more likely to be effective when:

- the vulnerability exploited involves a system program that runs with elevated (root) privileges;
- the vulnerability allows an attacker to take control of the execution of the system program, giving the attacker the ability to choose the system kernel calls or instructions that are issued by the system program;
- the attack does not cause the system program to behave anomalously (e.g. produce an error message in response to an invalid input supplied by the attacker) before the attack/attacker can take control of the execution of the system program;
- the system kernel calls occurring after the "point of seizure", i.e., the point in the data stream at which the attacker first takes control of the system program, include any or all of `execve`, or `open/write`, or `chmod`, or `chown`, or any other system kernel call that the attacker can use to effect the attack.

5.2 Choice of Normal Data

It is unreasonable to expect an attacker to be able to identify and access the precise segment, feature or attribute of the data that can be used to train an anomaly detector. The time, patience and system privileges required to do so may simply not be available to an attacker. However, since training data is vital to the function of an anomaly detector, the attacker has to construct an approximation of the training data that may have been used by the anomaly detector if he or she desires to exploit a blind spot of the detector.

For anomaly detectors like stide, i.e., anomaly detectors that monitor system programs, training data can be approximated more easily, because system programs typically behave in very set and regimented ways. For example, the `passwd` and `traceroute` system programs are limited in the number of ways that they can be used, and as a result it is possible to make reasonable assumptions about how these programs would be regularly invoked.

These assumptions may be aided by the wealth of easily accessible documentation that typically accompanies each system program, as well as by any general knowledge or experience already acquired by the attacker. It is important to note, however, that the success of this method for undermining stide relies on the attacker's being able to approximate normal usage of the system program.

To be successful at undermining stide, the attacker does not need to obtain every possible example of a system program's normal behavior. If the anomalous manifestation of an exploit can already be crafted by an extremely reduced subset of normal behavior, then it can only be expected that more examples of normal behavior contribute to an increased number of ways with which to construct the anomalous manifestation of the exploit.

For the `passwd` system program, normal data was obtained by executing the `passwd` system program with no arguments, and then by following the instructions displayed by the program to input the user's current password once, and then their new password twice. In other words passwd was invoked to expire an old password and install a new one.

For the traceroute system program, normal data was obtained by executing traceroute to acquire diagnostic information regarding the network connectivity between the local host and the Internet site nis.nsf.net. This site was chosen because it is the simplest example of using traceroute, based on the documentation provided with the program itself [7].

5.3 Establishing Attack Manifestations in Sensor Data

Two issues are addressed in this subsection. The first is whether the attacks embodied by the execution of the chosen exploits actually manifested in the sensor data, and the second is whether the manifestation is an anomalous event detectable by stide. Simply because an attack can be shown to manifest in the sensor data does not necessarily mean that the manifestation is automatically anomalous. It is necessary to establish that the manifestation of the exploits are initially detectable by stide in order to show that any modifications to the same exploits effectively render them undetectable by the same detector.

Before proceeding any further it is necessary to define what is meant by the term *manifestation* within the scope of this study. The manifestation of an attack is defined to be that sequence of system calls issued by the exploited/privileged system program, and due to the presence and activity of the exploit. The remainder of this section describes how the manifestations of the two exploits, passwd and traceroute, were obtained.

passwd. The passwd exploit was downloaded from [15]; then it was compiled and deployed. There were no parameters that needed to be set in order to execute the exploit. The successful execution of the exploit was confirmed by checking that elevated privileges were indeed conferred.

The manifestation of the passwd exploit was determined manually. An inspection of the source code for both the passwd exploit and that portion of the Linux kernel responsible for the race condition vulnerability identified the precise system calls that were attributable to the attack. The sequence of system calls that comprise the manifestation of the attack embodied by the passwd exploit is setuid, setgid, execve. Stide was then deployed, using the normal data described above as training data, plus the test data comprised of the data collected while the passwd exploit was executed. Stide was run with detector window sizes ranging from 1 to 15. It was found that the attack was detectable at all detector window sizes. More precisely, the attack was detectable by stide because setuid, setgid, and execve were all foreign symbols. From the detection map of stide in Figure 1, it can be seen that stide is capable of detecting size-1 foreign symbols at any detector window size.

traceroute. The traceroute exploit was downloaded from [8]; then it was compiled and deployed. The traceroute exploit expects values for two arguments. The first argument identifies the local platform, and the second argument is a hexadecimal number that represents the address of a specific function in memory. This address is overwritten to point to an attacker-specified function. The successful execution of the exploit was confirmed by checking that elevated privileges were indeed conferred.

The manifestation of the traceroute exploit was determined manually. An inspection of the source code for the traceroute exploit as well as for the traceroute system program, identified the precise system calls that were attributable to the attack. The sequence of system calls that comprise the manifestation of the attack embodied by the traceroute exploit is: brk, brk, brk, setuid, setgid, execve. Mirroring the deployment strategy for passwd, stide was trained on the previously collected traceroute normal data, and run with detector-window sizes 1-15. The attack was shown to be detectable at all window sizes, because setuid, setgid, and execve were all foreign symbols.

6 Manipulating the Manifestation; Modifying Exploits

Three vital items of knowledge have been established up to this point: the characteristics of the minimal-foreign-sequence event that stide is sometimes unable to detect; the conditions ensuring that stide does not detect such an event (the detector-window size must be smaller than the size of the minimal foreign sequence); and the fact that stide is completely capable of detecting the two chosen exploits when they were simply executed on the host system without any modifications. This means that the anomaly detector is completely effective at detecting these exploits should an attacker decide to deploy them.

How can these exploits be modified so that the anomaly detector does not sound the alarm when the modified exploits are deployed? How can an attacker provide his or her attack(s) with every opportunity to complete successfully and stealthily? This section shows how both exploits, guided by the detection map established for stide, can be modified to produce manifestations (or signatures) in the sensor data that are not visible to the detector.

6.1 Modifying passwd and traceroute

In aiming to replace the detectable anomalous manifestations of the exploits with manifestations that are undetectable by stide, there are two points that must be considered. Recall that each exploit embodies some goal to be attained by the attacker, e.g., elevation of privileges.

First, because the method for achieving the goal that is embodied by each exploit, passwd and traceroute, produces an anomalous event detectable by stide, namely a foreign symbol, another method for achieving the same goal must be found to replace it. Note that the goal of both exploits is the typical one of securing an interactive shell with elevated privileges. Interestingly, the

Table 2. The system calls that implement each of three methods that attempt to achieve the same goal of securing an interactive root account accessible to the attacker.

	Description of method	System calls that implement method
1	Changing the access rights to the `/etc/passwd` file in order to give the attacker permission to modify the file (write permission)	`chmod, exit`
2	Changing the *ownership* of the `/etc/passwd` file to the attacker	`chown, exit`
3	Opening the `/etc/passwd` file to append a new user with root privileges	`open, write, close, exit`

new means of achieving the same goal involves changing the value of only one variable in both exploit programs.

Second, the new method of achieving the same goal must not produce any manifestation that is detectable by stide. Although this could mean that both exploits are modified so that their manifestations appear normal, i.e., their manifestations match sequences that already exist in the normal data, it is typically more difficult to do this than to cause the exploits to manifest as foreign sequences. The difficulty lies in the fact that the kinds of normal sequences that can be used to effect an attack may be small. This makes it more likely that an attacker may require sequences that lie outside the normal vocabulary, i.e., foreign sequences.

6.2 New Means to Achieve Original Goals in Exploit Programs

In Section 5.3 it was shown that the execution of the `passwd` and `traceroute` exploits were detectable by stide because both exploits manifested anomalously as the foreign symbols `setuid`, `setgid`, and `execve`. Any attack that introduces a foreign symbol into the sensor data that is monitored by stide, will be detected. This is because foreign symbol manifestations lie in the visible region of stide's detection map. In order for the `traceroute` or `passwd` exploits to become undetectable by stide, they must not produce the system calls `setuid`, `setgid`, and `execve`. Instead an alternate method causing the exploits to manifest as minimal foreign sequences is required. Only system calls that are already present in the normal data can be the manifestation of the exploits.

For the `passwd` exploit, another method for achieving the same goal of securing an interactive shell with elevated privileges that does not involve the foreign symbols `setuid`, `setgid`, and `execve` would be to cause the exploit program to give the attacker permission to *modify* the `/etc/passwd` file. With such access, the attacker can then edit the accounts and give him or herself administrative privileges, to be activated upon his or her next login. The system calls required to implement this method are `chmod` and `exit`. These two calls are found in the normal data for the `passwd` system program.

There at least two other methods that will achieve the same goal. A second method would be to give the attacker *ownership* of the `/etc/passwd` file, and a third method would be to make the affected system program directly edit the `/etc/passwd` file to add a new administrative (root) account that is accessible to the attacker. The system calls that would implement all three methods respectively are listed in Table 2.

For the `traceroute` exploit, the other method for achieving the same goal of securing an interactive shell with elevated privileges that does not involve the foreign symbols `setuid`, `setgid`, and `execve`, is to make the affected system program directly edit the `/etc/passwd` file to add a new administrative (root) account that is accessible to the attacker. The system calls required to implement this method are `open`, `write`, `close`, and `exit`. All these system calls can be found in the normal data for the `traceroute` system program.

6.3 Making the Exploits Manifest as Minimal Foreign Sequences

In the previous subsection, the two exploits were made to manifest as system calls that can be found in the normal data for the corresponding `passwd` and `traceroute` system programs. This is still insufficient to hide the manifestations of the exploits from stide, because even though system calls that already exist in the normal data were used to construct the new manifestation of each exploits, the order of the system calls with respect to each other can still be foreign to the order of system calls that typically occur in the normal data. For example, even if `chmod` and `exit` both appear in the `passwd` normal data, both calls never appear sequentially. This means that the sequence `chmod`, `exit`, is a foreign sequence of size 2, foreign to the normal data. More precisely, this is a minimal foreign sequence of size 2, because the sequence does not contain within it any smaller foreign sequences or foreign symbols.

As a consequence, stide with a detector window of size 2 or larger would be fully capable of detecting such a manifestation. In order to make the manifestation invisible to stide, it is necessary to increase the size of the minimal foreign sequence. Increasing the size raises the chances of falling into stide's blind spot. Referring to Figure 1, it can be seen that the larger the size of the minimal foreign sequence, the larger the size of the blind spot.

To increase the size of the minimal foreign sequence, the short minimal foreign sequences that are the manifestations of both exploits (`chmod`, `exit` for the `passwd` exploit, and `open`, `write`, `close`, and `exit` for the `traceroute` exploit) must be padded with system calls from the normal data that would result in larger minimal foreign sequences with common subsequences. For example, for `passwd` the short minimal foreign sequence that is the manifestation of the new method described in the previous section is `chmod`, `exit`. This is a minimal foreign sequence of size 2. To increase this minimal foreign sequence it can be seen that in the normal data for `passwd`, the system call `chmod` is followed by the sequence utime, close, munmap, and elsewhere in the normal data, munmap is followed by exit. These two sequences

1. chmod, utime, close, munmap
2. munmap, exit
   can be concatenated to create a third sequence
3. chmod, utime, close, munmap, exit.

A method of attack can be developed which manifests as this concatenated sequence. This method is functionally equivalent to the method developed in the previous subsection; it gives the attacker permission to modify /etc/passwd with the chmod system call and exits with the exit system call. The three system calls utime, close, and munmap are made in such a way that they do not alter the state of the system.

If stide employed a detector window of size 2, and the detector window slid over the manifestation of the exploit that is the sequence chmod, utime, close, munmap, exit, no anomalies would result; no alarms would be generated because the manifestation no longer contains any foreign sequences of size 2. However, if stide employed a detector window of size 3, a single anomaly would be detected, namely the minimal foreign sequence of size 3, close, munmap, exit, which would result in an alarm.

The simple example given above describes the general process for creating the larger minimal foreign sequences required to fool stide. By performing an automated search of the normal data it is possible to find all sequences that can be used by an attacker as padding for the manifestation of a particular exploit. The general process for creating larger minimal foreign sequences was automated and used to modify both the passwd and traceroute exploits.

It is important to note that because stide only analyzes system calls and not their arguments, it is possible to introduce system calls to increase the size of minimal foreign sequences without affecting the state of the system. Executing system calls introduced by the attacker that are aimed at exploiting stide's blind spot need not cause any unintended side-effects on the system because the arguments for each system call is ignored. It is therefore possible to introduce system calls that do nothing, such as reading and writing to an empty file descriptor, or opening a file that cannot exist. This point argues for using more diverse data streams in order to provide more effective intrusion detection. Analyzing only the system call stream may be a vulnerability in anomaly detectors.

7 Evaluating the Effectiveness of Exploit Modifications

A small experiment is performed to show that the modified exploits were indeed capable of fooling stide. As shown in the previous section, a single deployment of a modified exploit is accompanied by a parameter that determines the size of the minimal foreign sequence that will be the manifestation of the exploit. Each exploit was deployed with parameter values that ranged between 2 and 7. A minimum value of 2 was chosen, because it is the smallest size possible for a minimal foreign sequence. The maximum value chosen was 7, because a minimal foreign sequence of size 7 would be invisible to stide employing a detector window of size 6. In the literature, stide is often used with a detector window of size 6 [6, 21]. 6 has been referred to as the "magic" number that has caused stide to begin detecting anomalies in intrusive data [6,11]. Using a detector window of size 6 in this experiment serves to illustrate a case where 6 may not be the best size to use because it will miss detecting exploits that manifest as minimal foreign sequences of size 7 and higher.

Each of the two exploits were deployed 6 times, one for each minimal foreign sequence size from 2 to 7. For each execution of an exploit, stide was deployed with detector window sizes 1 to 15. 1 was chosen as the minimum value simply because it is the smallest detector window size that the detector can be deployed with, and 15 was chosen as the maximum arbitrarily.

7.1 Results

The x-axis for the graph in Figure 2 represents the size of the minimal foreign sequence anomaly, and the y-axis represents the size of the detector window. Each star marks the size of the detector window that successfully detected a minimal foreign sequence whose corresponding size is marked on the x-axis. The term detect for stide means that the manifestation of the exploit must have registered as at least one sequence mismatch. Only the results for `traceroute` are presented. The results for `passwd` are very similar and have been omitted due to space limitations.

The graph in Figure 2 mirrors the detection map for stide, showing that the larger the minimal foreign sequence that is the manifestation of an exploit, the larger the detector window required to detect that exploit. Each circle marks the intersection between the size of the minimal foreign sequence that is the manifestation of the exploit and the size of the detector window used by stide, namely 6. Within each circle the presence of the star indicates that the manifestation of the exploit was detected by stide with a window size of 6.

Each successive circle along the x-axis at $y = 6$ depicts a shift in the manifestation of the exploit in terms of the increasing size of the minimal foreign sequence. These shifts are due to having modified the exploit. The arrows indicate a succession of modifications. For example, without any modification the exploit will naturally manifest as a foreign symbol in the data stream; this is represented by the circle at $x = 1, y = 6$. The first modification of the exploit resulted in a minimal foreign sequence of size 2; this is represented by the circle at $x = 2, y = 6$ pointed to by the arrow from the circle at $x = 1, y = 6$. The second modification yields a size-3 foreign sequence, and so forth. There is no circle at $x = 7$ because it was impossible to modify the exploit to shift its manifestation to a size-7 minimal foreign sequence, given the normal data for `traceroute`.

To summarize, if stide were deployed with a detector window of size 6, then it is possible to modify the `traceroute` exploit incrementally, so that it manifests as successively larger minimal foreign sequences, until a size is reached (size 7) at which the manifestation falls out of stide's visible detection range, and into its

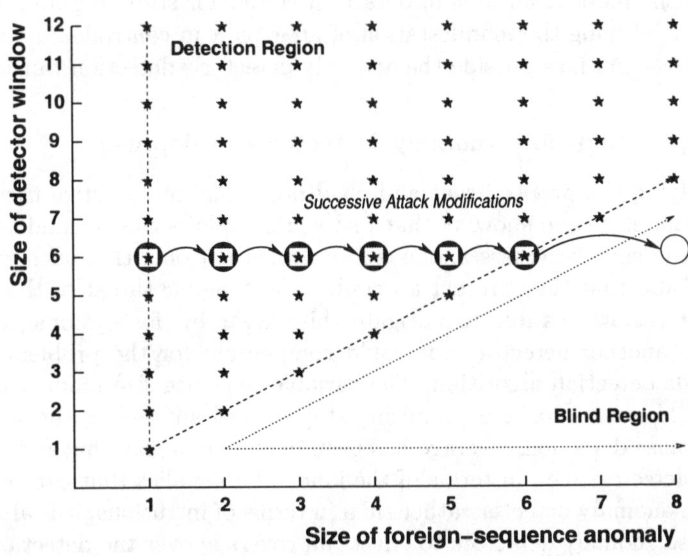

Fig. 2. The manifestation of each version of the traceroute exploit plotted on the detector coverage map for stide, assuming that stide has been configured with a detector window size of 6. Each version of the exploit can be detected by fewer and fewer configurations of stide until the last is invisible.

blind spot. This shows that it is possible to exert control over a common exploit so that its manifestation is moved from an anomaly detector's detection region, to its region of complete blindness. Such movement of an exploit's manifestation effectively hides the exploit from the detector's view.

8 Discussion

The results show that it is possible to hide the presence of the `passwd` and `traceroute` common exploits from stide by modifying those exploits so that they manifest only within stide's detection blind spot. Achieving an attack's objectives is not affected by the modifications to the exploit programs; neither is the training data tampered with in order to render an anomaly detector blind to the attacks. Note that, at present, these results can be said to be relevant only to other anomaly-based intrusion detection systems that employ anomaly detectors operating on sequences of categorical data. Although the results make no claims about other families of anomaly detectors that, for example, employ probabilistic concepts, it is possible that the methods described in this study may be applicable to a broader range of anomaly detectors.

The results presented in this paper show that it is also possible to control the manifestation of an attack so that the manifestation moves from an area of detection blindness to an area of detection clarity for stide. Figure 2 shows the results of modifying the manifestation of an exploit in controlled amounts until the manifestation falls outside the anomaly detector's detection range.

8.1 Implications for Anomaly Detector Development

By identifying the precise event and conditions that characterize the detection blindness for stide and showing that real-world exploits can be modified to take advantage of such weaknesses, one is forewarned not only that such weaknesses exist, but also that they present a possible and tangible threat to the protected system. It is now possible to mitigate this threat by, for example, combining stide with another detector that could compensate for the problems inherent in the stide detection algorithm. The variable sequence size model explored by Marceau [11] seems to be a promising step toward addressing the weakness in stide. Because detection coverage has been defined in a way that is pertinent to anomaly detectors, i.e., in terms of the kinds of anomalies that can be detected by a given anomaly detector rather than in terms of intrusions, it is also possible to compose anomaly detectors to effect full coverage over the detection space.

It is interesting to note that the ease with which an attacker can introduce sequences into the system call data suggests that sequences of system calls may not be a sufficiently expressive form of data to allow an anomaly detector to more effectively monitor and defend an information system. Increasing the number of different kinds of data analyzed, or changing the kind of data analyzed by an anomaly detector, may make an impact on the effectiveness of the intrusion-detection capabilities of an anomaly detector.

8.2 Implications for Anomaly Detector Evaluation

There are a few benefits of an anomaly-based evaluation method, an evaluation method focused on how well anomaly detectors detect anomalies. First, the results of an anomaly-based evaluation increases the scope of the results. In other words, what was established to be true with respect to the anomaly-detection performance of the anomaly detector on synthetic data will also be true of the anomaly detector on real-world data sets. This cannot be said of current evaluation procedures for anomaly detectors because current evaluation schemes evaluate an anomaly detector in terms of how well it detects attacks. This constrains the scope of the results to the data set used in the evaluation because an attack that manifests as a specific kind of anomaly in one data set may no longer do so in another data set due to changing normal behavior.

Second, the results of an anomaly-based evaluation can only contribute to increasing the accuracy of anomaly detectors performing intrusion detection. Current evaluation methods do not establish the detection capabilities of an anomaly detector with regard to the detection of the anomalous manifestations of attacks. The fact that attacks may manifest as different types of anomalies also means that different types of anomaly detectors may be required to detect them. If anomaly detectors are not evaluated on how well they detect anomalies, it is difficult to determine which anomaly detector would best suit the task of detecting a given type of attack. The events that anomaly detectors directly detect are anomalies, but typically anomaly detectors are evaluated on how well they detect attacks, the events that anomaly detectors do not detect except by making the assumption that attacks manifests as those anomalies detected by an anomaly detector.

9 Related Work

The concept of modifying an attack so that it successfully accomplishes its goal while eluding detection is not a new one. Ptacek and Newsham [14] highlighted a number of weaknesses that the network intrusion detection community needed to address if they were to defeat a wily attacker using network signature-based intrusion detection systems. Although the work presented in this paper differs in that it focuses on anomaly-based intrusion detection, it strongly reiterates the concern that an unknown weakness in an intrusion detection system creates a "dangerously false sense of security." In the case of this work, it was shown that a weakness in an anomaly detector could realistically be exploited to compromise the system being protected by that detector.

Wagner and Dean [20] introduced a new class of attacks against intrusion detection systems which they called the "mimicry attack". A "mimicry attack" is where an attacker is able to "develop malicious exploit code that mimics the operation of the application, staying within the confines of the model and thereby evading detection ..." Wagner and Dean studied this class of attack theoretically, but until this present paper, no study has shown if it were possible to create and deploy a mimicry attack in the real-world that truly affected the performance of an intrusion-detection system.

The present study confirms that the class of mimicry attacks does pose a serious threat to an anomaly-based intrusion detection system. By modifying common real-world exploits to create examples of this class of attack, this study also shows how mimicry attacks are able to undermine the protection offered by an anomaly-based intrusion detection system.

10 Conclusion

This study has shown how an anomaly-based intrusion detection system can be effectively undermined by modifying common real-world exploits. It presented a method that identified weaknesses in an anomaly-based intrusion detector, and showed how an attacker can effectively modify common exploits to take advantage of those weaknesses in order to craft an offensive mechanism that renders an anomaly-based intrusion detector blind to the on-going presence of those attacks.

The results show that it is possible to hide the presence of the `passwd` and `traceroute` common exploits from stide by modifying those exploits so that they manifest only within stide's detection blind spot. The results also show that it is possible to control the manifestation of an attack such that the manifestation moves from an area of detection clarity to one of detection blindness for stide.

References

1. Herve Debar, Marc Dacier, and Andreas Wespi. Towards a taxonomy of intrusion-detection systems. *Computer Networks*, 31(8):805–822, April 1999.
2. Stephanie Forrest, Steven A. Hofmeyr, Anil Somayaji, and Thomas A. Longstaff. A sense of self for unix processes. In *Proceedings of the 1996 IEEE Symposium on Security and Privacy*, 6–8 May 1996, Oakland, California, pages 120–128, IEEE Computer Society Press, Los Alamitos, California, 1996.
3. Cristian Gafton. passwd(1). Included in `passwd` version 0.64.1-1 software package, January 1998.
4. Anup K. Ghosh, Aaron Schwartzbard, and Michael Schatz. Learning program behavior profiles for intrusion detection. In *Proceedings of the 1st Workshop on Intrusion Detection and Network Monitoring*, 9–12 April 1999, Santa Clara, California, pages 51–62, The USENIX Association, Berkeley, California, 1999.
5. Anup K. Ghosh, James Wanken, and Frank Charron. Detecting anomalous and unknown intrusions against programs. In *Proceedings of the 14th Annual Computer Security Applications Conference*, 7–11 December 1998, Phoenix, Arizona, pages 259–267, IEEE Computer Society Press, Los Alamitos, 1998.
6. Steven A. Hofmeyr, Stephanie Forrest, and Anil Somayaji. Intrusion detection using sequences of system calls. *Journal of Computer Security*, 6(3):151–180, 1998.
7. Van Jacobson. Traceroute(8). Included in `traceroute` version 1.4a5 software package, April 1997.
8. Michel "MaXX" Kaempf. Traceroot2: Local root exploit in LBNL traceroute. Internet: `http://packetstormsecurity.org/0011-exploits/traceroot2.c`, March 2002.
9. Sandeep Kumar. *Classification and Detection of Computer Intrusions*. PhD thesis, Purdue University, West Lafayette, Indiana, August 1995.
10. Teresa Lunt. Automated audit trail analysis and intrusion detection: A survey. In *Proceedings of the 11th National Computer Security Conference*, Baltimore, Maryland, pages 65–73, October 1988.
11. Carla Marceau. Characterizing the behavior of a program using multiple-length N-grams. In *New Security Paradigms Workshop*, 18–22 September 2000, Ballycotton, County Cork, Ireland, pages 101–110, ACM Press, New York, New York, 2001.
12. Roy A. Maxion and Kymie M. C. Tan. Anomaly detection in embedded systems. *IEEE Transactions on Computers*, 51(2):108–120, February 2002.
13. Andrew P. Moore. CERT/CC vulnerability note VU#176888, July 2002. Internet: `http://www.kb.cert.org/vuls/id/176888`.
14. Thomas H. Ptacek and Timothy N. Newsham. Insertion, evasion, and denial of service: Eluding network intrusion detection. Secure Networks, Inc., Calgary, Alberta, Canada, January 1998.
15. Wojciech Purczynski (original author) and "lst" (author of improvements). Epcs2: Exploit for execve/ptrace race condition in Linux kernel up to 2.2.18. Internet: `http://www.securiteam.com/exploits/5NP061P4AW.html`, March 2002.
16. SecurityFocus Vulnerability Archive. LBNL Traceroute Heap Corruption Vulnerability, Bugtraq ID 1739. Internet: `http://online.securityfocus.com/bid/1739`, March 2002.

17. SecurityFocus Vulnerability Archive. Linux PTrace/Setuid Exec Vulnerability, Bugtraq ID 3447. Internet: `http://online.securityfocus.com/bid/3447`, March 2002.
18. Anil Somayaji and Geoffrey Hunsicker. IMMSEC Kernel-level system call tracing for Linux 2.2, Version 991117. Obtained through private communication. Previous version available on the Internet:
`http://www.cs.unm.edu/~immsec/software/`, March 2002.
19. Kymie M. C. Tan and Roy A. Maxion. "Why 6?" Defining the operational limits of stide, an anomaly-based intrusion detector. In *Proceedings of the 2002 IEEE Symposium on Security and Privacy*, 12–15 May 2002, Berkeley, California, pages 188–201, IEEE Computer Society Press, Los Alamitos, California, 2002.
20. David Wagner and Drew Dean. Intrusion detection via static analysis. In *Proceedings of the 2001 IEEE Symposium on Security and Privacy*, 14–16 May 2001, Berkeley, California, IEEE Computer Society Press, Los Alamitos, California, 2001.
21. Christina Warrender, Stephanie Forrest, and Barak Pearlmutter. Detecting intrusions using system calls: Alternative data models. In *Proceedings of the 1999 IEEE Symposium on Security and Privacy*, 9–12 May 1999, Oakland, California, pages 133–145, IEEE Computer Society Press, Los Alamitos, California, 1999.

Analyzing Intensive Intrusion Alerts via Correlation

Peng Ning, Yun Cui, and Douglas S. Reeves

Department of Computer Science
North Carolina State University
Raleigh, NC 27695-7534
ning@csc.ncsu.edu, ycui4@eos.ncsu.edu, reeves@csc.ncsu.edu

Abstract. Traditional intrusion detection systems (IDSs) focus on low-level attacks or anomalies, and raise alerts independently, though there may be logical connections between them. In situations where there are intensive intrusions, not only will actual alerts be mixed with false alerts, but the amount of alerts will also become unmanageable. As a result, it is difficult for human users or intrusion response systems to understand the alerts and take appropriate actions. Several complementary alert correlation methods have been proposed to address this problem. As one of these methods, we have developed a framework to correlate intrusion alerts using *prerequisites of intrusions*. In this paper, we continue this work to study the feasibility of this method in analyzing real-world, intensive intrusions. In particular, we develop three utilities (called *adjustable graph reduction, focused analysis*, and *graph decomposition*) to facilitate the analysis of large sets of correlated alerts. We study the effectiveness of the alert correlation method and these utilities through a case study with the network traffic captured at the DEF CON 8 Capture the Flag (CTF) event. Our results show that these utilities can simplify the analysis of large amounts of alerts, and also reveals several attack strategies that were repeatedly used in the DEF CON 8 CTF event.

Keywords: Intrusion Detection, Alert Correlation, Attack Scenario Analysis

1 Introduction

Intrusion detection has been considered the second line of defense for computer and network systems along with the prevention-based techniques such as authentication and access control. Intrusion detection techniques can be roughly classified as *anomaly detection* (e.g., NIDES/STAT [1]) and *misuse detection* (e.g., NetSTAT [2]).

Traditional intrusion detection systems (IDSs) focus on low-level attacks or anomalies; they cannot capture the logical steps or attacking strategies behind these attacks. Consequently, the IDSs usually generate a large amount of alerts, and the alerts are raised independently, though there may be logical connections between them. In situations where there are intensive intrusions, not only will actual alerts be mixed with false alerts, but the amount of alerts will also become unmanageable. As a result, it is difficult for human users or intrusion response systems to understand the intrusions behind the alerts and take appropriate actions.

To assist the analysis of intrusion alerts, several alert correlation methods (e.g., [3, 4,5,6]) have been proposed recently to process the alerts reported by IDSs. (Please see

A. Wespi, G. Vigna, and L. Deri (Eds.): RAID 2002, LNCS 2516, pp. 74–94, 2002.

Section 2 for details.) As one of these proposals, we have developed a framework to correlate intrusion alerts using *prerequisites of intrusions* [6].

Our approach is based on the observation that most intrusions are not isolated, but related as different stages of series of attacks, with *the early stages preparing for the later ones*. Intuitively, the prerequisite of an intrusion is the necessary condition for the intrusion to be successful. For example, the existence of a vulnerable service is the prerequisite of a remote buffer overflow attack against the service. Our method identifies the prerequisites (e.g., existence of vulnerable services) and the consequences (e.g., discovery of vulnerable services) of each type of attacks, and correlate the corresponding alerts by matching the consequences of some previous alerts and the prerequisites of some later ones. For example, if we find a port scan followed by a buffer overflow attack against one of the scanned ports, we can correlate them into the same series of attacks. We have developed an intrusion alert correlator based on this framework [7]. Our initial experiments with the 2000 DARPA intrusion evaluation datasets [8] have shown that our method can successfully correlate related alerts together [6,7].

This paper continues the aforementioned work to study the effectiveness of this method in analyzing real-world, intrusion intensive data sets. In particular, we would like to see how well the alert correlation method can help human users organize and understand intrusion alerts, especially when IDSs report a large amount of alerts. We argue that this is a practical problem that the intrusion detection community is facing. As indicated in [9], "encountering 10-20,000 alarms per sensor per day is common."

In this paper, we present three utilities (called *adjustable graph reduction, focused analysis*, and *graph decomposition*) that we have developed to facilitate the analysis of large sets of correlated alerts. These utilities are intended for human users to analyze and understand the correlated alerts as well as the strategies behind them. We study the effectiveness of these utilities through a case study with the network traffic captured at the DEF CON 8 Capture the Flag (CTF) event [10]. Our results show that they can effectively simplify the analysis of large amounts of alerts. Our analysis also reveals several attack strategies that appeared in the DEF CON 8 CTF event.

The remainder of this paper is organized as follows. Section 2 reviews the related work. Section 3 briefly describes our model for alert correlation. Section 4 introduces three utilities for analyzing hyper-alert correlation graphs. Section 5 describes our case study with the DEF CON 8 CTF dataset. Section 6 concludes this paper and points out some future work.

2 Related Work

Intrusion detection has been studied for about twenty years. An excellent overview of intrusion detection techniques and related issues can be found in a recent book [11].

Several alert correlation techniques have been proposed to facilitate analysis of intrusions. In [3], a probabilistic method was used to correlate alerts using similarity between their features. However, this method is not suitable for fully discovering causal relationships between alerts. In [12], a similar approach was applied to detect stealthy portscans along with several heuristics. Though some such heuristics (e.g., feature separation heuristics [12]) may be extended to general alert correlation problem, the approach cannot fully recover the causal relationships between alerts, either.

Techniques for aggregating and correlating alerts have been proposed by others [4]. In particular, the correlation method in [4] uses a *consequence mechanism* to specify what types of alerts may follow a given alert type. This is similar to the specification of misuse signatures. However, the consequence mechanism only uses alert types, the probes that generate the alerts, the severity level, and the time interval between the two alerts involved in a consequence definition, which do not provide sufficient information to correlate all possibly related alerts. Moreover, it is not easy to predict how an attacker may arrange a sequence of attacks. In other words, developing a sufficient set of consequence definitions for alert correlation is not a solved problem.

Another approach has been proposed to "learn" alert correlation models by applying machine learning techniques to training data sets embedded with known intrusion scenarios [5]. This approach can automatically build models for alert correlation; however, it requires training in every deployment, and the resulting models may overfit the training data, thereby missing attack scenarios not seen in the training data sets. The alert correlation techniques that we present in this paper address this same problem from a novel angle, overcoming the limitations of the above approaches.

Our method can be considered as a variation of JIGSAW [13]. Both methods try to uncover attack scenarios based on specifications of individual attacks. However, our method also differs from JIGSAW. First, our method allows partial satisfaction of prerequisites (i.e., required capabilities in JIGSAW [13]), recognizing the possibility of undetected attacks and that of attackers gaining information through non-intrusive ways (e.g., talking to a friend working in the victim organization), while JIGSAW requires all required capabilities be satisfied. Second, our method allows aggregation of alerts, and thus can reduce the complexity involved in alert analysis, while JIGSAW currently does not have any similar mechanisms. Third, we develop a set of utilities for interactive analysis of correlated alerts, which is not provided by JIGSAW.

An alert correlation approach similar to ours was proposed recently in the MIRADOR project [14]. The MIRADOR approach also correlates alerts using partial match of prerequisites (pre-conditions) and consequences (post-conditions) of attacks. However, the MIRADOR approach uses a different formalism than ours. In particular, the MIRADOR approach treats alert aggregation as an individual stage before alert correlation, while our method allows alert aggregation during and after correlation. As we will see in the later sections, this difference leads to the three utilities for interactive alert analysis.

GrIDS uses activity graphs to represent the causal structure of network activities and detect propagation of large-scale attacks [15]. Our method also uses graphs to represent correlated alerts. However, unlike GrIDS, in which nodes represent hosts or departments and edges represent network traffic between them, our method uses nodes to represent alerts, and edges the relationships between the alerts.

Several languages have been proposed to represent attacks, including STAT [2,16], Colored-Petri Automata (CPA), LAMBDA [17], and MuSig [18] and its successor [19]. In particular, LAMBDA uses a logic-based method to specify the precondition and postcondition of attack scenarios, which is similar to our method. (See Section 3.) However, all these languages specify entire attack scenarios, which are limited to known scenarios.

In contrast, our method (as well as JIGSAW) describes prerequisites and consequences of individual attacks, and correlate detected attacks (i.e., alerts) based on the relationship between these prerequisites and consequences. Thus, our method can potentially correlate alerts from unknown attack scenarios.

Alert correlation has been studied in the context of network management (e.g., [20], [21], and [22]). In theory, alert correlation methods for network management are applicable to intrusion alert correlation. However, intrusion alert correlation faces more challenges that its counter part in network management: While alert correlation for network management deals with alerts about natural faults, which usually exhibits regular patterns, intrusion alert correlation has to cope with less predictable, malicious intruders.

3 Preliminary: Alert Correlation Using Prerequisites of Intrusions

In this section, we briefly describe our model for correlating alerts using prerequisites of intrusions. Please read [6] for further details.

The alert correlation model is based on the observation that in series of attacks, the component attacks are usually not isolated, but related as different stages of the attacks, with the early ones preparing for the later ones. For example, an attacker needs to install the Distributed Denial of Service (DDOS) daemon programs before he can launch a DDOS attack.

To take advantage of this observation, we correlate alerts using prerequisites of the corresponding attacks. Intuitively, the *prerequisite* of an attack is the necessary condition for the attack to be successful. For example, the existence of a vulnerable service is the prerequisite of a remote buffer overflow attack against the service. Moreover, an attacker may make progress (e.g., discover a vulnerable service, install a Trojan horse program) as a result of an attack. Informally, we call the possible outcome of an attack the *(possible) consequence* of the attack. In a series of attacks where attackers launch earlier ones to prepare for later ones, there are usually strong connections between the consequences of the earlier attacks and the prerequisites of the later ones.

Accordingly, we identify the prerequisites (e.g., existence of vulnerable services) and the consequences (e.g., discovery of vulnerable services) of each type of attacks and correlate detected attacks (i.e., alerts) by matching the consequences of some previous alerts and the prerequisites of some later ones.

Note that an attacker does not have to perform early attacks to prepare for later ones. For example, an attacker may launch an individual buffer overflow attack against the service blindly. In this case, we cannot, and should not correlate it with others. However, if the attacker does launch attacks with earlier ones preparing for later ones, our method can correlate them, provided the attacks are detected by IDSs.

3.1 Prerequisite and Consequence of Attacks

We use predicates as basic constructs to represent prerequisites and (possible) consequences of attacks. For example, a scanning attack may discover UDP services vulnerable to certain buffer overflow attacks. We can use the predicate *UDPVulnerableToBOF* (*VictimIP, VictimPort*) to represent this discovery. In general, we use a logical formula, i.e., logical combination of predicates, to represent the prerequisite of an attack. Thus, we may have a prerequisite of the form *UDPVulnerableToBOF* (*VictimIP, VictimPort*) $\wedge$ *UDPAccessibleViaFirewall* (*VictimIP, VictimPort*). To simplify the discussion, we restrict the logical operators to $\wedge$ (conjunction) and $\vee$ (disjunction).

We use a set of logical formulas to represent the (possible) consequence of an attack. For example, an attack may result in compromise of the root account as well as modification of the .rhost file. We may use the following set of logical formulas to represent the consequence of this attack: {*GainRootAccess* (*IP*), *rhostModified* (*IP*)}. This example says that as a result of the attack, the attacker may gain root access to the system and the .rhost file may be modified.

Note that the consequence of an attack is the *possible* result of the attack. In other words, the consequence of an attack is indeed the worst consequence. (Please read [6] for details.) For brevity, we refer to *possible consequence* simply as *consequence* throughout this paper.

3.2 Hyper-alerts and Hyper-alert Correlation Graphs

With predicates as basic constructs, we use a *hyper-alert type* to encode our knowledge about each type of attacks.

Definition 1 A *hyper-alert type T* is a triple (*fact, prerequisite, consequence*) where (1) *fact* is a set of attribute names, each with an associated domain of values, (2) *prerequisite* is a logical formula whose free variables are all in *fact*, and (3) *consequence* is a set of logical formulas such that all the free variables in *consequence* are in *fact*.

Intuitively, the *fact* component of a hyper-alert type gives the information associated with the alert, *prerequisite* specifies what must be true for the attack to be successful, and *consequence* describes what could be true if the attack indeed succeeds. For brevity, we omit the domains associated with attribute names when they are clear from context.

Example 1 Consider the buffer overflow attack against the *sadmind* remote administration tool. We may have the following hyper-alert type for such attacks: *SadmindBufferOverflow* = ({*VictimIP, VictimPort*}, *ExistHost* (*VictimIP*) $\wedge$ *VulnerableSadmind* (*VictimIP*), {*GainRootAccess*(*VictimIP*)}). Intuitively, this hyper-alert type says that such an attack is against the host running at IP address *VictimIP*. (We expect the actual values of *VictimIP* are reported by an IDS.) As the prerequisite of a successful attack, there must exist a host at the IP address *VictimIP* and the corresponding *sadmind* service should be vulnerable to buffer overflow attacks. The attacker may gain root privilege as a result of the attack. □

Given a hyper-alert type, a *hyper-alert instance* can be generated if the corresponding attack is reported by an IDS. For example, we can generate a hyper-alert instance of type *SadmindBufferOverflow* from an alert that describes such an attack.

Definition 2 Given a hyper-alert type *T* = (*fact, prerequisite, consequence*), a *hyper-alert (instance) h of type T* is a finite set of tuples on *fact*, where each tuple is associated with an interval-based timestamp [*begin_time, end_time*]. The hyper-alert h implies that *prerequisite* must evaluate to True and all the logical formulas in *consequence* might evaluate to True for each of the tuples.

The *fact* component of a hyper-alert type is essentially a relation schema (as in relational databases), and a hyper-alert is a relation instance of this schema. One may point out that an alternative way is to represent a hyper-alert as a record, which is equivalent to a single tuple on *fact*. However, such an alternative cannot accommodate certain alerts possibly reported by an IDS. For example, an IDS may report an IPSweep attack along with multiple swept IP addresses, which cannot be represented as a single record. Thus, we believe the current notion of a hyper-alert is a more appropriate choice.

A hyper-alert *instantiates* its *prerequisite* and *consequence* by replacing the free variables in *prerequisite* and *consequence* with its specific values. Note that *prerequisite* and *consequence* can be instantiated multiple times if *fact* consists of multiple tuples. For example, if an IPSweep attack involves several IP addresses, the *prerequisite* and *consequence* of the corresponding hyper-alert type will be instantiated for each of these addresses.

In the following, we treat timestamps implicitly and omit them if they are not necessary for our discussion.

Example 2 Consider the hyper-alert type *SadmindBufferOverflow* defined in example 1. We may have a hyper-alert $h_{SadmindBOF}$ that includes the following tuples: $\{$(*VictimIP* = 152.141.129.5, *VictimPort* = 1235), (*VictimIP* = 152.141.129.37, *VictimPort* = 1235)$\}$. This implies that if the attack is successful, the following two logical formulas must be True as the prerequisites of the attack: *ExistHost* (152.141.129.5) $\wedge$ *VulnerableSadmind* (152.141.129.5), *ExistHost* (152.141.129.37) $\wedge$ *VulnerableSadmind* (152.141.129.37), and the following two predicates might be True as consequences of the attack: *GainRootAccess* (152.141.129.5), *GainRootAccess* (152.141.129.37). This hyper-alert says that there are buffer overflow attacks against *sadmind* at IP addresses 152.141.129.5 and 152.141.129.37, and the attacker may gain root access as a result of the attacks. □

To correlate hyper-alerts, we check if an earlier hyper-alert *contributes* to the prerequisite of a later one. Specifically, we decompose the prerequisite of a hyper-alert into parts of predicates and test whether the consequence of an earlier hyper-alert makes some parts of the prerequisite True (i.e., makes the prerequisite easier to satisfy). If the result is positive, then we correlate the hyper-alerts. This approach is specified through the following definitions.

Definition 3 Consider a hyper-alert type T = (*fact, prerequisite, consequence*). The *prerequisite set (or consequence set, resp.) of* T, denoted $P(T)$ (or $C(T)$, resp.), is the set of all such predicates that appear in *prerequisite* (or *consequence*, resp.). Given a hyper-alert instance h of type T, the *prerequisite set (or consequence set, resp.) of* h, denoted $P(h)$ (or $C(h)$, resp.), is the set of predicates in $P(T)$ (or $C(T)$, resp.) whose arguments are replaced with the corresponding attribute values of each tuple in h. Each element in $P(h)$ (or $C(h)$, resp.) is associated with the timestamp of the corresponding tuple in h.

Definition 4 Hyper-alert h_1 *prepares for* hyper-alert h_2 if there exist $p \in P(h_2)$ and $C \subseteq C(h_1)$ such that for all $c \in C$, $c.end\_time < p.begin\_time$ and the conjunction of all the logical formulas in C implies p.

Given a sequence S of hyper-alerts, a hyper-alert h in S is a *correlated hyper-alert* if there exists another hyper-alert h' such that either h prepares for h' or h' prepares for h. Otherwise, h is called an *isolated hyper-alert*.

Let us further explain the alert correlation method with the following example.

Example 3 Consider the *Sadmind Ping* attack with which an attacker discovers possibly vulnerable *sadmind* services. The corresponding hyper-alert type can be represented by *SadmindPing* = ({*VictimIP, VictimPort*}, *ExistsHost* (*VictimIP*), {*VulnerableSadmind* (*VictimIP*)}). It is easy to see that $P(SadmindPing) = \{ExistHost(VictimIP)\}$, and $C(SadmindPing) = \{VulnerableSadmind(VictimIP)\}$.

Suppose a hyper-alert $h_{SadmindPing}$ of type *SadmindPing* has the following tuples: {(*VictimIP* = 152.141.129.5, *VictimPort* = 1235)}. Then the prerequisite set of $h_{SadmindPing}$ is $P(h_{SadmindPing})$ = {*ExistsHost* (152.141.129.5)}, and the consequence set is $C(h_{SadmindPing})$ = {*VulnerableSadmind* (152.141.129.5)}.

Now consider the hyper-alert $h_{SadmindBOF}$ discussed in Example 2. Similar to $h_{SadmindPing}$, we can easily get $P(h_{SadmindBOF})$ = {*ExistsHost* (152.141.129.5), *ExistsHost* (152.141.129.37), *VulnerableSadmind* (152.141.129.5), *VulnerableSadmind* (152.141.129.37)}, and $C(h_{SadmindBOF})$ = {*GainRootAccess* (152.141.129.5), *GainRootAccess* (152.141.129.37)}.

Assume that all tuples in $h_{SadmindPing}$ have timestamps earlier than every tuple in $h_{SadmindBOF}$. By comparing the contents of $C(h_{SadmindPing})$ and $P(h_{SadmindBOF})$, it is clear that the element *VulnerableSadmind* (152.141.129.5) in $P(h_{SadmindBOF})$ (among others) is also in $C(h_{SadmindPing})$. Thus, $h_{SadmindPing}$ prepares for, and should be correlated with $h_{SadmindBOF}$. □

The prepare-for relation between hyper-alerts provides a natural way to represent the causal relationship between correlated hyper-alerts. We also introduce the notion of a *hyper-alert correlation graph* to represent a set of correlated hyper-alerts.

Definition 5 A *hyper-alert correlation graph* $CG = (N, E)$ is a connected graph, where the set N of nodes is a set of hyper-alerts and for each pair $n_1, n_2 \in N$, there is a directed edge from n_1 to n_2 in E if and only if n_1 prepares for n_2.

A hyper-alert correlation graph is an intuitive representation of correlated alerts. It can potentially reveal intrusion strategies behind a series of attacks, and thus lead to better understanding of the attacker's intention. We have performed a series of experiments with the 2000 DARPA intrusion detection evaluation datasets [7]. Figure 1 shows one of the hyper-alert correlation graphs discovered from these datasets. Each node in Figure 1 represents a hyper-alert. The numbers inside the nodes are the alert IDs generated by the IDS. This hyper-alert correlation graph clearly shows the strategy behind the sequence of attacks. (For details please refer to [7].)

4 Utilities for Analyzing Intensive Alerts

As demonstrated in [7], the alert correlation method is effective in analyzing small amount of alerts. However, our experience with intrusion intensive datasets (e.g., the DEF CON 8 CTF dataset [10]) has revealed several problems.

First, let us consider the following scenario. Suppose an IDS detected an *SadmindPing* attack, which discovered the vulnerable *Sadmind* service on host *V*, and later an *SadmindBufferOverlfow* attack against the *Sadmind* service. Assuming that they were launched from different hosts, should we correlate them? On the one hand, it is possible that one or two attackers coordinated these two attacks from two different hosts, trying to

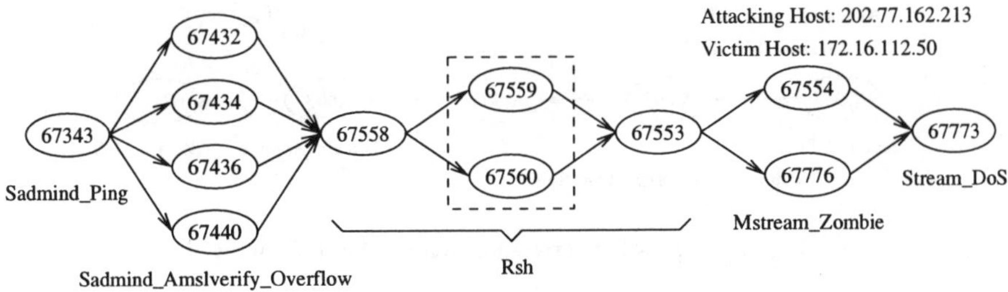

Fig. 1. A hyper-alert correlation graph discovered in the 2000 DARPA intrusion detection evaluation datasets

avoid being correlated. On the other hand, it is also possible that these attacks belonged to two separate efforts. Such a scenario clearly introduces a dilemma, especially when there are a large amount of alerts.

One may suggest to use time to solve this problem. For example, we may correlate the aforementioned attacks if they happened within t seconds. However, knowing this method, an attacker may introduce delays between attacks to bypass correlation.

The second problem is the overwhelming information encoded by hyper-alert correlation graphs when intensive intrusions trigger a large amount of alerts. Our initial attempt to correlate the alerts generated for the DEF CON 8 CTF dataset [10] resulted in 450 hyper-alert correlation graphs, among which the largest hyper-alert correlation graph consists of 2,940 nodes and 25,321 edges. Such a graph is clearly too big for a human user to comprehend in a short period of time.

Although the DEF CON 8 dataset involves intensive intrusions not usually seen in normal network traffic, the actual experience of intrusion detection practitioners indicates that "encountering 10-20,000 alarms per sensor per day is common [9]." Thus, it is necessary to develop techniques or tools to deal with the overwhelming information.

In this section, we propose three utilities, mainly to address the second problem. Regarding the first problem, we choose to correlate the alerts when it is possible, leaving the final decision to the user. We would like to clarify that these utilities are intended for human users to analyze alerts interactively, not for computer systems to draw any conclusion automatically, though some of the utilities may be adapted for automatic systems. These utilities are summarized as follows.

1. *Adjustable graph reduction.* Reduce the complexity (i.e., the number of nodes and edges) of hyper-alert correlation graphs while keeping the structure of sequences of attacks. The graph reduction is adjustable in the sense that users are allowed to control the degree of reduction.
2. *Focused analysis.* Focus analysis on the hyper-alerts of interest according to user's specification. This may generate hyper-alert correlation graphs much smaller and more comprehensible than the original ones.
3. *Graph decomposition.* Cluster the hyper-alerts in a hyper-alert correlation graph based on the common features shared by the hyper-alerts, and decompose the graph

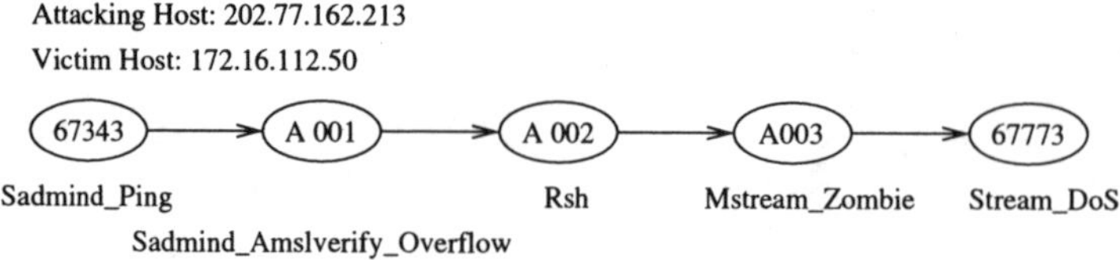

Fig. 2. A hyper-alert correlation graph reduced from Fig. 1

into smaller graphs according to the clusters. This can be considered to combine a variation of the method proposed in [3] with our method.

4.1 Adjustable Reduction of Hyper-alert Correlation Graphs

A natural way to reduce the complexity of a hyper-alert correlation graph is to reduce the number of nodes and edges. However, to make the reduced graph useful, any reasonable reduction should maintain the structure of the corresponding attacks.

We propose to aggregate hyper-alerts of the same type to reduce the number of nodes in a hyper-alert correlation graph. Due to the flexible definition of hyper-alerts, the result of hyper-alert aggregation will remain valid hyper-alerts. For example, in Figure 1, hyper-alerts 67432, 67434, 67436, and 67440 are all instances of hyper-alert type *Sadmind_Amslverify_Overflow*. Thus, we may aggregate them into one hyper-alert. As another example, hyper-alerts 67558, 67559, 67560, and 67553 are all instances of *Rsh*, and can be aggregated into a single hyper-alert.

Edges are reduced along with the aggregation of hyper-alerts. In Figure 1, the edges between the *Rsh* hyper-alerts are subsumed into the aggregated hyper-alert, while the edges between the *Sadmind_Ping* hyper-alert and the four *Sadmind_Amslverify_Overflow* hyper-alerts are merged into a single edge. As a result, we have a reduced hyper-alert correlation graph as shown in Figure 2.

Reduction of a hyper-alert correlation graph may lose information contained in the original graph. Indeed, hyper-alerts that are of the same type but belong to different sequences of attacks may be aggregated and thus provide overly simplified results. Nevertheless, our goal is to lose as little information of the structure of attacks as possible.

Depending on the actual alerts, the reduction of a hyper-alert correlation graph may be less simplified, or over simplified. We would like to give a human user more control over the graph reduction process. In the following, we use a simple mechanism to control this process, based on the notion of an interval constraint [6].

Definition 6 Given a time interval I (e.g., 10 seconds), a hyper-alert h satisfies *interval constraint of* I if (1) h has only one tuple, or (2) for all t in h, there exist another t' in h such that there exist $t.begin\_time < T < t.end\_time$, $t'.begin\_time < T' < t'.end\_time$, and $|T - T'| < I$.

We allow hyper-alert aggregation only when the resulting hyper-alerts satisfy an interval constraint of a given threshold I. Intuitively, we allow hyper-alerts to be aggregated only when they are close to each other. The larger a threshold I is, the more a hyper-alert correlation graph can be reduced. By adjusting the interval threshold, a user can control the degree to which a hyper-alert correlation graph is reduced.

4.2 Focused Analysis

Focused analysis is implemented on the basis of focusing constraints. A *focusing constraint* is a logical combination of comparisons between attribute names and constants. (In our work, we restrict logical operations to AND ($\wedge$), OR ($\vee$), and NOT ($\neg$).) For example, we may have a focusing constraint $SrcIP = 129.174.142.2 \vee DestIP = 129.174.142.2$. We say a focusing constraint C_f is *enforceable w.r.t. a hyper-alert type* T if when we represent C_f in a disjunctive normal form, at least for one disjunct C_{fi}, all the attribute names in C_{fi} appear in T. For example, the above focusing constraint is enforceable w.r.t. $T = (\{SrcIP, SrcPort\}, NULL, \emptyset)$, but not w.r.t. $T' = (\{VictimIP, VictimPort\}, NULL, \emptyset)$. Intuitively, a focusing constraint is enforceable w.r.t. T if it can be evaluated using a hyper-alert instance of type T.

We may *evaluate* a focusing constraint C_f with a hyper-alert h if C_f is enforceable w.r.t. the type of h. A focusing constraint C_f evaluates to True for h if there exists a tuple $t \in h$ such that C_f is True with the attribute names replaced with the values of the corresponding attributes of t; otherwise, C_f evaluates to False. For example, consider the aforementioned focusing constraint C_f, which is $SrcIP = 129.174.142.2 \vee DestIP = 129.174.142.2$, and a hyper-alert $h = \{(SrcIP = 129.174.142.2, SrcPort = 80)\}$, we can easily have that C_f = True for h.

The idea of focused analysis is quite simple: we only analyze the hyper-alerts with which a focusing constraint evaluates to True. In other words, we would like to filter out irrelevant hyper-alerts, and concentrate on analyzing the remaining hyper-alerts. We are particularly interested in applying focusing constraints to *atomic hyper-alerts*, i.e., hyper-alerts with only one tuple. In our framework, atomic hyper-alerts correspond to the alerts reported by an IDS directly.

Focused analysis is particularly useful when we have certain knowledge of the alerts, the systems being protected, or the attacking computers. For example, if we are interested in the attacks against a critical server with IP address *Server_IP*, we may perform a focused analysis using *DestIPAddress* = *Server_IP*. However, focused analysis cannot take advantage of the intrinsic relationship among the hyper-alerts (e.g., hyper-alerts having the same IP address). In the following, we introduce the third utility, graph decomposition, to fill in this gap.

4.3 Graph Decomposition Based on Hyper-alert Clusters

The purpose of graph decomposition is to use the inherent relationship between (the attributes of) hyper-alerts to decompose a hyper-alert correlation graph. Conceptually, we cluster the hyper-alerts in a large correlation graph based on the "common features" shared by hyper-alerts, and then decompose the original correlation graphs into subgraphs on the basis of the clusters. In other words, hyper-alerts should remain in the same graph only when they share certain common features.

We use a *clustering constraint* to specify the "common features" for clustering hyper-alerts. Given two sets of attribute names A_1 and A_2, a *clustering constraint* $C_c(A_1, A_2)$

is a logical combination of comparisons between constants and attribute names in A_1 and A_2. (In our work, we restrict logical operations to AND ($\wedge$), OR ($\vee$), and NOT ($\neg$).) A clustering constraint is a constraint for two hyper-alerts; the attribute sets A_1 and A_2 identify the attributes from the two hyper-alerts. For example, we may have two sets of attribute names $A_1 = \{SrcIP, DestIP\}$ and $A_2 = \{SrcIP, DestIP\}$, and $C_c(A_1, A_2) = (A_1.SrcIP = A_2.SrcIP) \wedge (A_1.DestIP = A_2.DestIP)$. Intuitively, this is to say two hyper-alerts should remain in the same cluster if they have the same source and destination IP addresses.

A clustering constraint $C_c(A_1, A_2)$ is *enforceable w.r.t. hyper-alert types* T_1 *and* T_2 if when we represent $C_c(A_1, A_2)$ in a disjunctive normal form, at least for one disjunct C_{ci}, all the attribute names in A_1 appear in T_1 and all the attribute names in A_2 appear in T_2. For example, the above clustering constraint is enforceable w.r.t. T_1 and T_2 if both of them have $SrcIP$ and $DestIP$ in the $fact$ component. Intuitively, a focusing constraint is enforceable w.r.t. T if it can be evaluated using two hyper-alerts of types T_1 and T_2, respectively.

If a clustering constraint $C_c(A_1, A_2)$ is enforceable w.r.t. T_1 and T_2, we can *evaluate* it with two hyper-alerts h_1 and h_2 that are of type T_1 and T_2, respectively. A clustering constraint $C_c(A_1, A_2)$ evaluates to True for h_1 and h_2 if there exists a tuple $t_1 \in h_1$ and $t_2 \in h_2$ such that $C_c(A_1, A_2)$ is True with the attribute names in A_1 and A_2 replaced with the values of the corresponding attributes of t_1 and t_2, respectively; otherwise, $C_c(A_1, A_2)$ evaluates to False. For example, consider the clustering constraint $C_c(A_1, A_2) : (A_1.SrcIP = A_2.SrcIP) \wedge (A_1.DestIP = A_2.DestIP)$, and hyper-alerts $h_1 = \{(SrcIP = 129.174.142.2, SrcPort = 1234, DestIP = 152.1.14.5, DestPort = 80)\}$, $h_2 = \{(SrcIP = 129.174.142.2, SrcPort = 65333, DestIP = 152.1.14.5, DestPort = 23)\}$, we can easily have that $C_c(A_1, A_2)$ = True for h_1 and h_2. For brevity, we write $C_c(h_1, h_2)$ = True if $C_c(A_1, A_2)$ = True for h_1 and h_2.

Our clustering method is very simple with a user-specified clustering constraint $C_c(A_1, A_2)$. Two hyper-alerts h_1 and h_2 are in the same cluster if $C_c(A_1, A_2)$ evaluates to True for h_1 and h_2 (or h_2 and h_1). Note that $C_c(h_1, h_2)$ implies that h_1 and h_2 are in the same cluster, but h_1 and h_2 in the same cluster do not always imply $C_c(h_1, h_2)$ = True or $C_c(h_2, h_1)$ = True. This is because $C_c(h_1, h_2) \wedge C_c(h_2, h_3)$ does not imply $C_c(h_1, h_3)$, nor $C_c(h_3, h_1)$.

4.4 Discussion

The alert correlation method is developed to uncover the high-level strategies behind a sequence of attacks, not to replace the original alerts reported by an IDS. However, as indicated by our initial experiments [7], alert correlation does provide evidence to differentiate between alerts. If an alert is correlated with some others, it is more possible that the alert corresponds to an actual attack.

It is desirable to develop a technique which can comprehend a hyper-alert correlation graph and generate feedback to direct intrusion detection and response processes. We consider such a technique as a part of our future research plan. However, given the current status of intrusion detection and response techniques, it is also necessary to allow human users to understand the attacks and take appropriate actions.

Table 1. General statistics of the initial analysis

# total hyper-alert types	115	# total hyper-alerts	65054
# correlated hyper-alert types	95	# correlated	9744
# uncorrelated hyper-alert types	20	# uncorrelated	55310
# partially correlated hyper-alert types	51	% correlated	15%

The three utilities developed in this section are intended to help human users analyze attacks behind large amounts of alerts. They can make attack strategies behind intensive alerts easier to understand, but cannot improve the performance of alert correlation.

5 Analyzing DEF CON 8 CTF Dataset: A Case Study

To study the effectiveness of the alert correlation method and the utilities proposed in Section 4, we performed a series of experiments on the network traffic collected at the DEF CON 8 CTF event [10]. In our experiments, we used NetPoke[1] to replay the network traffic in an isolated network monitored by a RealSecure Network Sensor 6.0 [23]. In all the experiments, the Network Sensor was configured to use the *Maximum_Coverage* policy with a slight change, which forced the Network Sensor to save all the reported alerts. Our alert correlator [7] was then used to process the alerts to discover the hyper-alert correlation graphs. The hyper-alert correlation graphs were visualized using the GraphViz package [24]. For the sake of readability, transitive edges are removed from the graphs.

In these experiments, we mapped each alert type reported by the RealSecure Network Sensor to a hyper-alert type (with the same name). The prerequisite and consequence of each hyper-alert type were specified according to the descriptions of the attack signatures provided with the RealSecure Network Sensor 6.0.

It would be helpful for the evaluation of our method if we could identify false alerts, alerts for sequences of attacks, and alerts for isolated attacks. Unfortunately, due to the nature of the dataset, we are unable to obtain any of them. Thus, in this study, we focus on the analysis of the attack strategies reflected by hyper-alert correlation graphs, but only discuss the uncorrelated alerts briefly.

5.1 Initial Attempt

In our initial analysis of the DEF CON 8 CTF dataset, we tried to correlate the hyper-alerts without reducing the complexity of any hyper-alert correlation graphs. The statistics of the initial analysis are shown in Table 1.

Table 1 shows that only 15% alerts generated by RealSecure are correlated. In addition, 20 out of 115 hyper-alert types that appear in this data set do not have any instances correlated. Among the remaining 95 hyper-alert types, 51 types have both correlated and uncorrelated instances.

[1] NetPoke is a utility to replay packets to a live network that were previously captured with the tcpdump program.
`http://www.ll.mit.edu/IST/ideval/tools/tools_index.html`

Table 2. Statistics of top 10 uncorrelated hyper-alert types.

Hyper-alert type	# uncorrelated alerts	# correlated alerts	Hyper-alert type	# uncorrelated alerts	# correlated alerts
IPHalfScan	33745	958	Windows_Access_Error	11657	0
HTTP_Cookie	2119	0	SYNFlood	1306	406
IPDuplicate	1063	0	PingFlood	1009	495
SSH_Detected	731	0	Port_Scan	698	725
ServiceScan	667	2156	Satan	593	280

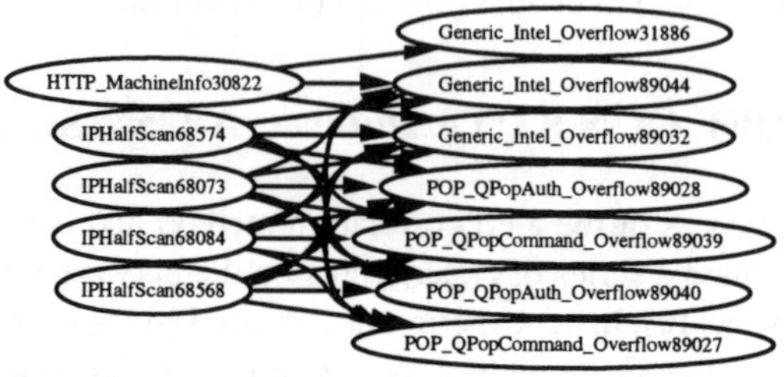

Fig. 3. A small hyper-alert correlation discovered in initial analysis

Table 2 shows the statistics of the top 10 uncorrelated hyper-alert types (in terms of the number of uncorrelated hyper-alerts). Among these hyper-alert types, uncorrelated *IPHalfScan* counted 61% of all uncorrelated hyper-alerts. *Windows_Access_Error* counted 21% of all uncorrelated alerts. According to the description provided by RealSecure, a *Windows_Access_Error* represents an unsuccessful file sharing connection to a Windows or Samba server, which usually results from an attempt to brute-force a login under another account's privileges. It is easy to see that the corresponding attacks could hardly prepare for any other attacks (since they failed). The third largest hyper-alert type *HTTP_Cookie* counted for 3.3% of the total alerts. Though such alerts have certain privacy implications, we do not treat them as attacks, considering the nature of the DEF CON CTF events. These three hyper-alert types counted for 74.5% of all the alerts. We omit the discussion of the other uncorrelated hyper-alerts.

Figure 3 shows one of the small hyper-alert correlation graphs. The text in each node is the type followed by the ID of the hyper-alert. All the hyper-alerts in this figure were destined to the host at 010.020.001.024. All the *IPHalfScan* attacks were from source IP 010.020.011.240 at source port 55533 or 55534, and destined to port 110 at the victim host. After these attacks, all the attacks in the second stage except for 31886 were from 010.020.012.093 and targeted at port 110 of the victim host. The only two hyper-alerts that were not targeted at port 110 are hyper-alert 30882, which was destined to port 80 of the victim host, and hyper-alert 31886, which was destined to port 53. Thus, it is very possible that all the hyper-alerts except for 30882 and 31886 were related.

Not all of the hyper-alert correlation graphs are as small and comprehensible as Figure 3. In particular, the largest graph (in terms of the number of nodes) has 2,940 nodes and 25,321 edges, and on average, each graph has 21.75 nodes and 310.56 edges.

Obviously, most of the hyper-alert correlation graphs are too big to understand for a human user.

5.2 Graph Reduction

We further analyzed the hyper-alert correlation graphs with the three utilities proposed in Section 4. Due to space reasons, we only report our analysis results about the largest hyper-alert correlation graph in this section.

We first applied graph reduction utility to the hyper-alert correlation graphs. Figure 4 shows the fully reduced graph. Compared with the original graph, which has 2,940 nodes and 25,321 edges, the fully reduced graph has 77 nodes and 347 edges (including transitive edges).

The fully reduced graph in Figure 4 shows 7 stages of attacks. The layout of this graph was generated by GraphViz [24], which tries to reduce the number of cross edges and make the graph more balanced. As a result, the graph does not reflect the actual stages of attacks. Nevertheless, Figure 4 provides a much clearer outline of the attacks.

The hyper-alerts in stage 1 and about half of those in stage 2 correspond to scanning attacks or attacks to gain information of the target systems (e.g., *ISS, Port_Scan*). The upper part of stage 2 include attacks that may lead to execution of arbitrary code on a target system (e.g., *HTTP_WebSite_Sample*). Indeed, these hyper-alerts directly prepare for some hyper-alerts in stage 5, but GraphViz arranged them in stage 2, possibly to balance the graph. Stages 3 consists of a mix of scanning attacks (e.g., *Nmap_Scan*), attacks that reveal system information (e.g,, *HTTP_PHP_Read*), and attacks that may lead to execution of arbitrary code (e.g., *HTTP_Campas*). Stage 4 mainly consists of buffer overflow attacks (e.g., *POP_QPopCommand_Overflow*), detection of backdoor programs (e.g., *BackOrifice*), and attacks that may lead to execution of arbitrary code. The next 3 stages are much cleaner. Stage 5 consists of attacks that may be used to copy programs to target hosts, stage 6 consists of detection of two types of DDOS (Distributed Denial of Service) daemon programs, and finally, stage 7 consists of the detection of an actual DDOS attack.

Note that the fully reduce graph in Figure 4 is an approximation to the strategies used by the attackers. Hyper-alerts for different, independent sequences of attacks may be aggregated together in such a graph. For example, if two individual attackers use the sequence of attacks (e.g., using the same script downloaded from a website) to attack the same target, the corresponding hyper-alerts may be correlated and aggregated in the same fully reduced graph. Nevertheless, a fully reduced graph can clearly outline the attack strategies, and help a user understand the overall situation of attacks.

As we discussed earlier, the reduction of hyper-alert correlation graphs can be controlled with interval constraints. Figure 5 shows the numbers of nodes and edges of the reduced graphs for different interval sizes. The shapes of the two curves in Figure 5 indicate that most of the hyper-alerts that are of the same type occurred close to each other in time. Thus, the numbers of nodes and edges have a deep drop for small interval thresholds and a flat tail for large ones. A reasonable guess is that some attackers tried the same type of attacks several times before they succeeded or gave up. Due to space reasons, we do not show these reduced graphs.

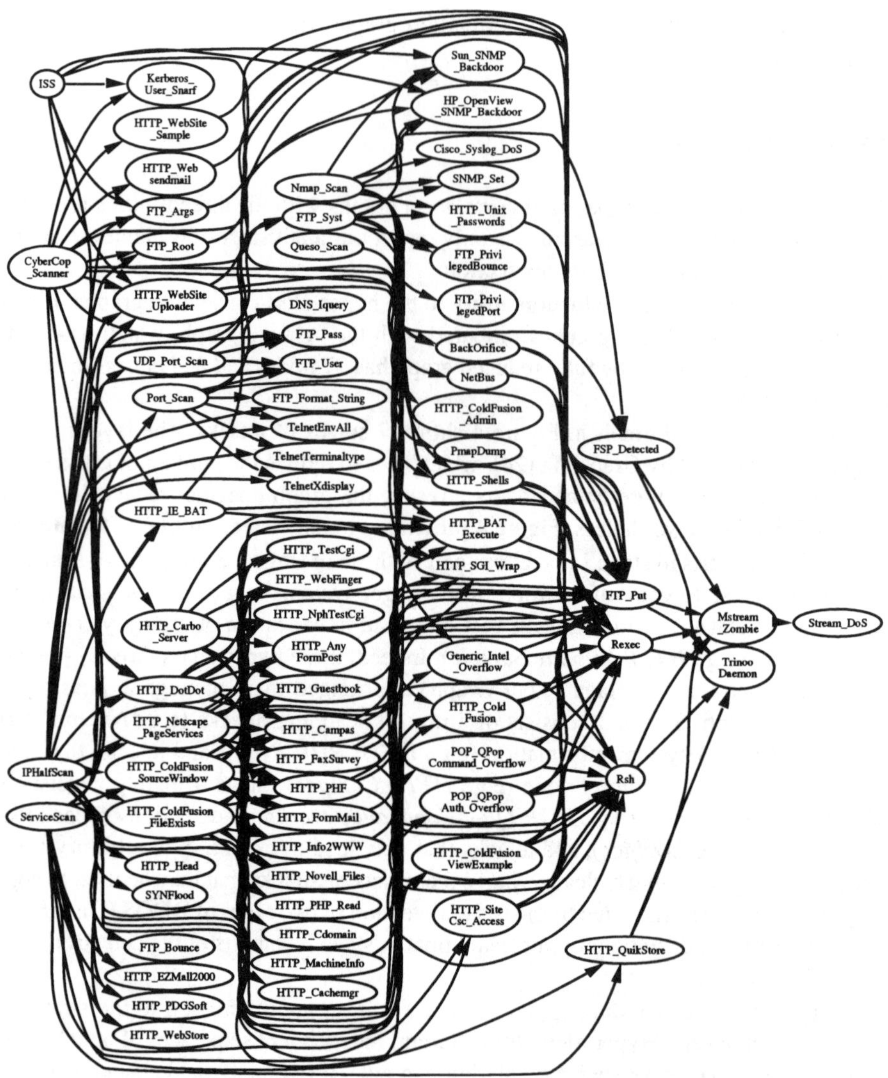

Fig. 4. The fully reduced graph for the largest aggregated hyper-alert correlation graph.

5.3 Focused Analysis

Focused analysis can help filter out the interesting parts of a large hyper-alert correlation graph. It is particularly useful when a user knows the systems being protected or the potential on-going attacks. For example, a user may perform a focused analysis with focusing constraint $DestIP = ServerIP$, where $ServerIP$ is the IP address of a critical server, to find out attacks targeted at the server. As another example, he/she may

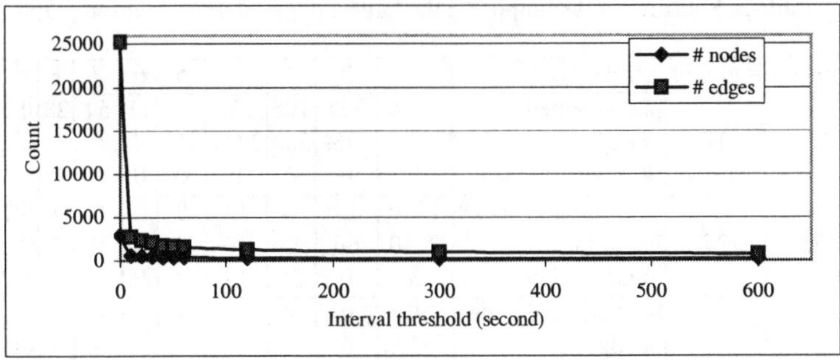

Fig. 5. Sizes of the reduced graphs w.r.t. the interval threshold for the largest hyper-alert correlation graph

use $SrcIP = ServerIP \vee DestIP = ServerIP$ to find out attacks targeted at or originated from the server, suspecting that the server may have been compromised.

In our experiments, we tried a number of focusing constraints after we learned some information about the systems involved in the CTF event. Among these focusing constraints are (1) C_{f1} : ($DestIP$ = 010.020.001.010) and (2) C_{f2} : ($SrcIP$ = 010.020.011.251 $\wedge DestIP$ = 010.020.001.010). We applied both focusing constraints to the largest hyper-alert correlation graph. The results consist of 2154 nodes and 19423 edges for C_{f1}, and 51 nodes and 28 edges for C_{f2}. The corresponding fully reduced graphs are shown in Fig. 6 and Fig. 7, respectively. (Isolated nodes are shown in gray.) These two graphs also appear in the results of graph decomposition (Section 5.4). We defer the discussion of these two graphs to the next subsection.

Focused analysis is an attempt to approximate a sequence of attacks that satisfy the focusing constraint. Its success depends on the closeness of focusing constraints to the invariants of the sequences of attacks. A cunning attacker would try to avoid being correlated by launching attacks from different sources (or stepping stones) and introducing delays in between attacks. Thus, this utility should be used with caution.

5.4 Graph Decomposition

We applied three clustering constraints to decompose the largest hyper-alert correlation graph discovered in Section 5.1. In all these clustering constraints, we let $A_1 = A_2 = \{SrcIP, DestIP\}$.

1. $C_{c1}(A_1, A_2)$: $A_1.DestIP = A_2.DestIP$. This is to cluster all hyper-alerts that share the same destination IP addresses. Since most of attacks are targeted at the hosts at the destination IP addresses, this is to cluster hyper-alerts in terms of the victim systems.
2. $C_{c2}(A_1, A_2)$: $A_1.SrcIP = A_2.SrcIP \wedge A_1.DestIP = A_2.DestIP$. This is to cluster all the hyper-alerts that share the same source and destination IP addresses.
3. $C_{c3}(A_1, A_2)$: $A_1.SrcIP = A_2.SrcIP \vee A_1.DestIP = A_2.DestIP \vee A_1.SrcIP = A_2.DestIP \vee A_1.DestIP = A_2.SrcIP$. This is to cluster all the hyper-alerts that

Table 3. Statistics of decomposing the largest hyper-alert correlation graph.

	# clusters	# graphs	cluster ID	1	2	3	4	5	6	7	8	9	10	11	12
			# connected nodes	2154	224	105	227	83	11	54	28	0	23	6	0
C_{c1}	12	10	# edges	19423	1966	388	2741	412	30	251	51	0	26	5	0
			# isolated nodes	0	0	0	0	0	0	0	0	1	0	0	4
			# correlated nodes	1970	17	0	12	0	0	0	3	0	29	0	0
C_{c2}	185	37	# edges	2240	66	0	10	0	0	0	2	0	28	0	0
			# isolated nodes	3	0	21	17	35	26	15	12	4	22	13	26
			# connected nodes	2935	0	–	–	–	–	–	–	–	–	–	–
C_{c3}	2	1	# edges	25293	0	–	–	–	–	–	–	–	–	–	–
			# isolated nodes	4	1	–	–	–	–	–	–	–	–	–	–

are connected via common IP addresses. Note that with this constraint, hyper-alerts in the same cluster may not share the same IP address directly, but they may connect to each other via other hyper-alerts.

Table 3 shows the statistics of the decomposed graphs. C_{c1} resulted in 12 clusters, among which 10 clusters contain edges. C_{c2} resulted in 185 clusters, among which 37 contain edges. Due to space reasons, we only show the first 12 clusters for C_{c2}. C_{c3} in effect removes one hyper-alert from the original graph. This hyper-alert is *Stream_DoS*, which does not share the same IP address with any other hyper-alerts. This is because the source IPs of *Stream_DoS* are spoofed and the destination IP is the target of this attack. This result shows that all the hyper-alerts except for *Stream_DoS* share a common IP address with some others.

The isolated nodes in the resulting graphs are the hyper-alerts that prepare for or are prepared for by those that do not satisfy the same clustering constraints. Note that having isolated hyper-alerts in a decomposed graph does not imply that the isolated hyper-alerts are correlated incorrectly. For example, an attacker may hack into a host with a buffer overflow attack, install a DDOS daemon, and start the daemon program, which then tries to contact its master program. The corresponding alerts (i.e., the detection of the buffer overflow attack and the daemon's message) will certainly not have the same destination IP address, though they are related.

Figures 6 and 7 show a decomposed graph for C_{c1} and C_{c2}, respectively. Both graphs are fully reduced to save space. All the hyper-alerts in Figure 6 are destined to 010.020.001.010. Figure 6 shows several possible attack strategies. The most obvious ones are those that lead to the *Mstream_Zoombie* and *TrinooDaemon*. However, there are multiple paths that lead to these two hyper-alerts. Considering the fact that multiple attackers participated in the DEF CON 8 CTF event, we cannot conclude which path caused the installation of these daemon programs. Indeed, it is possible that none of them is the actual way, since the IDS may have missed some attacks.

Figure 6 involves 75 source IP addresses, including IP address 216.136.173.152, which does not belong to the CTF subnet. We believe that these attacks belong to different sequences of attacks, since there were intensive attacks from multiple attackers who participated in the CTF event.

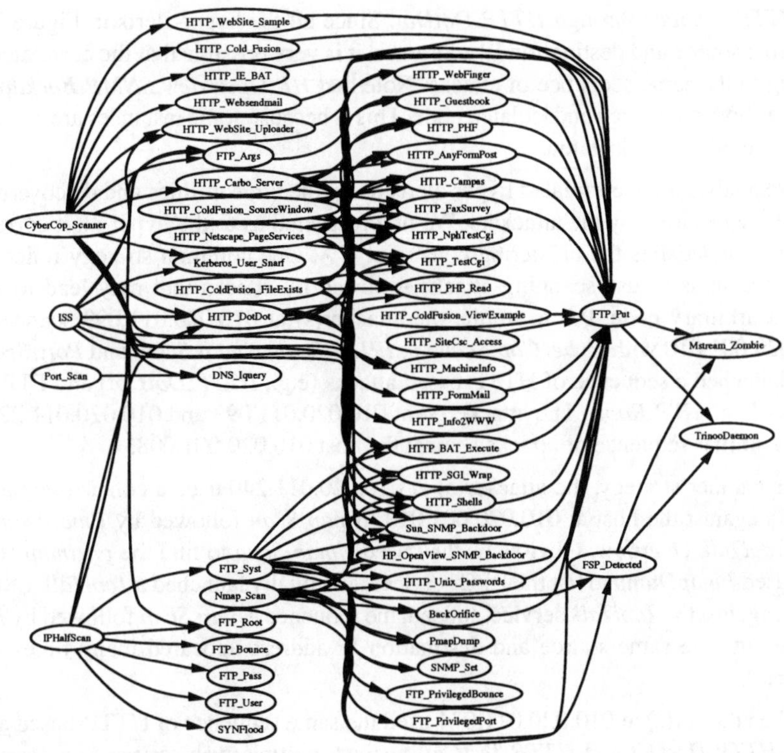

Fig. 6. A fully reduced hyper-alert correlation graph resulting from graph decomposition with C_{c1}. (Cluster ID = 1; DestIP = 010.020.001.010.)

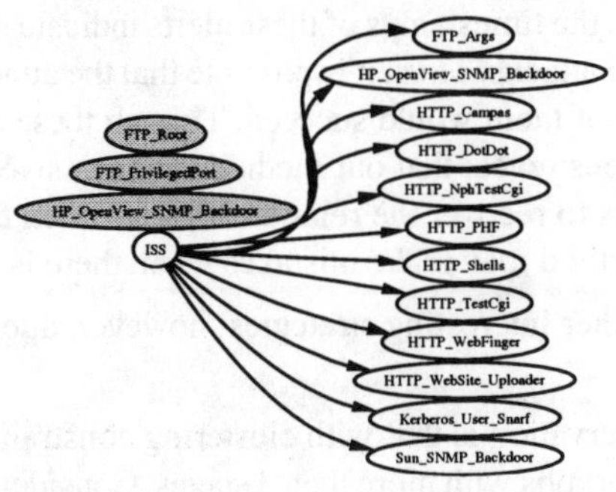

Fig. 7. A fully reduced hyper-alert correlation graph resulting from graph decomposition with C_{c2}. (Cluster ID = 10; SrcIP = 010.020.011.251; DestIP = 010.020.001.010.)

Figure 7 is related to Figure 6, since they both are about destination IP address 010.020.001.010. Indeed, Figure 7 is a part of Figure 6, though in Figure 6, *ISS* prepares for *HTTP_Campas* through *HTTP_DotDot*. Since all the hyper-alerts in Figure 7 have the same source and destination IP addresses, it is very possible that the correlated ones belong to the same sequence of attacks. Note that *HP_OpenView_SNMP_Backdoor* appears as both connected and isolated node. This is because some instances are correlated, while the others are isolated.

We analyzed the correlated hyper-alerts using the three utilities and discovered several strategies used by the attackers. We first restricted us to the hyper-alert correlation graphs that satisfies the clustering constraint C_{c2}. One common strategy reflected by these graphs is to use scanning attacks followed by attacks that may lead to execution of arbitrary code. For example, the attacker(s) at 010.020.011.099 scanned host 010.020.001.010 with *CyberCop_Scanner, IPHalfScan, Nmap_Scan*, and *Port_Scan* and then launched a sequence of HTTP-based attacks (e.g., *HTTP_DotDot*) and FTP based attacks (e.g., *FTP_Root*). The attacker(s) at 010.020.011.093 and 010.020.011.227 also used a similar sequence of attacks against the host 010.020.001.008.

As another strategy, the attacker(s) at 010.020.011.240 used a concise sequence of attacks against the host at 010.020.001.013: *Nmap_Scan* followed by *PmapDump* and then *ToolTalk_Overflow*. Obviously, they used *Nmap_Scan* to find the *portmap* service, then used *PmapDump* to list the RPC services, and finally launched a *ToolTalk_Overflow* attack against the *ToolTalk* service. Indeed, the sequence *Nmap_Scan* followed by *PmapDump* with the same source and destination IP address appeared many times in this dataset.

The attacker(s) at 010.020.011.074 used the same sequence of HTTP-based attacks (e.g., *HTTP_DotDot* and *HTTP_TestCgi*) against multiple web servers (e.g., servers at 010.020.001.014, 010.020.001.015, 010.020.001.019, etc.). Our hyper-alert correlation graphs shows that *HTTP_DotDot* prepares for the following HTTP-based attacks. However, our further analysis of the dataset shows that this may be an incorrect correlation. Though it is possible that the attacker used *HTTP_DotDot* to collect necessary information for the later attacks, the timestamps of these alerts indicate that the attacker(s) used a script to launch all these attacks. Thus, it is possible that the attacker(s) simply launch all the attacks, hoping one of them would succeed. Though these alerts are indeed related, these prepare-for relations reveal that our method is aggressive in correlating alerts. Indeed, alert correlation is to recover the relationships between the attacks behind alerts; any alert correlation method may make mistakes when there is not enough information.

There are several other interesting strategies; however, due to space reasons, we do not list them here.

One interesting observation is that with clustering constraint C_{c2}, there is not many hyper-alert correlation graphs with more than 3 stages. Considering the fact that there are many alerts about *BackOrifice* and *NetBus* (which are tools to remotely manage hosts), we suspect that many attackers used multiple machines during their attacks. Thus, their strategies cannot be reflected by the restricted hyper-alert correlation graphs.

When relax the restriction to allow hyper-alert correlation graphs involving different source IP but still with the same destination IP addresses (i.e., with clustering constraint C_{c1}), we have graphs with more stages. Figure 6 is one such fully reduced hyper-alert correlation graph. However, due to the amount of alerts and source IP addresses involved in this graph, it is difficult to conclude which hyper-alerts belong to the same sequences of attacks.

In summary, during the analysis of the DEF CON 8 CTF dataset, the utilities have greatly simplified the analysis process. We have discovered several attack strategies that were possibly used during the attacks. However, there are a number of situations where we could not separate multiple sequences of attacks. This implies that additional work is necessary to address this problem.

6 Conclusion and Future Work

In this paper, we presented three utilities, *adjustable graph reduction*, *focused analysis*, and *graph decomposition*, which were developed to facilitate the analysis of large sets of correlated alerts. We studied the effectiveness of these utilities through a case study with the DEF CON 8 CTF dataset [10]. Our results show that these utilities can simplify the analysis of large amounts of alerts. In addition, our analysis reveals several attack strategies that are repeatedly used in the DEF CON 8 CTF event.

Due to the nature of the DEF CON 8 CTF dataset, we were unable to evaluate the successful rate of the alert correlation method. Although we have found several attack strategies, we also encountered several situations where it is difficult to conclude about attack strategies. Indeed, a hyper-alert correlation graph is an approximation to a real attack strategy. Thus, the hyper-alert correlation method and the three utilities should be used with caution.

Our future work includes several problems, including seeking more techniques to improve alert correlation, automating the analysis process, further refinement of our toolkit, and systematic development of hyper-alert types. We also plan to analyze alert streams from operational networks such as the NC State University campus network.

Acknowledgement. This work is partially supported by the U.S. Army Research Office under grant DAAD19-02-1-0219, by the National Science Foundation under grant 0207297, and by NCSU Center for Advanced Computing & Communication. The authors would like to thank the anonymous reviewers for their valuable comments.

References

1. Javits, H., Valdes, A.: The NIDES statistical component: Description and justification. Technical report, SRI International, Computer Science Laboratory (1993)
2. Vigna, G., Kemmerer, R.A.: NetSTAT: A network-based intrusion detection system. Journal of Computer Security **7** (1999) 37–71
3. Valdes, A., Skinner, K.: Probabilistic alert correlation. In: Proceedings of the 4th International Symposium on Recent Advances in Intrusion Detection (RAID 2001). (2001) 54–68
4. Debar, H., Wespi, A.: Aggregation and correlation of intrusion-detection alerts. In: Recent Advances in Intrusion Detection. LNCS 2212 (2001) 85 – 103
5. Dain, O., Cunningham, R.: Fusing a heterogeneous alert stream into scenarios. In: Proceedings of the 2001 ACM Workshop on Data Mining for Security Applications. (2001) 1–13

6. Ning, P., Reeves, D.S., Cui, Y.: Correlating alerts using prerequisites of intrusions. Technical Report TR-2001-13, North Carolina State University, Department of Computer Science (2001)
7. Ning, P., Cui, Y.: An intrusion alert correlator based on prerequisites of intrusions. Technical Report TR-2002-01, North Carolina State University, Department of Computer Science (2002)
8. MIT Lincoln Lab: 2000 DARPA intrusion detection scenario specific datasets. http://www.ll.mit.edu/IST/ideval/data/2000/2000_data_index.html (2000)
9. Manganaris, S., Christensen, M., Zerkle, D., Hermiz, K.: A data mining analysis of RTID alarms. Computer Networks **34** (2000) 571–577
10. DEFCON: Def con capture the flag (CTF) contest. http://www.defcon.org/html/defcon-8-post.html (2000) Archive accessible at http://wi2600.org/mediawhore/mirrors/shmoo/.
11. Bace, R.: Intrusion Detection. Macmillan Technology Publishing (2000)
12. Staniford, S., Hoagland, J., McAlerney, J.: Practical automated detection of stealthy portscans. To appear in Journal of Computer Security (2002)
13. Templeton, S., Levit, K.: A requires/provides model for computer attacks. In: Proceedings of New Security Paradigms Workshop, ACM Press (2000) 31 – 38
14. Cuppens, F., Miege, A.: Alert correlation in a cooperative intrusion detection framework. In: Proceedings of the 2002 IEEE Symposium on Security and Privacy. (2002)
15. Staniford-Chen, S., Cheung, S., Crawford, R., Dilger, M., Frank, J., Hoagland, J., Levitt, K., Wee, C., Yip, R., Zerkle, D.: GrIDS - a graph based intrusion detection system for large networks. In: Proceedings of the 19th National Information Systems Security Conference. Volume 1. (1996) 361–370
16. Ilgun, K., Kemmerer, R.A., Porras, P.A.: State transition analysis: A rule-based intrusion detection approach. IEEE Transaction on Software Engineering **21** (1995) 181–199
17. Cuppens, F., Ortalo, R.: LAMBDA: A language to model a database for detection of attacks. In: Proc. of Recent Advances in Intrusion Detection (RAID 2000). (2000) 197–216
18. Lin, J., Wang, X.S., Jajodia, S.: Abstraction-based misuse detection: High-level specifications and adaptable strategies. In: Proceedings of the 11th Computer Security Foundations Workshop, Rockport, MA (1998) 190–201
19. Ning, P., Jajodia, S., Wang, X.S.: Abstraction-based intrusion detection in distributed environments. ACM Transactions on Information and System Security **4** (2001) 407–452
20. Gruschke, B.: Integrated event management: Event correlation using dependency graphs. In: Proceedings of the 9th IFIP/IEEE International Workshop on Distributed Systems: Operations & Management. (1998)
21. Ricciulli, L., Shacham, N.: Modeling correlated alarms in network management systems. In: In Western Simulation Multiconference. (1997)
22. Gardner, R., Harle, D.: Pattern discovery and specification translation for alarm correlation. In: Proceedings of Network Operations and Management Symposium (NOMS'98). (1998) 713–722
23. ISS, Inc.: RealSecure intrusion detection system. (http://www.iss.net)
24. AT & T Research Labs: Graphviz - open source graph layout and drawing software. (http://www.research.att.com/sw/tools/graphviz/)

A Mission-Impact-Based Approach to INFOSEC Alarm Correlation[1]

Phillip A. Porras, Martin W. Fong, and Alfonso Valdes

SRI International
333 Ravenswood Avenue
Menlo Park, CA 94025-3493
{porras,mwfong,valdes}@sdl.sri.com

Abstract. We describe a mission-impact-based approach to the analysis of security alerts produced by spatially distributed heterogeneous information security (INFOSEC) devices, such as firewalls, intrusion detection systems, authentication services, and antivirus software. The intent of this work is to deliver an automated capability to reduce the time and cost of managing multiple INFOSEC devices through a strategy of topology analysis, alert prioritization, and common attribute-based alert aggregation. Our efforts to date have led to the development of a prototype system called the EMERALD Mission Impact Intrusion Report Correlation System, or M-Correlator. M-Correlator is intended to provide analysts (at all experience levels) a powerful capability to automatically fuse together and isolate those INFOSEC alerts that represent the greatest threat to the health and security of their networks.

Keywords: Network security, intrusion report correlation, alert management, alert prioritization.

1 Introduction

Among the most visible areas of active research in the intrusion detection community is the development of technologies to manage and interpret security-relevant alert streams produced from an ever-increasing number of INFOSEC devices. While the bulk of the work in security alert management and intrusion report correlation has spawned from the intrusion detection community, this paper takes a much broader definition of alert stream contributors. Over recent years, the growing number of security enforcement services, access logs, intrusion detection systems, authentication servers, vulnerability scanners, and various operating system and applications logs have given administrators a potential wealth of information to gain insight into security-relevant activities occurring within their systems. We broadly define these various security-relevant log producers as INFOSEC devices, and recognize them as having potential contributions to the problems of security incident detection and confidence reinforcement in discerning the credibility of INFOSEC alarms.

[1] Supported by DARPA through Air Force Research Laboratory, contract F30602-99-C-0187.

A. Wespi, G. Vigna, and L. Deri (Eds.): RAID 2002, LNCS 2516, pp. 95-114, 2002.

Unfortunately, this broader view of alert stream contributors adds to the complexity facing intrusion report correlation systems. INFOSEC devices range greatly in function, even within a single technology. For example, within the intrusion detection space, the variety of analysis methods that may be employed, the spatial distribution of sensors, and their target event streams (network traffic, host audit logs, other application logs), increases the difficulty in understanding the semantics of what each sensor is reporting, as well as the complexity of determining equivalence among the intrusion reports from different sensors.

The motivation for our work is straightforward: as we continue to incorporate and distribute advanced security services into our networks, we need the ability to understand the various forms of hostile and fault-related activity that our security services observe as they help to preserve the operational requirements of our systems. Today, in the absence of significant fieldable technology for security-incident correlation, there are several challenges in providing effective security management for mission-critical network environments:

- Domain expertise is not widely available that can interpret and isolate high threat operations within active and visible Internet-connected networks. Also not widely available are skills needed to understand the conditions under which one may merge INFOSEC alerts from different sources (e.g., merging firewall and OS syslogs with intrusion detection reports). In an environment where thousands (or tens of thousands) of INFOSEC alarms may be produced daily, it is important to understand redundancies in alert production that can simplify alert interpretation. Equally important are algorithms for prioritizing which security incidents pose the greatest administrative threats.
- The sheer volume of INFOSEC device alerts makes security management a time-consuming and therefore expensive effort [13]. There are numerous examples of organizations that have found even small deployment of IDS sensors to be an overwhelming management cost. As a result, these IDS components are often tuned down to an extremely narrow and ad hoc selection of a few detection heuristics, effectively minimizing the coverage of the IDS tool.
- In managing INFOSEC devices, it is difficult to leverage potentially complementary information produce from heterogeneous INFOSEC devices. For example, is captured in a firewall log, is typically manually analyzed in isolation from potentially relevant alert information captured by an IDS, syslog, or other INFOSEC alert source.

The remainder of this paper describes the design, implementation, and provides illustrative experiments developed during a two-year research study of IDS interoperability and intrusion report management that address the above issues.

2 EMERALD M-Correlator Algorithm Overview

M-Correlator is designed to consolidate and rank a stream of security incidents relative to the needs of the analyst, given the topology and operational objectives of the protected network. The first phase of INFOSEC alert processing involves dynamically controllable filters, which provide remote subscribers with an ability to eliminate low-interest alerts, while not preventing INFOSEC devices from producing these alerts that may be of interest to other analysts. Next, the alerts are vetted against the known topology of the target network. A *relevance score* (Section 2.2) is produced through a comparison of the alert target's known topology against the known vulnerability requirements of the incident type (i.e., incident vulnerability dependencies). Vulnerability dependency information is provided to M-Correlator through an *Incident Handling Fact Base* (Section 2.1). Next, a *priority calculation* (Section 2.3) is performed per alert to indicate (a) the degree to which an alert is targeting a critical asset or resource, and (b) the amount of interest the user has registered for this class of security alert. Last, an overall *incident rank* (Section 2.4) is assigned to each alert, which provides a combined assessment of the degree to which the incident appears to impact the overall mission of the network, and the probability that the activity reported in this alert was successful.

M-Correlator next attempts to combine related alerts with an attribute-based *alert clustering algorithm* (Section 3). The resulting correlated incident stream represents a filtered, lower-volume, content rich security-incident stream, with an incident-ranking scheme that allows analysts to identify those incidents that pose the greatest risk to the currently specified mission objectives of the monitored network.

2.1 An Incident Handling Fact Base

M-Correlator includes an *Incident Handling Fact Base* that provides the necessary information to optimally interpret alert content against the mission specification and relevance analysis. The incident handling fact base provides critical information needed to

- Augment terse INFOSEC device alerts with meaningful descriptive information, and associate alerts with M-Correlator-specific incident codes and classifications
- Understand the dependencies of incident types to their required OS versions, hardware platform, network services, and applications
- Understand which incident types can be merged by the M-Correlator alert clustering algorithm

Table 1 enumerates the field definitions of entries in the M-Correlator incident handling fact base. Entries in this fact base are referenced in subsequent sections, which describe topology vetting, prioritization, incident ranking, and alert clustering. The current M-Correlator fact base provides incident definitions for more than 1,000 intrusion report types from ISS's Realsecure, Snort [20], the EMERALD [19] suite of host and network-based intrusion detection sensors, and Checkpoint's Firewall-1 product line. Incident types that are not represented in this fact base can still be managed and aggregated by the M-Correlator; however, the advanced alert clustering and relevance calculations are not performed on alerts that are absent from this fact base.

Table 1. Incident-Handling Fact-Base Field Definitions

Field Type	Description
Incident Code	A unique code to indicate incident type. These codes have been derived from the original Boeing/NAI IDIP incident codes that were used by the Common Intrusion Detection Framework CISL specification [10]. A mapping between this incident code and other well-known attack code specifications such as Bugtraq ID, CERT ID, and MITRE CVE codes is available using the References field.
COTS Codes	An equivalent code listing of well-known commercial off-the-shelf (COTS) incident name or numeric code value that expresses this incident.
Incident Class	An M-Correlator general categorization scheme used for abstractly registering interest in an incident that represents a common impact to the system. Incident types are associated with only one incident class (see Section 2.3 for details).
Description	Human-readable incident description.
Vulnerable OS and Hardware	OS type(s) and version(s), and hardware architectures required for the successful invocation of the incident.
Bound Ports and Applications	The list of required network services and applications that must be enabled on the target of an alert for this incident type to succeed.
Cluster List	One or more index values that may be associated with incident types. Two alerts that share a common cluster name may be candidates for merger should other attributes be aligned.
References	Bugtraq ID [3], CERT ID [4], Common Vulnerabilities and Exposures (CVE) ID [2,6], available descriptive URL.

2.2 Relevance Formulation

M-Correlator maintains an internal topology map of the protected network, which is dynamically managed by the analyst. Automated topology map generation is supported using *Nmap* [16], through which M-Correlator can identify the available assets on the network, IP address to hostname mappings, OS type and version information, active TCP and UDP network services per host, and hardware type. Nmap can be run on intervals to maintain an updated topology database, and this database can be dynamically inserted into the M-Correlator runtime process. Given the topology database and the vulnerable OS, hardware, and bound ports fields of the incident-handling knowledge (Section 2.1), M-Correlator develops a relevance score that assesses per alert, the likelihood of successful intrusion.

As each alert is processed by M-Correlator, the associated known dependencies for that alert, as indicated within the incident handling fact base, are compared against the configuration of the target machine. Positive and negative matches against these required dependencies result in increased or decreased weighting of the relevance score, respectively. Our model for calculating asset relevance may identify as many as five attributes that match the known topology of the target host:

- OS type and version
- Hardware type
- Service suite
- Enabled network service
- Application

The relevance score is calculated on a scale from 0 to 255. 0 indicates that the incident vulnerabilities required for the successful execution of the reported security incident were *not* matched to the known topology of the target host. An unknown alert, incompletely specified dependency information in the fact base, or incomplete topology information regarding the target host, results in a neutral relevance score of 127 (i.e., the score does not contribute positively or negatively to the overall incident rank for that security incident). Scores nearer to 255 indicate that the majority of required dependencies of the reported security incident were matched to the known topology of the target host.

2.3 Priority Formulation

The objective of mission impact analysis is to fuse related alerts into higher-level security incidents, and rank them based on the degree of threat each incident poses to the mission objectives of the target network. A *mission* is defined with respect to an administrative network domain. Mission-impact analysis seeks to isolate the highest threat security incidents together, providing the analyst with an ability to reduce the total number of incidents that must be reviewed. Abstractly, we define security incident prioritization as follows:

$$\text{Let} \quad \underline{Stream = \{e_1, e_2, e_3, \ldots, e_n\}}$$

$$\text{Find} \quad HighImpact = \{e_\alpha, e_\beta, \ldots, e_\psi\} \subseteq Stream$$

$$\underset{e_i \in \text{HighImpact}}{\forall} \text{Threat\_Rank}(e_i, \text{Mission}) > T_{\text{acceptable}}$$

The *mission* is the underlying objective for which the computing resources and data assets of the monitored network are brought together and used. We express this concept of mission through a *mission specification,* which is defined by the analyst. A mission specification is defined in two parts: (1) an enumeration by the analyst of those data assets and services that are most critical to the client users of the network, and (2) an identification of which classes of intrusion incidents are of greatest concern to the analyst. With respect to the critical assets and services of the protected network, the analyst must register the following items within the mission specification:

- Critical computing assets (such as file servers on which the user community depends)
- Critical network services (such as web server, a DBMS)
- Sensitive data assets (these are primarily files and directories considered highly sensitive or important to the mission of the network)
- Administrative and untrusted user accounts such as might be used by consultants

Next, the analyst can specify those intrusion incidents, or classes of incident, of greatest concern given the analyst's responsibilities within the organization. This portion of the mission specification is referred to as the *interest profile*. Interest profiles may be user specific, just as the responsibilities of analysts may be distinct. Each alert processed by M-Correlator is associated with a unique incident class type. Each incident signature listed in the incident handling knowledge base is associated with one of the following incident classes, which were derived, in part, from a review of previous work in incident classifications and vulnerability analysis [15, 2, 11]:

- Privilege Violation — Theft or escalation of access rights to that of system or administrative privileges.
- User Subversion — An attempt to gain privileges associated with a locally administered account. This may include reports of user masquerading.
- Denial of Service — An attempt to block or otherwise prevent access to an internal asset, including host, application, network service, or system resource, such as data or a device.
- Probe - An attempt to gain information on assets or services provided within the monitored domain.
- Access Violation — An attempt to reference, communicate with, or execute data, network traffic, OS services, devices, or executable content, in a manner deemed inconsistent with the sensor's surveillance policy.
- Integrity Violation — An attempt to alter or destroy data or executable content that is inconsistent with the sensor's surveillance policy.
- System Environment Corruption — An unauthorized attempt to alter the operational configuration of the target system or other system asset (e.g., network service configuration).
- User Environment Corruption — An unauthorized attempt to alter the environment configuration of a user account managed within the monitored domain.
- Asset Distress — Operational activity indicating a current or impending failure or significant degradation of a system asset (e.g., host crash, lost service, destroyed system process, file system, or processtable exhaustion).
- Suspicious Usage — Activity representing significantly unusual or suspicious activity worthy of alert, but not directly attributable to another alert class.
- Connection Violation — A connection attempt to a network asset that occurred in violation of the network security policy.

Incident Rank: An assessment and ranking of events {e1,. ..., en, ...}

with respect to $\begin{cases} \textit{Mission Profile} = \{CR_{assets}, CR_{resources}, CR_{users}, Incident_{weight}\} \\ \textit{Probability of Success} \dashrightarrow \{\text{Alert Outcome, Relevance}\} \end{cases}$

Fig. 1. Incident Rank Calculation

- Binary Subversion — Activity representing the presence of a Trojan horse or virus.
- Action Logged — A security relevant event logged for potential use in later forensic analyses.
- Exfiltration — An attempt to export data or command interfaces through an unexpected or unauthorized communication channel.

M-Correlator allows analysts to specify low, medium-low, medium, medium-high, and high interest in a particular incident type.

2.4 Incident Rank Calculation

Incident ranking represents the final assessment of each security incident with respect to (a) the incident's impact on the mission profile as reflected by the priority calculation, and (b) the probability that the security incident reported by the INFOSEC device(s) has succeeded. Most sensors provide little if any indication regarding the outcome of an observed security incident, providing strong motivation for the production of a relevance score, where possible. It should be noted that the concept of outcome is decoupled here from that of the relevance analysis, in that outcome represents a sensor provided conclusion produced from a method unknown to the correlation engine. Relevance represents an assessment of the target system's susceptibility to an attack given vulnerability dependencies and the attack target's configuration. While both outcome and relevance may reinforce each other in increasing an overall incident rank score, so too can they neutralize each other in the face of disagreement.

Once a mission profile is specified, security incidents may be assessed and ranked against the profile. We concisely define incident ranking as illustrated in Figure 1.

2.4.1 The Bayes Calculation

Mathematically, relevance, priority, and incident rank calculations are formulated using an adaptation of the Bayes framework for belief propagation in trees, described in [17] and [21]. Our adaptation of the Bayes framework employs simultaneous observable attributes (cf., per attribute updates of a network). Additionally, this Bayes network produces values for relevance, priority, and incident rank even when only a limited set of observed attributes are available — a behaviour that is unachievable with continuous variable calculations. In this framework, belief in hypotheses at the root node is related to propagated belief at other nodes and directly observed evidence at leaf nodes by means of conditional probability tables (CPTs). At each node, "prior" probabilities $\pi(parent)$ are propagated from the parent, and "likelihoods" $\lambda(child)$ are propagated to the parent. The branch and node structure of the tree expresses the three major aspects of the calculation, namely, outcome, relevance, and priority.

Bayes networks compute belief in a number of hypothesis states. In our adaptation, the root node considers the hypothesis "criticality" and states "low", "medium", and "high". A mapping function transforms this to a single value on a scale of 0 to 255. The predefined CPTs encode the mathematical relationship between observable evidence and derived intermediate node values to the overall criticality of the alert with respect to the mission. The predefined conditional probability tables were created by interactively training the network with exemplar attribute sets. Although all such calculations are ultimately subjective, we tuned the CPTs to provide "reasonable" trade-offs and balance between relevance-, priority-, and outcome-related attributes to best match our (human) expert's intuitive alert ranking. In effect, the process of tuning the Bayes network CPTs is similarly to tuning an expert system via knowledge engineering.

While CPT initialization begins with expert tuning, we recognize the need to adapt the framework for specific environments. To this end, we include an adaptive mode wherein the analyst presents simulated alerts, which are ranked by the system. At this time the analyst either accepts the outcome or enters a desired ranking. This causes the CPTs to adapt slightly in order to more accurately reflect the administrator's preference. The adaptation occurs with no knowledge of the underlying Bayes formalism on the part of the administrator. The analyst may optionally revert to the original CPT values as well.

2.4.2 The Rank Tree

Figure 2 represents the complete incident rank tree, which brings together the contributions of alert outcome (when provided by the INFOSEC device), relevance score, and security incident priority score. These three contributors are represented by the three major branches of the incident rank tree. The priority subtree represents a merger of the incident class importance, as defined by the analyst, and the criticality of the attack target with respect to the mission of the network. The elements of the respective CPTs reflect $P(criticality = c|priority = p)$. Each of these matrices represents two values of criticality by three values of priority. Therefore, the local knowledge base consists of a set of CPTs linking the attribute to the appropriate node on its main branch. If the attribute is not observed in a given alert, the state of the corresponding node is not changed, and thus this attribute does not influence the result one way or the other. If this attribute is observed in a subsequent update for the same alert, our system adjusts the previous prioritization for the new information.

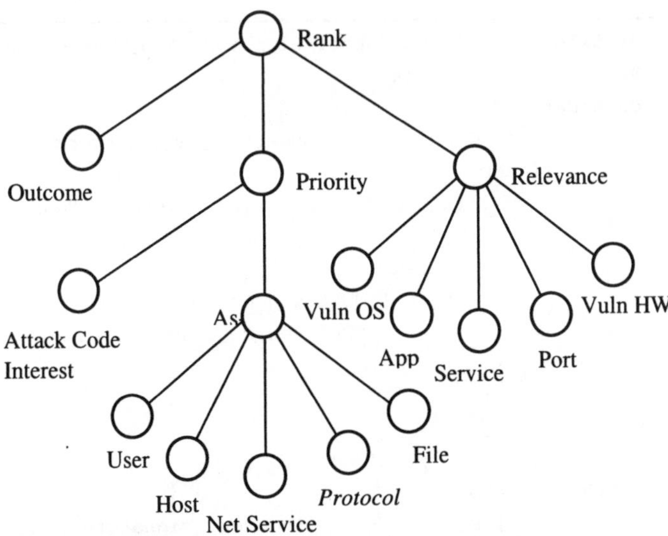

Fig.2. Incident Rank Calculation

As discussed in Section 2.2, our model identifies five equally weighted contributing attributes that formulate the relevance score: vulnerable OS, vulnerable hardware, service suite, bound ports, and application. The relevance subtree in Figure 2 illustrates these elements. Again, the Bayes net is robust in cases where the alert does not provide values for all these attributes.

3 Alert Clustering Algorithm

M-Correlator employs an alert clustering algorithm, which is used to consolidate both network and host-based INFOSEC alerts that occur in close (dynamically adjustable) temporal proximity into correlated security-incident reports. INFOSEC alerts regarding network communications are merged through an analysis of common network session, as defined by port and IP address matches, and common observer, alert type, or, more liberally, by common alert classification as defined in the incident handling fact base. INFOSEC alerts regarding host activity are merged through an analysis of common session, as defined through user session attributes such as process ID or user ID, common observer, alert type, or more liberally by common alert classification.

Figure 3, shows an example M-Correlator clustering policy. (Note that we are deliberately restricting this discussion to the more straightforward properties of these policies.) Given a new security-incident report, the M-Correlator first determines if the report is a candidate for the policy. In this example, if the report originates from either a network sensor or a host-based sensor in which the source process ID and source user names are known, then the report is a candidate for further processing.

```
Profile         Cross_Sensor_Signature_And_Session_Match
Policy          Liberal
Candidate_If     [ OR
        [ IN_GROUP          observer_name Network_Sensor ]
        [ AND
            [ IN_GROUP      observer_name Host_Sensor ]
            [ NOT
                [ AND
                    [ NULL source_pid ]
                    [ NULL source_username ]
                ]
            ]
        ]
    ]
Match_If [ AND
           [ EQ              incident_signature ]
           [ NULL_OR_EQ      observer_stream ]
           [ OR
              [ AND
                  [ IN_GROUP observer_name Network_Sensor ]
                  [ ELEMENTS_EQ            source_IParray ]
                  [ ELEMENTS_EQ            target_IParray ]
              ]
              [ AND
                  [ IN_GROUP observer_name Host_Sensor ]
                  [ ELEMENTS_EQ            target_IParray ]
                  [ NULL_OR_EQ             source_pid ]
                  [ NULL_OR_EQ             source_username ]
              ]
           ]
         ]
Delay_Until_Expire        600
Delay_Until_Flush         90
Initial_Flush_Delay       90

Enable                    true
Unique_Match              true
Merge_Action              fuse
```

Fig. 3. Example Alert Cluster Policy Specification

The clustering policy's Match_If clause defines the criteria by which reports are clustered. Thus, in this case, all clustered report incident signatures and their observer stream identifiers (if extant) must match. Also, if the sensor is network-based, the reports must have matching source and target IP addresses; while if host-based, the reports must have matching target IP addresses, and, if extant, matching source process IDs and source user names.

A clustering policy also specifies the longevity of a clustered report when there are no subsequent candidate reports; the delay before a clustered report is initially issued, and the refresh interval between clustered report updates, again whose purpose is to reduce report traffic.

The incident-handling fact base also supports the specification of a set of attributes that represent loose relationships among alerts of different classes. For example, consider a firewall that reports a series of connection violations between external host A and internal host B, and suppose this activity temporally overlaps an internal network IDS report of a port sweep on several ports. That is, the port sweep manifested itself by two sets of activity: (1) connection attempts that were blocked by the firewall filtering policy, and (2) connections that were allowed through the firewall, but in aggregate caused the network IDS to consider the flurry of connections as a potential port scan. Alert clustering tags are established by the incident handling fact base maintainer, and allow M-Correlator a greater ability to leverage associations unique to specific known scenarios. In this example, a shared cluster name within the incident-handling fact base allows M-Correlator to merge the connection violation reports with the port scan alerts.

4 An Example Mission Specification

A brief example mission specification is subsequently used here to illustrate mission-based impact analysis. This example is based on a simulated heterogeneous network, illustrated in Figure 4. The network consists of hosts employing four different operating systems and protected by several distributed INFOSEC devices. Four Sun Solaris systems are protected by host-based intrusion detection sensors (SRI's EMERALD eXpert-BSM [14]), and three network intrusion detection systems (eBayes-TCP [21], eXpert-Net, and RealSecure). Traffic entering the LAN is filtered

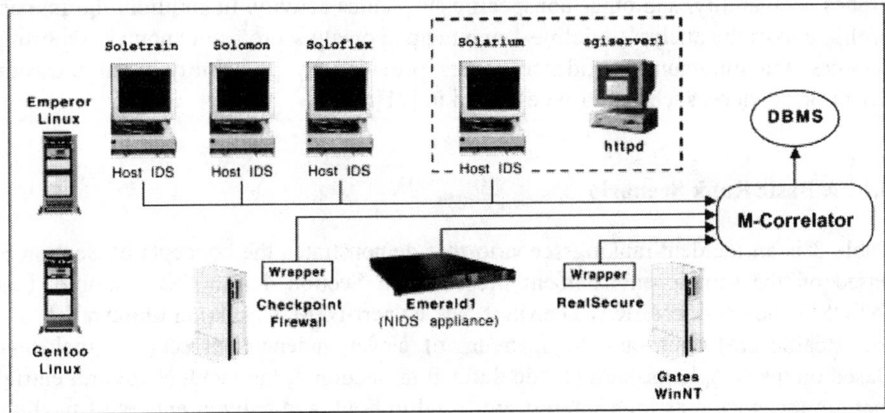

Fig. 4. An Experimental Simulation Network

through a Checkpoint firewall, whose alert reporting mechanism is wrapped to forward INFOSEC alerts to the M-Correlator.

Within this experimental LAN, there is one mission-critical server, Solarium, which

Host_specifier ::= (<IPaddress> | <host_name>) ['/' <num_mask_bits>]
Port_list ::= '[' <port> {<port>} ']'
Service_specifier ::= [<proto>] ':' [Host_specifier] ':' [(<port> | Port_list)]
Resource_specifier ::= [(<user_name> | <uid>)] ':'
[<path>] ':' [Host_specifier]

operates as a fileserver for the rest of the network, and one mission-critical SGI HTTP server. There are two administrative accounts for this LAN, em_admin1 and em_admin2, and three untrusted consultant accounts that the administrators want to monitor closely. A highly sensitive and valuable source code directory is located on host Solomon. With respect to administrative responsibilities, the analyst in this case study is most concerned with rapidly responding to potential subversion of system control, through either privilege theft or modification of the systems themselves. The analyst is also concerned with attacks against users, but to a lesser extent than system-directed attacks that may have direct impact on the security of all user accounts. The analyst, who is responsible for maintaining availability and watching for suspicious probes and behaviors, feels that at this time alerts of these types should receive a slightly lower priority than direct attacks on the system and its users. Table 2 provides an example mission specification based on the experimental LAN in Figure 4.

The top left quadrant defines the list of host and network services that are critical to the mission of the example network. The middle left quadrant enumerates both critical user accounts and critical host files or directories. The syntax for critical asset and resource specifications is as follows:

The right quadrant defines the analyst's level of interest in the various classes of security alerts. In this scenario, the analyst is primarily concerned with direct threats of privilege and access violations, while being much less concerned with issues of probes, availability, and other nonspecific suspicious activity. In addition, the interest profile allows the analyst to define a minimum anomaly score from anomaly reporting services and minimum confidence scores provided by probability-based intrusion detection services, such as those described in [21].

4.1 A Basic Rank Scenario

Table 3 is an incident-ranking scenario that demonstrates the concepts of Section 3, based on the sample environment presented in Section 4. In this example, four INFOSEC devices (EMERALD eXpert-Net, eXpert-BSM, Checkpoint Firewall-1, and ISS RealSecure) contribute to a stream of eleven independent security incidents. Based on the sample mission profile defined in Section 4, the incident column entries that are recognized as high interest are listed in bold, and column entries of medium interest are underlined. Critical servers and untrusted source users, as defined in the sample mission profile, are also listed in bold.

Table 2. Sample Mission Specification

<table>
<tr><td><pre>Critical_Assets [
   :solarium:
   TCP:sgiserver1:http
]</pre></td><td><pre>Interest_Policy [
   PRIVILEGE_VIOLATION   High
   SYSTEM_ENV_CORRUPTION High
   ACCESS_VIOLATION   High
   BINARY_SUBVERSION
High
   CONNECTION_VIOLATION Medium
   USER_ENV_CORRUPTION   Medium
   USER_SUBVERSION   Medium
   INTEGRITY_VIOLATION   Medium
   EXFILTRATION   Medium
   PROBE
Low
   SUSPICIOUS_USAGE   Low
   ACTION_LOGGED   Low
   DENIAL_OF_SERVICE
Low
   ASSET_DISTRESS   Low
   MinAnomalyScore   90
   MinConfidenceScore
30
 ]</pre></td></tr>
<tr><td><pre>Critical_Resources [
   em_admin1:192.12.34.0/24
   em_admin1:192.12.34.0/24
   consult1:192.12.34.0/24
   consult2:192.12.34.0/24
   consult3:192.12.34.0/24
  :/proprietary/src/:solomon
  ]</pre></td><td></td></tr>
<tr><td><pre>Alert_Filters [
      __empty__
]</pre></td><td></td></tr>
</table>

The Active Port and OS/Architecture columns identify contributions of the relevance scores. “Yes” in the Active Port and OS/Architecture columns indicates that the dependent network services, operating system, and hardware required for the successful execution of the given alert were indeed found to be present and enabled on the target machine.

Table 4 illustrates the view of the Table 3 dataset from the perspective of the incident rank calculation process, and the resulting ranking produced by M-Correlator (1 represents the highest threat ranking and 11 represents the lowest threat ranking). Alerts of highest threat share the property that their combined contributing elements are of high priority, relevance, and positive outcome.

Table 3. Basic Experiment Incident Rank Example Set

Observer	Incident	Source IP	Target IP	Src user	Active Port	Outcome	OS/Arch
eBSM-Solarium	Root_Core_Creat	200.55.19.143.100	solarium.23	consult1	N/A	success	Yes
eBSM-Soloflex	Root_Core_Creat	200.55.19.144.3467	gates.23	consult1	N/A	fail	N/A
eXpert-Net	FTP_Cwd_Probe	200.55.19.145.4125	gentoo.21	consult1	No	success	N/A
eXpert-Net	FTP_Cwd_Probe	200.55.19.146.3341	gentoo.21	anonymous	No	fail	N/A
eXpert-Net	FTP_Cwd_Probe	200.55.19.147.5143	emperor.21	anonymous	Yes	success	N/A
Checkpoint	TCP_Conn_Denied	200.55.19.148.1657	gates.53		No	fail	No
ISS-realsecure	NFSMKNOD	200.55.19.149.1235	sgiserver.53		No	success	No
eBSM-Solarium	Private_File_Alt	200.55.19.150.1809	solarium.53	consult1	No	fail	No
eBSM-Solarium	Private_File_Alt	200.55.19.151.5413	solarium.53	consult1	No	success	No
ISS-realsecure	HTTP_SGI_Wrap	200.55.19.152.1243	sgiserver1.80	nobody	Yes	fail	Yes
ISS-realsecure	HTTP_SGI_Wrap	200.55.19.151.3467	sgiserver.80	consult1	Yes	success	Yes

4.2 A Multi-sensor Alert Scenario

As a further example of incident ranking in combination with M-Correlator's alert clustering algorithm, the following is a multi-INFOSEC device alert scenario. Like the basic scenario, this scenario is based on the experimental environment discussed in Section 4. Table 5 provides a sampling of alerts produced from five INFOSEC devices distributed in the M-Correlator experimental LAN: EMERALD eBayes-TCP and eXpert-Net network intrusion detection sensors, Checkpoint Firewall-1, ISS Realsecure, and eXpert-BSM.

In this scenario, 79 alerts are forwarded to M-Correlator for analysis and database storage. Table 5 shows the time at which each INFOSEC alert was generated, the INFOSEC device that generated the alert, the alert type, the source and destination of the attack, and alert outcome and relevance calculation. The alert is identified as relevant if M-Correlator was able to confirm that at least one of the alert dependencies in the relevance calculation was found to match. The alert is identified as nonrelevant if no alert dependencies are found to match, and N/A indicates that the relevance cal culation could not be performed because of lack of information from the incident-handling fact base or regarding the target host. Bold text is used in Table 5 to indicate that the value represents a critical item in the sample mission specification of Section 4.

Table 4. Basic Experiment Incident Rank Resolution

Ranking	Incident	Target IP	SrcUser	Active Port	OS/Arch	Outcome
	priority			relevance		outcome
4	Root_Core_Creat	**solarium.23**	consult1	N/A	Yes	success
10	Root_Core_Creat	gates.23	consult1	N/A	N/A	fail
7	FTP_CWD_Probe	gentoo.21	consult1	No	N/A	success
11	FTP_CWD_Probe	gentoo.21	anonymous	No	N/A	fail
6	FTP_CWD_Probe	emperor.21	anonymous	Yes	N/A	success
9	TCP_Conn_Denied	gates.53		No	No	fail
5	NFSMKNOD	sgiserver.53		No	No	success
8	Private_File_Alt	**solarium.53**	consult1	No	No	fail
3	Private_File_Alt	**solarium.53**	consult1	No	No	success
2	**HTTP_SGI_Wrap**	**sgiserver1.80**	nobody	Yes	Yes	fail
1	**HTTP_SGI_Wrap**	**sgiserver.80**	consult1	Yes	Yes	success

Table 6 presents the alert stream represented in Table 5 ranked and aggregated by the M-Correlator prototype. In this example, Table 5's 79 original INFOSEC alerts were processed by M-Correlator and stored in an Oracle database. The security incidents shown in Table 6 were merged and ranked as follows:

1. Entry 1: This incident represents the three EMERALD eXpert-BSM reports of malicious activity from the same user session. The aggregate priority of these alerts was high because this incident included a high-interest privilege subversion attack on the critical fileserver, Solarium. The attacks were also successful, and all were found relevant to the target machine.

2. Entry 2: The second-highest-ranked incident represents ISS Realsecure's Intel-Buffer-Overflow and EMERALD eXpert-Net's Imapd-Overflow alerts. These alerts both shared the common alert class privilege-subversion and common net

work session attributes. While this security incident was registered in the mission specification as high interest, and found both successful and relevant, the alert was performed against a lower-interest target host.

3. Entry 3: The M-Correlator used its alert clustering algorithm to merge the eBayes-TCP Port-Scan alert with the Checkpoint Firewall Connection-Violation alerts. These alerts pertain to the same source and target addresses, and share a common alert cluster tag called Generic-Probe. The alerts are ranked low because of the analyst's low interest in probing activity, and because the alert did not target a critical asset or resource.
4. Entry 4: The lowest-ranked security incident is a Realsecure Kerberos-User-Snarf probe. This alert represents a distinct security incident, was evaluated overall as low interest because the attack targeted a noncritical asset, and apparently represented a false alarm, as there is no Kerberos server present on the target machine.

4.3 Operational Exposures

In recent months the EMERALD development group has begun operational experiments with M-Correlator on operational data sets provided my multiple organizations inside the United States Department of Defense.

5 Related Research

The broad problem of information security alert management and post-sensor analysis is an area that has been undergoing a great deal of activity. In the commercial space, one issue receiving particular attention is that of sensor overload through alert fusion and better methods for visualization. Some systems offer a form of severity measurement that attempts to prioritize alerts. However, unlike M-Correlator, which attempts to formulate a security incident ranking based on a multi-attribute mission profile specification, the vast majority of severity metric services rely on hard-coded mappings of attributes to fixed severity indexes.

There is also a growing momentum of research activity to explore intrusion report correlation as a separable layer of analysis above intrusion detection. In the space of alert aggregation and false positive reduction there are a number of ongoing projects. These include work by Honeywell, which is developing Argus, a qualitative Bayesian estimation technology to combine results from multiple intrusion detection systems [9]. The work to date has emphasized false-positive reduction through *a priori* assessments of sensor reporting behavior. M-Correlator also attempts to reduce false positives, but through an emphasis on the relevance of an alert to the target system's configuration.

Table 5. Cross Sensor Incident Rank and Aggregation Dataset

9:41am	eBayes-TCP	Port_Scan	200.55.19.100 --> gates.[21 22 23]	Success, N/A
9:41am	Checkpoint	TCP_Connect_Violation	200.44.19.100 --> gates	Failed, N/A
		x 70 TCP_Connect_Violations		
9:45am	Realsecure	Kerberos_User_Snarf	195.16.19.56 --> emperor	Unknown, Non-Relevant
9:48am	eBayes-TCP	Port_Scan	200.55.19.100 --> gates.[21 22 23 79 80]	Success, N/A
9:51am	Realsecure	**Intel_Buffer_Overflow**	200.55.19.149.3450 --> gentoo.143	Unknown, **Relevant**
9:51am	eXpert-Net	**Imap_Overflow**	200.55.19.149.3450 --> gentoo.143	Success, **Relevant**
9:52am	eBayes-TCP	Port_Scan	200.55.19.100 --> Gates.[21 22 23 79 80 514]	Success, N/A
10:02am	eXpert-BSM	**Buffer_Overflow**	console --> **solarium**	Success, **Relevant**
10:04am	eXpert-BSM	Illegal_File_Alteration	console --> **solarium**	Success, **Relevant**
10:05am	eXpert-BSM	Illegal_File_Alteration	console --> **solarium**	Success, **Relevant**

IBM Zurich [7] is also exploring common attribute-based alert aggregation in the Tivoli Enterprise Console, as are Columbia University and Georgia-Tech using association rules [12]. Onera Toulouse is using an expert-system-based approach for similarity formulation [5], and SRI International is using a probabilistic-based approach to attribute similarity recognition [22]. M-Correlator's alert clustering algorithm is very similar in purpose and outcome to these systems as it, too, attempts to perform alert fusion in the presence of incident type disagreement, feature omission, and competing incident clustering opportunities.

Another major thrust of activity involves the development of complex (or multistage) attack modeling systems, capable of recognizing multistage (or multi-event) attack scenarios within the stream of distributed intrusion detection sensors. Stanford has developed extensions to its ongoing research in Complex Event Processing (CEP) in the domain of intrusion report correlation [18]. In this work, complex multistage scenarios can be represented in concept abstraction hierarchies using the CEP language. This approach provides methods for efficient event summarization and complex pattern recognition. The Stanford work differs significantly from M-Correlation in that it emphasizes scenario recognition and content summarization, whereas M-Correlation emphasizes impact analysis, where the impact is defined with respect to the mission objectives of the target network. In addition, IET Incorporated is involved in research to develop situation-aware visualization software for managing security information and recognizing composite attack models [1]. There is an emphasis on large-scale attack recognition using a probabilistic domain-modeling algorithm. The effort differs significantly from M-Correlation in that it too emphasizes model recognition. Finally, UC Santa Barbara's STAT-based web of sensors work is exploring the use of the STATL language as a basis for complex attack modeling [8].

Table 6. Ranked and Aggregated Security Incidents

Rank	Time	Incident	Connection	Observers	Other
1	10:02am - 10:05am	**Buffer_Overflow:** **- buffer_overflow** **- illegal_file_alteration** **- illegal_file_alteration**	Console --> **Solarium**	eXpert-BSM	**Success, Relevant**
2	9:51am	**Generic_Priv_Subv:** **- imapd_overflow** **- intel_buff_overflow**	200.55.19.149 --> gento.143	RealSecure, eXpert-Net	**Success, Relevant**
3	9:41am - 9:52am	Generic_Probe: - Port_Scan - TCP_Connect_Violation	200.55.19.100 --> gates.[21 22 23 79 80 514]	eBayes-TCP Checkpoint	**Success**
4	9:45am	**Kerberos_User_Snarf**	195.16.19.56 --> emperor	RealSecure	Unknown Irrelevant

6 Conclusion

We have discussed a mission-impact-based approach to alert prioritization and aggregation. This research has led to the development of a prototype system called the EMERALD M-Correlator, which is capable of receiving security alert reports from a variety of INFOSEC devices. Once translated to an internal incident report format, INFOSEC alerts are augmented, and, where possible, fused together through a chain of processing. A *relevance score* is produced through a comparison of the alert target's known topology against the vulnerability requirements of the incident type, which is provided to M-Correlator by an *Incident Handling Fact Base*. Next, a *priority calculation* is performed per alert to indicate (a) the degree to which the alert is targeted at critical assets, and (b) the amount of interest the user has registered for this alert type. Last, an overall *incident rank* is assigned to each alert, which brings together the priority of the alert with the likelihood of success.

Once ranked, the M-Correlator attempts to combine related incident alarms with an attribute-based *alert clustering algorithm*. The resulting correlated incident stream represents a filtered, lower-volume, content-rich security-incident stream, with an incident-ranking scheme that allows the analyst to identify those incidents that pose the greatest risk to the monitored network.

M-Correlator has reached a maturity level where trial releases of the system have occurred in several computing environments with the U.S. Department of Defense. These analyses have included live testing in a DoD Information Warfare exercise, as well as multiple batch analyses of INFOSEC alert streams from several DoD organizations. These activities are contributing to the process of validating the efficacy of mission-based correlation, as well as providing valuable insight into future extensions of the incident rank model. We believe that mission-impact based analyses will prove themselves extremely useful both to human analysts, and to other consumers, such as automated response technology that must sift through thousands of alerts daily in search of alarms worthy of proactive response. Extension of the basic M-Correlator algorithm is already underway to incorporate its logic into a real-time response engine.

References

[1] D'Ambrosio, B, M. Takikawa, D. Upper, J. Fitzgerald, and S. Mahoney, "Security Situation Assessment and Response Evaluation," *Proceedings (DISCEX II) DARPA Information Survivability Conference and Exposition,* Anaheim, CA, Vol. I, June 2001.

[2] D.W. Baker, S.M. Christey, W.H. Hill, and D.E. Mann, "The Development of a Common Enumeration of Vulnerabilities and Exposures," *Proceedings of the Second International Workshop on Recent Advances in Intrusion Detection (RAID),* September 1999.

[3] Bugtraq. Security Focus Online. http://online.securityfocus.com/archive/1

[4] CERT Coordination Center. Cert/CC Advisories Carnegie Mellon, Software Engineering Institute. Online. http://www.cert.org/advisories/

[5] F. Cuppens, "Managing Alerts in a Multi-Intrusion Detection Environment," *Proceedings 17th Computer Security Applications Conference,* New Orleans, LA, December 2001.

[6] Common Vulnerabilities and Exposures. The MITRE Corporation. http://cve.mitre.org/

[7] H. Debar and A. Wespi, "Aggregation and Correlation of Intrusion-Detection Alerts," *Proceedings 2001 International Workshop on Recent Advances in Intrusion Detection (RAID)*, Davis, CA, October 2001.

[8] G. Vigna, R.A. Kemmerer, and P. Blix, "Designing a Web of Highly-Configurable Intrusion Detection Sensors," *Proceedings 2001 International Workshop on Recent Advances in Intrusion Detection (RAID)*, Davis, CA, October 2001. C.W. Geib and R.P Goldman, "Probabilistic Plan Recognition for Hostile Agents," *Proceedings of FLAIRS 2001 Special Session on Uncertainty* - May 2001.

[10] C. Kahn, P.A. Porras, S. Staniford-Chen, and B. Tung, "A Common Intrusion Detection Framework," http://www.gidos.org.

[11] K. Kendall, "A Database of Computer Attacks for the Evaluation of Intrusion Detection Systems," Master's Thesis, Massachusetts Institute of Technology, June 1999.

[12] W. Lee, R.A. Nimbalkar, K.K. Yee, S.B. Patil, P.H. Desai, T.T. Tran, and S.J. Stolfo, "A Data Mining and CIDF-Based Approach for Detecting Novel and Distributed Intrusions", *Proceedings 2000 International Workshop on Recent Advances in Intrusion Detection (RAID),* Toulouse, France, October 2000.

[13] D. Levin, Y. Tenney, and H. Henri, "Issues in Human Interaction for Cyber Command and Control," *Proceedings (DISCEX II) DARPA Information Survivability Conference and Exposition,* Anaheim, CA, Vol. I, June 2001.

[14] U. Lindqvist and P.A. Porras, "eXpert-BSM: A Host-based Intrusion Detection Solution for Sun Solaris," *Proceedings 17th Computer Security Applications Conference,* New Orleans, LA, December 2001.

[15] U. Lindqvist, D. Moran, P.A. Porras, and M. Tyson, "Designing IDLE: The Intrusion Detection Library Enterprise," *Proceedings 1998 International Workshop on Recent Advances in Intrusion Detection (RAID)*, Louvain-la-Neuve, Belgium, September 1998.

[16] NMAP Network Mapping tool.
http://www.insecure.org/nmap/

[17] Pearl, J. "Probabilistic Reasoning in Intelligent Systems," Morgan-Kaufmann (1988).

[18] L. Perrochon, E. Jang, and D.C. Luckham.: Enlisting Event Patterns for Cyber Battlefield Awareness. *DARPA Information Survivability Conference & Exposition* (DISCEX'00), Hilton Head, South Carolina, January 2000.

[19] P.A. Porras and P.G. Neumann, "EMERALD: Event Monitoring Enabling Responses to Anomalous Live Disturbances," *Proceedings National Information Systems Security Conference*, NSA/NIST, Baltimore, MD, October 1997.

[20] M. Roesch, "Lightweight Intrusion Detection for Networks," *Proceedings of the 13th Systems Adminstration Conference — LISA 1999*, November, 1999.

[21] Valdes and K. Skinner, "Adaptive, Model-based Monitoring for Cyber Attack Detection", *Proceedings 2000 International Workshop on Recent Advances in Intrusion Detection (RAID),* Toulouse, France, October 2000.

[22] Valdes and K. Skinner, "Probabilistic Alert Correlation," *Proceedings 2001 International Workshop on Recent Advances in Intrusion Detection (RAID)*, Davis, CA, October 2001.

M2D2: A Formal Data Model for IDS Alert Correlation

Benjamin Morin[1], Ludovic Mé[2], Hervé Debar[1], and Mireille Ducassé[3]

[1] France Télécom R&D, Caen, France
{benjamin.morin, herve.debar}@rd.francetelecom.com,
[2] Supélec, Rennes, France
Ludovic.Me@supelec.fr,
[3] IRISA/INSA, Rennes, France
ducasse@irisa.fr

Abstract. At present, alert correlation techniques do not make full use of the information that is available. We propose a data model for IDS alert correlation called M2D2. It supplies four information types: information related to the characteristics of the monitored information system, information about the vulnerabilities, information about the security tools used for the monitoring, and information about the events observed. M2D2 is formally defined. As far as we know, no other formal model includes the vulnerability and alert parts of M2D2. Three examples of correlations are given. They are rigorously specified using the formal definition of M2D2. As opposed to already published correlation methods, these examples use more than the events generated by security tools; they make use of many concepts formalized in M2D2.

1 Introduction

Intrusion detection today faces four major challenges. Firstly, the overall number of alerts generated is overwhelming for operators. Secondly, too many of these alerts falsely indicate security issues and require investigation from these operators. Thirdly, the diagnosis information associated with the alerts is so poor that it requires the operator to go back to the original data source to understand the diagnosis and assess the real severity of the alert. Finally, the number of undetected attacks (a.k.a false negatives) must be reduced in order to provide an appropriate coverage for the monitoring of the environment.

Alarm correlation is the key to the first three challenges: it consists in reducing and conceptually interpreting multiple alarms such that new meanings are assigned to these alarms [1]. In order to reduce false negatives, heterogeneous analyzers are often used to get multiple points of view of the attacks. Multiplying analyzers also multiplies alerts. Alert correlation is therefore all the more essential.

At present, alert correlation techniques do not make full use of the information that is available. For example, they tend to only use the events generated

A. Wespi, G. Vigna, and L. Deri (Eds.): RAID 2002, LNCS 2516, pp. 115–137, 2002.

by security tools. We argue that alert correlation must take at least four information types into account: information related to the characteristics of the monitored information system, information about the vulnerabilities, information about the security tools used for the monitoring, and information about the events observed. Indeed, an organization information system (IS) is made of a network of hosts which potentially exhibit vulnerabilities. These vulnerabilities are exploited by attackers. Security tools are required to monitor the IS in order to locate events of interest, identify attacks and generate alerts.

In this paper, we propose a model, called M2D2, which supplies the above four types of information. It provides concepts and relations relevant to the security of an information system. M2D2 is formally defined, following Mc Hugh's suggestion that further intrusion detection significant progress will depend on the development of an underlying theoretical basis for the field [2].

The model has nevertheless a practical ground. Indeed, a relational database implements parts of it, and the corresponding tables are automatically generated from data sources. IDS and vulnerability scanners fill the database with events, vulnerabilities and products tables come from the ICAT [3] database.

Three examples of correlations are given. The first one aggregates alerts referring to the same host, even when the host name is not provided by the alert attributes. The second example identifies the hosts vulnerable to an attack occurrence. The third example aims at reducing false positives by analyzing the reaction or the absence of reaction of the security tools which are able to detect the cause of the alert. These examples are rigorously specified using the formal bases of M2D2. Moreover, they illustrate the need to work on the four types of information.

Two sets of related work address the modeling of information for intrusion detection : the NetStat Network Fact Base of Vigna and Kemmerer [4,5] and the Intrusion Reference Model (IRM) of Goldman et al. [6]. The Network Fact Base is formally defined but it is not comprehensive, only the topology of information systems and some of the events are modeled. IRM is quite comprehensive but we are not aware of any formalization.

As it is illustrated in the examples of correlations, the model must be both comprehensive and formally defined. Indeed, if it is not comprehensive enough, all the useful properties cannot be expressed; if it is not formally defined, the properties are not as rigorous as they must be. In the following, we reuse the concepts of Vigna's formal model [7], and extend the formalization to encompass the four types of information needed. The main contribution of this article is therefore a formal model of data relevant for alert correlation. This model is richer than the existing formal models.

Section 2 presents the concepts of the model and the relations and functions which link concepts together. Section 3 presents some correlation methods based on the model. Section 4 discusses related work. The last section concludes on the contributions of this article.

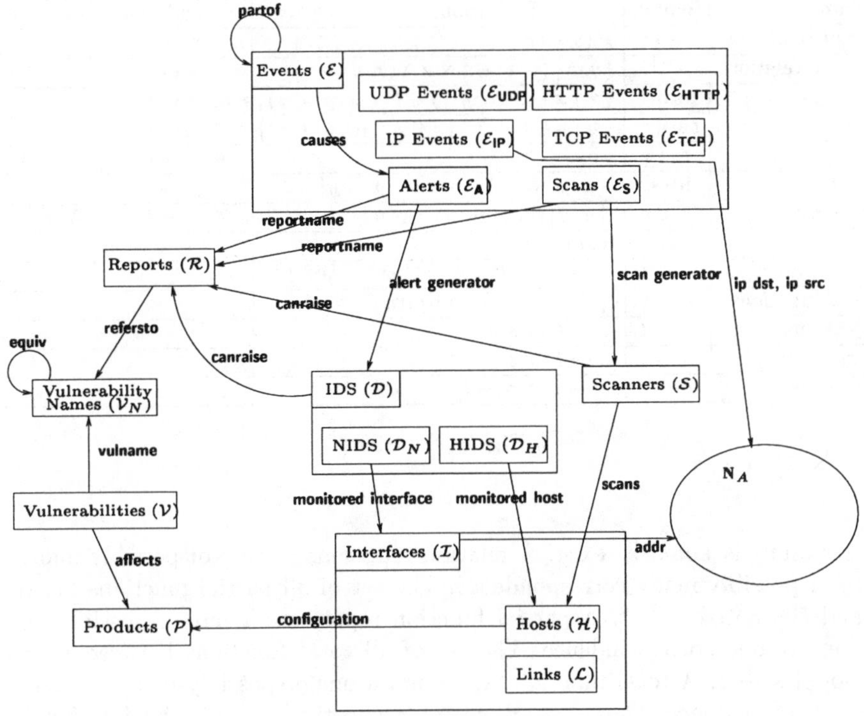

Fig. 1. M2D2 scheme

2 M2D2: Concepts and Relations

M2D2 integrates the four main types of information relevant to security identified in the introduction: IS characteristics, vulnerabilities, security tools, and events. The global schema of M2D2 can be found in figure 1. It summarizes the main concepts and relations described thereafter.

In the remainder of this section the four types of information are described in turn. For each type of information, the formal definition can be found in framed figures, and an informal description of the concepts can be found in the text. Every time a relation is informally discussed, its identifier is indicated in bold (**equiv** for example) so that its formal definition can be easily found in the corresponding frame.

2.1 Notations

For formalization, notations of the Z and B formal methods are used [8]. The precise definitions of all the operators used in this article can be found in figure 2. Let us briefly recall here the basic ones. The set of all the relations between two

Name	Syntax	Definition	Conditions
Set of relations	$s \leftrightarrow t$	$\mathcal{P}(s \times t)$	
Inverse relation	r^{-1}	$\{y, x \mid (y, x) \in t \times s \wedge (x, y) \in r\}$	$r \in s \leftrightarrow t$
Domain	$\mathbf{dom}(r)$	$\{x \mid x \in s \wedge \exists y \cdot (y \in t \wedge (x, y) \in r)\}$	$r \in s \leftrightarrow t$
Range	$\mathbf{ran}(r)$	$\{y \mid y \in t \wedge \exists x \cdot (x \in s \wedge (x, y) \in r)\}$	$r \in s \leftrightarrow t$
Composition	$r \; ; q$	$q \circ r$	$r \in s \leftrightarrow t \wedge q \in t \leftrightarrow u$
Identity	$\mathbf{id}(s)$	$\{x, y \mid (x, y) \in s \times s \wedge x = y\}$	
Restriction	$u \lhd r$	$\{x, y \mid (x, y) \in r \wedge x \in u\}$	$r \in s \leftrightarrow t \wedge u \subseteq s$
Image	$r[w]$	$\mathbf{ran}(w \lhd r)$	$r \in s \leftrightarrow t \wedge w \subseteq s$
Partial functions	$s \nrightarrow t$	$\{r \mid r \in s \leftrightarrow t \wedge (r^{-1} ; r) \subseteq \mathbf{id}(t)\}$	
Total functions	$s \rightarrow t$	$\{f \mid f \in s \nrightarrow t \wedge \mathbf{dom}(f) = s\}$	
Injections	$s \rightarrowtail t$	$s \rightarrow t \cap s \rightarrowtail\!\!\!\!\!\nrightarrow t$	
Surjections	$s \twoheadrightarrow t$	$s \nrightarrow\!\!\!\!\twoheadrightarrow t \cap s \rightarrow t$	
Bijections	$s \rightarrowtail\!\!\!\!\twoheadrightarrow t$	$s \rightarrowtail t \cap s \twoheadrightarrow t$	

Fig. 2. Notations

sets s and t is noted $s \leftrightarrow t$. A relation represents a zero-or-possibly-many to zero-or-possibly-many correspondence. The set of all partial functions between s and t is noted $s \nrightarrow t$. A partial function represents a zero-or-possibly-many to zero-or-one correspondence. The set of all total functions between s and t is noted $s \rightarrow t$. A total function represents a one-or-possibly-many to zero-or-one correspondence. The set of all partial (respectively total) injections between s and t is noted $s \rightarrowtail\!\!\!\!\!\nrightarrow t$ (resp. $s \rightarrowtail t$). A partial (respectively total) injection represents a zero-or-one to zero-or-one (resp. one to zero-or-one) correspondence. The set of all partial (respectively total) surjections between s and t is noted $s \nrightarrow\!\!\!\!\twoheadrightarrow t$ (resp. $s \twoheadrightarrow t$). A partial (respectively total) surjection represents a zero-or-possibly-many to one (resp. one-or-possibly-many to one) correspondence. The set of all bijections between s and t is noted $s \rightarrowtail\!\!\!\!\twoheadrightarrow t$. A bijection represents a one to one correspondence.

The distinction between relations and functions, as well as the injective property, have a major impact on implementation. An injective function can be implemented straightforwardly by an array. A non-injective function can be implemented by an array but the contents of the array may be a pointer to a list of values. In general, a relation which is not an injective function is best implemented by a database table. The partial versus total distinction has a major impact on the interpretation of the data. Indeed, if a function is partial some of the values may be undefined. A relation which is not a function may also be partial. It can be specified by the fact that the domain is only included in, and not necessarily equal to, the origin set ($\mathbf{dom}(r) \subseteq s$). Bijections are the easiest to implement and handle, unfortunately, very few relations are bijections in reality. In the following, we will therefore rigorously define the characteristics of the defined relations.

Definitions

$\mathcal{I}$ is the set of network interfaces.
A network is a hypergraph on $\mathcal{I}$, i.e is a family $N = (E_1, E_2, \ldots, E_m)$, where $E_i \subset \mathcal{I}$ is called an edge. Edges are divided in two subsets, $\mathcal{H}$ and $\mathcal{L}$
$\mathcal{H} \subseteq \mathcal{P}(\mathcal{I})$ is the set of hosts, and
$\mathcal{L} \subseteq \mathcal{P}(\mathcal{I})$ is the set of links.
$\mathcal{N}$ is the set of host names in the information system.
$\mathbb{N}_A = \{0, \ldots, 2^{32} - 1\}$ is the set of IPv4 address.

Functions & Relations

$\textbf{addr} \in \mathcal{I} \rightarrowtail \mathbb{N}_A$
$\textbf{hostname} \in \mathcal{H} \rightarrow \mathcal{N}$

Properties

Hosts and links both partition interfaces, i.e

$$\bigcup_{i=1}^{p} H_i = \mathcal{I} \text{ and } H_i \cap H_j = \emptyset \ (\forall i, j \in \{1, 2, \ldots, p\} \text{ and } i \neq j)$$
$$\bigcup_{i=1}^{q} L_i = \mathcal{I} \text{ and } L_i \cap L_j = \emptyset \ (\forall i, j \in \{1, 2, \ldots, q\} \text{ and } i \neq j)$$

Comments

A *hub* is a type of link, for example.
See figure 4 for a network example. Interfaces are dots, hosts are circles and links are lines between interfaces.
Hosts should uniquely be identified by their host name, however two distinct hosts can have the same host name, thus **hostname** is not a bijection.

Fig. 3. M2D2 IS topology model

2.2 Information System's Characteristics

Modeling information systems (IS) is a difficult task because IS are increasingly complex and dynamic. Alert correlation must take into account information system specificities. As an example, alert correlation requires evaluating the success probability of an attack, which can be achieved by comparing the target configuration with the vulnerability's known prerequisites.

As mentioned earlier, an IS characteristics depends on both the topology (network of hosts) and the products used.

Topology. The topology model is formally defined in figure 3. We adopt the TCP/IP networks topology formal model proposed by Vigna [7]. Vigna suggests to model a network with an hypergraph. Nodes of the graph are the network interfaces, edges are subsets of the interfaces set[1]. Edges are divided into two kinds of components, hosts and links. From that physical network model, an IP network is modeled with an injective function which maps interfaces with IP addresses (**addr**).

[1] in standard graphs, edges are nodes couples; in an hypergraph, edges are subsets of the nodes

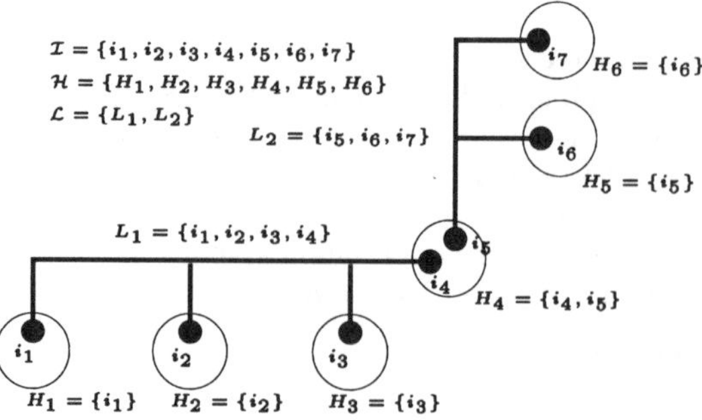

Fig. 4. A sample network

In addition to Vigna's model, M2D2 also includes other topological information, such as the mapping between hosts and host names (**hostname**). This information is useful for alert correlation purposes because IDS either give the former or the latter depending on their data source. We also rely on the IANA list for mapping services names to port numbers.

M2D2 information system model does not contain all relevant concepts yet. Missing concepts (users and files for example) must be integrated inside M2D2 in the future in order to reach comprehensiveness.

From a practical point of view, some automated tools exist to build and maintain the network topology and product database [2]. Network topology can also be manually built with graphical tools.

Products. Products are formally defined in figure 5. In M2D2, a product is simply a logical entity that is executed by a host. A product is a quadruplet (`vendor_id`, `product_id`, `version_id`, `type`). `vendor_id` and `product_id` respectively represent the vendor name (**vendor**) and the product name (**prodname**). The version number (**version**) is problematical because there is currently no canonical representation of software product versions, and version notations are heterogenous (generally an alphanumeric string). That is why **version** is a partial function. For alert correlation purposes, it is useful to compare the version number of two products, for example to evaluate the vulnerability of a host. There is currently no solution to this problem. The product set may be any set provided with a total order relation. We suggest to use $\mathbb{N}^p$. The total order relation ($<$) is given in figure 5. Each product belongs to a class (**prodtype**) which may be one of {`OperatingSystem`, `xxxServer`, `LocalApp`, `other`}, where `xxxServer` stands for any server-like application, i.e an application which listens to a port. `xxx` is to be replaced by a protocol name like `http`, `ftp`, `snmp`.

[2] SNMP protocol (`http://www.ietf.org/rfc/rfc1157.txt`), NetSaint tool for Unix (`http://www.netsaint.org`), System Management Server for Windows (`http://www.microsoft.com/smserver/`)

Definitions

$\mathcal{P}$ is the set of products,
VN is the set of product vendor names,
PN is the set of product names,
$PV = \mathbb{N}^p$ is the set of product version ($\mathbb{N}$ is the set of integers),
$PT = \{\texttt{OperatingSystem}, \texttt{LocalApp}, \texttt{httpServ}, \texttt{ftpServ}, \ldots, \texttt{other}\}$
is the set of product types.

Functions & Relations

$$
\begin{aligned}
\textbf{vendor} &\in \mathcal{P} \nrightarrow VN \\
\textbf{prodname} &\in \mathcal{P} \rightarrow PN \\
\textbf{version} &\in \mathcal{P} \rightarrow PV \\
\textbf{prodtype} &\in \mathcal{P} \nrightarrow PT \\
\textbf{configuration} &\in \mathcal{H} \rightarrow \mathscr{P}(\mathcal{P}) \\
< &\in PV \leftrightarrow PV
\end{aligned}
$$

Properties

$\forall v_1, v_2 \in \mathbb{N}^p, v_1 < v_2 \iff \exists j | \forall i, 1 \leq i \leq j, v_1[i] = v_2[i] \wedge v_1[j] < v_2[j]$

Comments

Not all products have a vendor attribute, thus **vendor** is a partial function.
Two distinct versions of the same product are modelized by two $\mathcal{P}$elements, i.e **version**is not a relation.

Fig. 5. M2D2 product model

Services, operating systems, local applications are products which run on hosts. The set of products running on a host is called a *configuration*. Formally speaking, a host configuration is a subset of products.

2.3 Vulnerabilities

Vulnerabilities are defined in figure 6. Most of current attacks consist in exploiting vulnerabilities that exist in entities of an IS. Shirey defines a vulnerability as a flaw or weakness in a system design, implementation, or management that could be exploited to violate the system security policy [9]. Shirey's vulnerable system corresponds to an M2D2 product. However, vulnerabilities sometimes do not refer to a single product: a host actual vulnerability may depend on the combined presence of several products (a web server on a given operating system for instance); a protocol design error may involve vulnerabilities in several implementations of the protocol (see SNMP's vulnerabilities `CAN-2002-0012` for instance (a.k.a BugTraq id 4088)). Arlat et al. define a vulnerability as a latent error present on a target [10], a target corresponding to a M2D2 host. However, it is not practical to model a vulnerable entity that way because each vulnerability would require a dedicated relation or function between the host set and the vulnerabilities set. In M2D2 a vulnerability affects a configuration (**affects**). In that way, a host is vulnerable if its configuration is a superset of one vulnerable configuration.

Definitions
$\mathcal{V}$ is the set of CVE and CAN vulnerabilities
$\mathcal{V}_N$ is the set of vulnerability names

Functions & Relations

$$
\begin{array}{rcl}
\textsf{affects} \in & \mathcal{V} & \leftrightarrow \mathcal{P}(\mathcal{P}) \\
\textsf{req} \in & \mathcal{V} & \rightarrow \{\texttt{Remote, RemoteUser, Local}\} \\
\textsf{con} \in & \mathcal{V} & \rightarrow \{\texttt{CodeExec, DoS, Info}\} \\
\textsf{vulname} \in & \mathcal{V} & \rightarrowtail \mathcal{V}_N \\
\textsf{equiv} \in & \textsf{vulname}[\mathcal{V}] & \leftrightarrow \mathcal{V}_N
\end{array}
$$

Properties
$\textsf{vulname}[\mathcal{V}]$ is the set of CVE and CAN vulnerability names.

Comments
affects is a relation because a vulnerability may affect several configurations and a configuration may be affected by many vulnerabilities.

Fig. 6. M2D2 vulnerability model

The vulnerability set is built from the ICAT vulnerability database. Moreover, ICAT uses the CVE/CAN[3] naming convention proposed by Mann [11] to uniquely identify the vulnerabilities. The Mitre proposes a list of name equivalences between CVE and other vulnerability names. This list is used inside M2D2 to cluster vulnerability names which refer to the same vulnerability (**equiv**). Ideally, **equiv** should be an equivalence relation, but it is not in reality. As Mann notices, the mapping between vulnerability names and CVE names is seldom one-to-one, so a non-CVE name may be equivalent to more than one CVE names. As a result, from one non-CVE vulnerability name, it is possible to get many CVE vulnerabilities.

Modeling vulnerabilities actually requires that a classification is established. As a matter of fact, no dictionary of terms for describing alerts exists [6], but alerts often refer to a known vulnerability. As a result, the information found in vulnerability databases remains the only source of information for describing alerts. One of the ICAT database strengths is that many attributes are used to classify vulnerabilities. For alert correlation purpose, we have chosen the two criteria which are described thereafter. Criteria can easily be extracted from the ICAT database, they are common to all classifications and are relevant for alert correlation, while remaining simple. For example, correlation may require to detect if a privilege elevation occurred, or if several kinds attacks have been attempted.

- Access requirements (**req**) for the vulnerability to be exploited
  - `Local`: the attacker needs a physical local access to the target,
  - `RemoteUser`: the attacker needs a remote access to the target with user-level privileges,
  - `Remote`: the attacker only needs a remote access to the target.

[3] `http://cve.mitre.org/`

Definitions

$\mathcal{D}$ is the set of IDS,
$\mathcal{S}$ is the set of vulnerability scanners,
$\mathcal{R}$ is the set of tool-specific report names (alert or scan),
$\mathcal{D}_N$ is the set of NIDS
$\mathcal{D}_H$ is the set of HIDS

Functions & Relations

$$\begin{array}{rcl}
\textbf{meth} \in & \mathcal{D} & \rightarrow \{\texttt{Misuse, Anomaly}\} \\
\textbf{data} \in & \mathcal{D} & \rightarrow \{\texttt{HostBased, NetworkBased}\} \\
\textbf{monitored_interface} \in & \mathcal{D}_N & \rightarrow \mathcal{I} \\
\textbf{monitored_host} \in & \mathcal{D}_H & \rightarrow \mathcal{H} \\
\textbf{scans} \in & \mathcal{S} & \leftrightarrow \mathcal{H} \\
\textbf{canraise} \in & \mathcal{D} \cup \mathcal{S} & \leftrightarrow \mathcal{R} \\
\textbf{refersto} \in & \mathcal{R} & \leftrightarrow \mathcal{V}_N
\end{array}$$

Properties

$\mathcal{D}_N = \textbf{data}^{-1}[\{\texttt{NetworkBased}\}]$
$\mathcal{D}_H = \textbf{data}^{-1}[\{\texttt{HostBased}\}]$

Fig. 7. M2D2 security tools model

- Consequence (**con**) of a successful exploit of the vulnerability:
  - `DoS`: the exploit results in a denial of service.
  - `InformationGathering`: the exploit results in information disclosure. For example, directory traversals and file reading fall under this category.
  - `CodeExecution`: the exploit results in the execution of arbitrary code. For example, buffer overflows and race conditions fall under this category.

2.4 Security Tools

Security tools model is formalized in figure 7. Vulnerabilities can be detected either when they are exploited by an attacker or simply when they are latent. IDS detect the former, vulnerability scanners detect the latter. IDS and vulnerability scanners are security tools.

Although they are part of the topology, security tools need to be explicitly described in a security model like M2D2. Security tools are event producers (see 2.5). Any security tool type should be integrated into M2D2. In the current model, IDS are qualified by their analysis method (**meth**), either misuse-detection or anomaly detection. Other security tools, like firewalls, will be included in future versions of M2D2.

IDS detect, identify and generate alerts about active attacks in real time. We distinguish two kinds of IDS according to their data source: HIDS (Host-based IDS) and NIDS (Network-based IDS) (**data**). This dichotomy is based on the *topological visibility* of an IDS: the *topological visibility* of an IDS refers to the set of network components an IDS can monitor. NIDS (Network-based IDS) read streams of events on a network interface (**monitored_interface**), thus they can monitor many hosts depending on their location in the network. Whereas a HIDS (should it be system-based or application based) monitor a single host (**monitored_host**), vulnerability scanners do vulnerability assessments on several hosts, thus their topological visibility is modeled with the relation between a scanner and a host (**scans**).

We also introduce the *operational visibility* of an IDS which refers to the ability of an IDS to detect vulnerability exploits according to its configuration[4]. This is modeled by a relation (**canraise**) between an IDS or a scanner and a report name. A report name is tool-specific, i.e the names of the reports are the ones that are used by the tool that generates the alerts. Report names are linked to one or more vulnerability identifiers (**refersto**). It is therefore transitively possible to know which vulnerability exploits can be detected by an IDS.

The correspondence between report names and vulnerabilities are found, for instance, in IDS signature files for misuse IDS.

It has to be noted that all report names do not refer to a vulnerability. As a matter of fact, all attacks are not vulnerability exploits (port scanning is such example). Some IDS can detect an attack but not recognize it as the exploit of a given vulnerability, either because the attack is not referenced yet or because the IDS cannot do so (case of the anomaly IDS).

2.5 Events, Alerts, and Scans

The event formal model is described in figure 8.

Both vulnerability scanners and IDS generate messages about the existence or about an ongoing exploit of a vulnerability. The former is called a *scan*, and the latter an *alert*. In the case of alerts, security tools should also generate the *events* which led them to generate alerts.

Arlat et al. define an event as the instantaneous image of a system [10]. In other words, an event is the manifestation of some activity the format of which is discussed in this section. In the intrusion detection field, event and alert are generally two distinct concepts [12]. An event is a low level entity (TCP packet, system call, syslog entry, for example) from which an analysis is performed by a security tool. An alert is a message from an analyzer signaling that one or more events of interest have been detected. We say that an alert is *a kind of* event, since it reflects the state of an IDS. Such a definition is compliant with the previous definitions and enables to elegantly model the fact that alerts produced by an IDS may be the events from other tools' point of view. In other words, current IDS are low level event consumers and alert producers but future IDS may be alert consumers and high level alerts producers.

M2D2 models alerts, scans, and the following event types: IP events, TCP events, UDP events, HTTP events and HTTP log events. We restricted the event types to the set of event types understood by most current IDS. However, M2D2 has to be extended with other event types such as operating system level events and other network application protocol events, with their respective attributes.

[4] e.g the set of activated signatures for a misuse IDS

Definitions

$\mathcal{E}$ is the set of event instances,
$\mathcal{E}_{\mathsf{A}}$ is the set of alerts,
$\mathcal{E}_{\mathsf{S}}$ is the set of scans,
$\mathcal{E}_{\mathsf{IP}}$ is the set of IPv4 events,
$\mathcal{E}_{\mathsf{TCP}}$ is the set of TCP events,
$\mathcal{E}_{\mathsf{UDP}}$ is the set of UDP events,
$\mathcal{E}_{\mathsf{HTTP}}$ is the set of HTTP events,
$\mathcal{E}_{\mathsf{LOG}}$ is the set of web servers log events.

Properties

$\mathcal{E}_{\mathsf{A}}, \mathcal{E}_{\mathsf{IP}}, \mathcal{E}_{\mathsf{TCP}}, \mathcal{E}_{\mathsf{UDP}}, \mathcal{E}_{\mathsf{HTTP}}, \mathcal{E}_{\mathsf{LOG}}$ partition $\mathcal{E}$

Functions & Relations

$$\begin{array}{rcl}
\textsf{tstamp} \in & \mathcal{E} & \rightarrow \mathbb{N} \\
\textsf{partof} \in & \mathcal{E} & \leftrightarrow \mathcal{E} \\
\textsf{alert_generator} \in & \mathcal{E}_{\mathsf{A}} & \rightarrow \mathcal{D} \\
\textsf{reportname} \in & \mathcal{E}_{\mathsf{A}} \cup \mathcal{E}_{\mathsf{S}} & \rightarrow \mathcal{R} \\
\textsf{scan_generator} \in & \mathcal{E}_{\mathsf{S}} & \rightarrow \mathcal{S} \\
\textsf{scan_host_target} \in & \mathcal{E}_{\mathsf{S}} & \rightarrow \mathcal{H} \\
\textsf{scan_port_target} \in & \mathcal{E}_{\mathsf{S}} & \rightarrow \mathbb{N} \\
\textsf{causes} \in & \mathcal{E} & \leftrightarrow \mathcal{E}_{\mathsf{A}}
\end{array}$$

Comments

Alerts can cause other alerts, therefore **dom(causes)** $\subset \mathcal{E}$, not $\mathcal{E} - \mathcal{E}_{\mathsf{A}}$
ran(causes) $\subsetneq \mathcal{E}_{\mathsf{A}}$ because alerts may have no cause.

Fig. 8. M2D2 model of events

The common attribute of all events is the timestamp (**tstamp**). The IDMEF [12] uses three distinct timestamps: `detecttime` (attack occurrence date), `createtime` (attack detection date), `analysertime` (alert generation date). Those three distinct timestamps depend on the nature of the M2D2 events: a M2D2 IPEvent (cf p. 126) timestamp is a `detecttime`. An alert (cf p. 125) timestamp is a `createtime` or a `analysertime`.

To model the aggregative nature of events, we introduce the **partof** relation. This kind of relation models the fact that an alert can aggregate (encapsulate) several other alerts or that an HTTP request is part of an HTTP log entry. This relation may be refined in the future if necessary.

In the remainder of this section, we give some details on some of the event types (alerts, scans, and HTTP log events). Others are self-explanatory (the reader may refer to the relevant RFCs).

Alerts. Alerts are formalized in figure 8.

Each alert has a single report name (**reportname**). Alerts are generated by IDS (**alert_generator**).

As with IDMEF definition, alerts are caused by the occurrence of events which are the manifestation of an attack (**causes**). Note that alerts may not be linked to a *causal* event. This does not mean that an alert has no cause, but that the causal information is not available (the IDS generating the alert does not give this information, for instance). Alerts may also have more than one causes. Finally, an event can be the cause of several alerts because:

- it can be shared by several alerts generated by distinct IDS, or
- IDS may generate several alerts for one single event. For example, the HTTP request
  `GET /cgi-bin/phf?/etc/passwd` may cause three alerts: one concerning the presence of `/etc/passwd` in the URL, one concerning the presence of `/cgi-bin/phf` and one concerning the success/failure of the request.

In other existing alert formats, *source* and *target* concepts are attributes of the alerts [12,13]. Source and target implicitly refer to the source and target of an attack. As events reflect the steps of an attack, we claim that the source and the target concepts are hold by the events themselves. For IP events, the source and the target are respectively the source IP address and the destination IP address. Port numbers play that role for TCP events. The target information may also be deduced from the location where an event was detected. This is the case for all the events raised by HIDS, since IDS only monitor the host they are running on.

Scans. Scan is a shorthand for "vulnerability scan". A scan is a vulnerability assessment provided by a vulnerability scanner (Nessus[5] for instance) against a specific host.

Like alerts, scans have a tool-specific report name (**reportname**). Scans are generated by scanners (**scan_generator**).

There are two major differences between scans and alerts. Firstly, scans are not caused by events, since scans do not detect active vulnerabilities exploits. However, it should be noted that to perform security tests, scanners generate intrusive traffic which can cause IDS alerts. Alert correlation should take this information into account.

The second difference is that while alert target information is carried by low level events, the target is explicitly carried by the scan. Vulnerability assessments are made against a host (**scan_host_target**) on a given port (**scan_port_target**).

IPEvents, TCPEvents, UDPEvents, HTTPEvents. These event types are modeled in figure 9.

In Vigna's model, a message is a triple (i_s, i_d, p), where $i_s, i_d \in \mathcal{I}$ are respectively the source interface and the destination interface which belong to the same link. p is the payload of the message.

[5] `http://www.nessus.org`

Definitions

$\mathcal{U}$ is the set of URL which have been raised by security tools.

Functions & Relations

$$\begin{array}{rcll}
\textsf{ip_src}, \textsf{ip_dst}, \textsf{idt} \in & \mathcal{E}_{\textsf{IP}} & \rightarrow & \mathbb{N}_A \\
\textsf{ipayload} \in & \mathcal{E}_{\textsf{IP}} & \nrightarrow & \mathcal{E}_{\textsf{TCP}} \cup \mathcal{E}_{\textsf{UDP}} \\
\textsf{sp}, \textsf{dp} \in & \mathcal{E}_{\textsf{TCP}} \cup \mathcal{E}_{\textsf{UDP}} & \rightarrow & \mathbb{N} \\
\textsf{seq}, \textsf{ack} \in & \mathcal{E}_{\textsf{TCP}} & \rightarrow & \mathbb{N} \\
\textsf{tpayload} \in & \mathcal{E}_{\textsf{TCP}} \cup \mathcal{E}_{\textsf{UDP}} & \rightarrowtail & \mathcal{E}_{\textsf{HTTP}} \\
\textsf{httpmeth} \in & \mathcal{E}_{\textsf{HTTP}} & \rightarrow & \{\texttt{GET},\texttt{POST},\texttt{HEAD}\} \\
\textsf{status} \in & \mathcal{E}_{\textsf{HTTP}} & \rightarrow & \mathbb{N} \\
\textsf{url} \in & \mathcal{E}_{\textsf{HTTP}} & \rightarrow & \mathcal{U}
\end{array}$$

Comments

- **ip_src, ip_dst, idt** Resp. source address, destination address and identification number of an IPv4 event,
- **sp, dp** Resp. source port and destination port numbers of a TCP or UDP event,
- **seq, ack** Resp. sequence number and acknowledgement number of a TCP event,
- **httpmeth** Method of an HTTP request,
- **status** Status of an HTTP request,
- **url** URL of an HTTP request.
- **ipayload** is a partial function because IP Event payload may not be available and thus is not provided by the security tool.

Fig. 9. M2D2 IP Events, TCP Events, UDP Events and HTTP Events

An IP datagram is a sequence of messages whose payload is a triple (a_s, a_d, p'), where $a_s, a_d \in \mathbb{N}$ are respectively the source and destination addresses of the IP datagram. The messages in the sequence represent the route of the datagram.

An IP session is a sequence of IP datagrams and represents a conversation between two hosts.

An UDP datagram is an IP datagram whose payload is a couple (p_s, p_d), where p_s,$p_d \in \mathbb{N}$ are respectively the source port and the destination port of the UDP datagram. An UDP session is a couple of UDP datagrams (request and response).

Basically, a TCP segment is an IP datagram whose payload is a tuple (p_s,p_d,s,a,F,p''), where p_s,p_d,s,$a \in \mathbb{N}$ are respectively the source port, the destination port, the sequence number and the acknowledgement number, F is a set of flags and p'' is the payload of the TCP segment. A TCP session is an IP session with constraints over the datagram attributes which are described in Vigna's model.

In the same way, a M2D2 TCP event (resp. an UDP event) is a TCP segment (resp. an UDP datagram) in Vigna's model.

We do not model sessions in M2D2 yet because most current NIDS do not work at the session level. However, extending M2D2 with sessions is straightforward since the M2D2 events are conceptually identical to the concepts found in Vigna's model.

An IP event is linked to its content with the **ipayload** function. In the same way, TCP or UDP event are linked to their content (an application protocol like HTTP for instance) with the **tpayload** function.

HTTP Events. An HTTP event models an HTTP interaction between a client and a server. HTTP event attributes potentially reflect a complete HTTP interaction. Thus, we use the **httpmeth**, **url**, and **status** functions. Note that if only a part of the interaction may be analyzed (that is the case for NIDS, which may only have access to a part of the interaction), then a special value is assigned to the event attribute.

HTTP Log events. Log events correspond to entries in HTTP server logs. An HTTP server log entry contains an HTTP event, and an identifier of the requester (either its IP address or its host name). That mix of information coming from two different protocol layers raises a modeling problem. We thus had to introduce a specific kind of events, the HTTP Log events. That kind of events reflect the contents of a HTTP server log.

3 Alert Correlation Using M2D2

A first step towards alert correlation is alert aggregation. It consists in grouping alerts following various criteria. In this section, we propose three relevant examples of aggregation methods which use many M2D2 concepts. As these concepts are defined in an unambiguous way, the aggregation methods are rigorously expressed. M2D2 contribution to alert correlation resides in the use of relations between vulnerabilities and topology, between topology and security tools, as well as between security tools and vulnerabilities to model alert aggregation methods.

3.1 Aggregation of Alerts Referring a Single Host

The goal of the **common_target** aggregation function (cf figure 10) is to group alerts referring to a given target host. This set is made of two subsets: one contains the alerts generated by HIDS running on the host, and the other contains alerts generated by NIDS having detected malicious IP events destined to the host.

As HIDS monitor a single host (cf figure 7, **monitored_host**), the set of alerts referring to a host is equal to the set of alerts generated by the HIDS monitoring the host. We call this aggregation function **hids_target** (cf figure 10).

Functions & Relations

$\textsf{hids\_target} \in \mathcal{H} \leftrightarrow \mathcal{E}_\mathsf{A}$
$\textsf{nids\_target} \in \mathcal{H} \leftrightarrow \mathcal{E}_\mathsf{A}$
$\textsf{common\_target} \in \mathcal{H} \leftrightarrow \mathcal{E}_\mathsf{A}$
$\textsf{belongs\_to\_host} \in \mathcal{I} \rightarrow \mathcal{H}$

Properties

$\textsf{hids\_target} = \textsf{monitored\_host}^{-1}; \textsf{alert\_generator}^{-1}$
$\textsf{nids\_target} = \textsf{addr}; \textsf{ip\_dst}^{-1}; \textsf{causes}$
$i \ \textsf{belongs\_to\_host} \ h \iff i \in h$

$\textsf{common\_target} = \textsf{hids\_target} \cup \textsf{nids\_target}$

Fig. 10. The **common_target** aggregation function referring the same target host

Functions & Relations

$\textsf{sub\_config} \in ran(\textsf{affects}) \leftrightarrow ran(\textsf{configuration})$
$\textsf{relative\_vulns} \in \mathcal{E}_\mathsf{A} \leftrightarrow \mathcal{V}_N$
$\textsf{harmful} \in \mathcal{E}_\mathsf{A} \leftrightarrow \mathcal{H}$

Properties

$v \ \textsf{sub\_config} \ c \iff v \subset c$
$\textsf{relative\_vulns} = \textsf{reportname}; \textsf{refersto}; \textsf{equiv}$
$\textsf{harmful} = \textsf{relative\_vulns}; \textsf{vulname}^{-1}; \textsf{affects}; \textsf{sub\_config}; \textsf{configuration}^{-1}$

Fig. 11. The successful attacks identification function

An intermediate relation is required to uniquely identify a host from a network interface (this is a function because hosts partition the interface set, cf figure 10). We call this function **belongs_to_host** (cf 16).

NIDS monitor an interface which has access to a stream of network events, some of which being destined to the host under consideration. The aggregation function, **nids_target** (cf figure 10) consists in aggregating all the alerts caused by IP events whose IP destination address maps to one of the host interfaces.

3.2 Identification of the Hosts Vulnerable to an Attack Occurrence

The successful attack identification function, called **harmful** (cf figure 11), applies to a single alert and gives the set of hosts on which the attack may work. In other words, it gives information about the success of the attack.

An attack may be successful if one of the configurations affected by the vulnerability which is exploited by the attack really *exists* on some hosts in the information system. For example, an attack exploiting a ProFTP vulnerability against a WU-ftp server may not work. Thus the attack severity is low. That does not mean that the attack should not be reported (it is still an attack attempt), but its severity may be lowered.

A vulnerability $v \in \mathcal{V}$ exists on a host if the host configuration is a superset of at least one of v vulnerable configurations. Thus, we introduce the **sub_config** (cf figure 11) relation, which maps all the vulnerability configurations with all the host configurations. **sub_config** provides all the vulnerabilities exhibited by a given host, and $\textbf{sub\_config}^{-1}$ provides all the hosts which exhibit a given vulnerability. Building and updating the **sub_config** relation in batch mode is required for efficiency because evaluating existing vulnerabilities in real time would be prohibited.

We introduce the **relative_vulns** relation which applies to an alert and maps the vulnerability names the alert report name refers to. From an alert relative vulnerability set, we get the corresponding CVE vulnerability subset, and then the vulnerable configurations set is extracted. Thanks to the **sub_config** relation, we get the set of vulnerable host configurations. Lastly, applying $\textbf{configuration}^{-1}$ to the host configuration set provides the set of hosts which are vulnerable.

3.3 Detecting False Positives

In this section we present a correlation function which is a way to detect false positives. When an IDS generates an alert, it is relevant to check if all the IDS which were able to detect the events causing the alarm did generate an alert too. When an IDS generates an alert, a simple sanity check is to verify that others able to both process the events and create the alert did so. If this constraint is verified, then the alert likelihood is reinforced. If not, the proper behavior of all IDSes should be checked to decide whether the one sending the alert is providing the wrong diagnosis or whether the others are misbehaving.

Let us consider an alert $a \in \mathcal{E}_{\mathbf{A}}$. We first need to identify the alerts which are *similar* to a. That is, we need to know which alerts are caused by the same set of events. Once we have the set of similar alerts, we can know which set of IDS raised them. Finally, we have to compare this set with the set of IDS which were able to detect the events (the ability of an IDS to detect events and generate alerts depends on its topological visibility and its operational visibility (cf section 2.4)).

In the following, we first describe two methods to identify similar alerts. We then show how to get the set of IDS which did react. Lastly, we propose a function to get the *potentially reactive IDS* set (i.e the set of IDS which could react to an attack).

Alerts similarity. In Howard's classification [14], an *attack* is a series of *actions* taken by an attacker to achieve an unauthorized result. Events result from these actions. As it has been described in section 2, one or more alerts may result from every each event. Determining the likelyhood of an alert can be achieved by enumerating the IDS which also generated an alert for a single *action* (be it from an attack or from a legitimate action which is believed malicious). We call

Definitions

$\overset{ip}{=} \in \mathcal{E}_{\mathsf{IP}} \leftrightarrow \mathcal{E}_{\mathsf{IP}}$

$$\forall i_1, i_2 \in \mathcal{E}_{\mathsf{IP}},\ i_1 \overset{ip}{=} i_2 \iff \begin{cases} \mathbf{ip\_src}(i_1) = \mathbf{ip\_src}(i_2) \wedge \mathbf{ip\_dst}(i_1) = \mathbf{ip\_dst}(i_2) \\ \qquad \wedge\ \mathbf{idt}(i_1) = \mathbf{idt}(i_2) \\ \qquad \wedge\ |\mathbf{tstamp}(i_1) - \mathbf{tstamp}(i_2)| < \varepsilon \\ \qquad \text{OR} \\ \exists i_3 \in \mathcal{E}_{\mathsf{IP}}, \text{ such that } i_1 = i_3 \wedge i_2 = i_3 \end{cases}$$

$\mathbf{sim\_alert\_ip} \in \mathcal{E}_{\mathsf{A}} \leftrightarrow \mathcal{E}_{\mathsf{A}}$

Properties

$\overset{ip}{=}$ is an equivalence relation.

$\mathbf{sim\_alert\_ip} = (\mathbf{causes}^{-1}; \overset{ip}{=}; \mathbf{causes})$

$\mathbf{sim\_alert\_ip}[\{a\}]$ is the set of alerts which are caused by the same IP event as a.

Fig. 12. Alerts caused by the same IP event

alerts caused by a single action *similar alerts.* Several ways to aggregate alerts are possible, for example:

- aggregating alerts caused by the same event
- aggregating alerts referring to the same vulnerability
- aggregating alerts caused by events belonging to the same TCP/IP session
- aggregating alerts on a temporal relation basis

In the remainder of this section, we illustrate the first two examples of similar alerts aggregation methods.

Aggregating alerts through events. Several alerts can be generated from a single event either because more than one triggers within the same IDS matches the event (a trigger is a condition fulfillment leading an IDS to generate an alert), or because more than one IDS have access to the event itself. Thus, alerts caused by the same events should be aggregated. We call events causing an alert causal events. It should be noted that two alerts do not need to have the same set of causal events to be aggregated: alerts can be aggregated if they have at least one causal event in common. For example, a suspicious HTTP request can be detected by a NIDS (A) and by an application-based IDS (B). A alert is caused by an HTTP event and the underlying network layers event. B alert is caused by an HTTP event, which is part of a web server log occurrence. Thus, A and B do not have the same set of causal events; the HTTP event is the only one they have in common. However, A alert and B alert are aggregated because they have one causal event in common.

As every alert comes along with its own causal event set, we need to define a relation to compare events. Such a relation strongly depends on the timestamp

$$\textsf{sim\_alert\_vuln} \in \mathcal{E}_\textsf{A} \leftrightarrow \mathcal{E}_\textsf{A} \cup \mathcal{E}_\textsf{S}$$
$$\textsf{sim\_alert\_vuln} = \textsf{relative\_vulns}; \textsf{refersto}^{-1}; \textsf{reportname}^{-1}$$

Fig. 13. Alerts referring to the same vulnerability

$$\textsf{reactive\_ids} \in \mathcal{E}_\textsf{A} \leftrightarrow \mathcal{D}$$
$$\textsf{reactive\_ids} = (\textsf{sim\_alert\_ip} \cup \textsf{sim\_alert\_vuln}); \textsf{alert\_generator}$$

Fig. 14. Reactive IDS

of the events. Although distinct IDS alert timestamps (IDMEF `analysertime`) may differ, the timestamp of the causal events (IDMEF `detecttime`) should be very close. Two events cannot have the same timestamps as clock synchronization is impossible to achieve. This is why we impose two events to have close timestamps (in figure 12, ε is a constant whose value is close to the maximum gap existing between two synchronized clocks).

We recursively define the comparison relation between IP events, noted $\stackrel{ip}{=}$ (cf figure 12). Two IP events are equal if they have the same attributes (destination address, source address, identification number) and close timestamps or if there is another IP event which is equal to the two others. A recursive definition is required here otherwise $\stackrel{ip}{=}$ would not be an equivalence relation due to the time constraint. Equality relations for other types of events can be built in the same way.

Applying the **causes** (figure 8) relation to an $\stackrel{ip}{=}$ equivalence class gives the set of alerts which are caused by the same IP event. The **sim_alert_ip** (cf figure 12) is a function which applies to an alert, and gives a set of alerts which are caused by the same IP event (if the alert is caused by an IP event, otherwise this is the empty set).

Aggregating alerts through vulnerability. As a second example, we propose a way to aggregate alerts generated by IDS which not only use distinct data sources, but also use distinct alert naming conventions. Thus, we propose to group alerts which refer to the same vulnerability (figure 13).

Let us consider an alert $a \in \mathcal{E}_\textsf{A}$. The set of a similar alerts is composed of the alerts whose relative vulnerabilities (defined in figure 11) are the same as a.

Reactive IDS. Reactive IDS are the IDS which have generated an alert because of a single action. Actions are on the attacker's side so we do not have access to this concept. We only have access to the manifestations of actions, i.e events and alerts. Thus, for a given alert $a \in \mathcal{E}_\textsf{A}$, the reactive IDS are the IDS which have generated alerts similar to a. In the preceding paragraphs, we introduced two examples of similar alert aggregation functions. The reactive IDS can be

$\textsf{oper\_able} \in \mathcal{E}_{\mathsf{A}} \leftrightarrow \mathcal{D} \cup \mathcal{S}$
$\textsf{oper\_able} = \textsf{relative\_vulns}; \textsf{refersto}^{-1}; \textsf{canraise}^{-1}$

Fig. 15. Operationnaly-able IDS

$\textsf{topo\_able\_hids} \in \mathcal{E}_{\mathsf{A}} \leftrightarrow \mathcal{D}_H$
$\textsf{topo\_able\_hids} = \textsf{alert\_generator}; \textsf{monitored\_host}; \textsf{monitored\_host}^{-1}$
$\textsf{topo\_able\_nids} \in \mathcal{E}_{\mathsf{A}} \leftrightarrow \mathcal{D}_H$
$\textsf{topo\_able\_nids} = \textsf{causes}^{-1}; \textsf{ip\_dst}; \textsf{addr}^{-1}; \textsf{belongs\_to\_host}; \textsf{monitored\_host}^{-1}$
$\textsf{topo\_able} \in \mathcal{E}_{\mathsf{A}} \leftrightarrow \mathcal{D}_H$
$\textsf{topo\_able} = \textsf{topo\_able\_hids} \cup \textsf{topo\_able\_nids}$

Fig. 16. Topologically-able IDS

$\textsf{able} \in \mathcal{E}_{\mathsf{A}} \leftrightarrow \mathcal{D}_H$
$\textsf{able} = \textsf{topo\_able} \cap \textsf{oper\_able}$

Fig. 17. Potentialy reactive IDS

obtained by simply applying **alert_generator** (cf figure 8) to the similar alerts. We call this relation **reactive_ids** (cf figure 14).

Potentially reactive IDS. A potentially reactive IDS is an IDS which is both topologically and operationally able (see section 2.4) to detect an attack. Topological and operational ability both apply to alerts: given an alert $a \in \mathcal{E}_{\mathsf{A}}$, operationally-able IDS (resp. topologically-able IDS) are IDS which are operationally-able (resp. topologically-able) to generate an alert similar to a.

Given an alert $a \in \mathcal{E}_{\mathsf{A}}$, the operationally-able IDS set is composed of the IDS which can raise a report which refers to one of a relative vulnerabilities. This function is formalized in figure 15.

Evaluating the topologically-able IDS faces two major pitfalls. The first one is that the topological visibility of an HIDS is limited to a single host, whereas the topological visibility of an NIDS is not even limited to a static set of hosts. An NIDS having access to a network stream of events depends on the route followed by the datagrams. Thus, it is generally not possible to say that an NIDS has access to the datagrams of such attack without knowing *a priori* the route taken by the datagrams, and this information is not available.

The second pitfall is that an attack may concern more than a single host. In other words, a set of events may be considered malicious only because of the multiplicity of the targets (network probes are such examples). In such a case, although an IDS is operationally able and topologically able to detect some *part of* an attack, it may not generate an alert.

Moreover, if an alert $a \in \mathcal{E}_\mathsf{A}$ is generated by an HIDS, no reliable conclusion can be sketched about NIDS topological ability to detect the corresponding attack: if the attack is launched locally, it has no network-side effect ; if it is launched remotely, no information about the route taken by the datagrams is available

Thus, the set of topologically-able IDS is currently limited to the set of HIDS. We propose two functions to obtain topologically HIDS set for a given alert. The first one is called **topo_able_hids** and applies to alerts generated by HIDS and gives HIDS which monitor the same host. The second relation, called **topo_able_nids**, applies to alerts generated by an NIDS and gives HIDS which monitor the destination host of the underlying IP event which causes the alert (cf figure 16).

Lastly, potentially reactive IDS are both topologically-able and operationally-able IDS. As the we are currently limited to topologically-able HIDS, potentially reactive IDS is also limited to potentially reactive HIDS. The corresponding relation is in figure 17.

3.4 Discussion

M2D2 provides an expressive tool to build complex correlation methods. For example, the reduction of false positives described in section 3.3 could have been hardly designed without such a rich and formal model. Moreover, the three aggregations presented above have all illustrated how powerful and easy it is to compose relations. Aggregations themselves could be composed further. For example, the first two aggregations can be composed to create another relevant aggregation relation ($\mathbf{harmful}^{-1}$; **common_target**). Indeed, applying $\mathbf{harmful}^{-1}$ to a given host gives alerts whose related attacks may be successful. When used in conjunction with the **common_target** function, we know which attacks may work on a given host.

The formal specification also helps detect a number of problems. For example, the relation sub_config figure 11 has been introduced to ensure type consistency between the composed relations. This has a practical impact on the implementation: either an intermediate table has to be created or, at correlation time, the sub-configuration will have to be checked.

4 Related Work

Formal models. As already mentioned in the introduction, the closest related work are the NetStat Network Fact Base (NNFB) of Vigna and Kemmerer [4,5] and the Intrusion Reference Model (IRM) of Goldman et al. [6]. Both the NetStat Network Fact Base and the Network Entity Relationship Database (NERD) of IRM are based on Vigna's network formal model [7], and so are the topolopy and network events of M2D2. The concept of product in M2D2 is similar to the concept of services in NNFB and NERD. The security tools formalized in M2D2 are mentioned in NNFB and IRM but they are not precisely specified. As far as we know, no model includes the vulnerability and alert parts of M2D2. On the other hand, IRM includes a security goals database. It captures security policies set by the administrators. Evidence about violations of the security policy are then aggregated. This feature, not yet formalized, could be added in M2D2 in the future. Finally, as mentioned in section 2.5, there are no system level events modeled in M2D2. We are currently working on the addition of such events.

Another piece of work related to ours is the one of the IETF Intrusion Detection Working Group (IDWG) [12]. The Intrusion Detection Message Exchange Format (IDMEF) has been proposed to provide a common format of alerts [12]. The alert specification in M2D2 is compatible with IDMEF, IDMEF-compliant alerts can be received and processed within the M2D2 framework without loss of semantical information. The scope of M2D2 is larger than IDMEF, to take into account all sources of information relevant to information security, and as such simplifies the IDMEF model through the removal of classes with similar semantic content while conveying the same information through aggregation relationships. This is in particular the case of the SOURCE and TARGET classes, replaced by the notion of node while the *source* and *target* information are carried by aggregation relationships.

Implicit models for correlation. Another aspect of the related work is correlation. In order to perform fusion or correlation, researchers in the field have worked out implicit models dedicated to their correlation algorithms. The objective of M2D2 is to have a solid modeling foundation that allows multiple correlation tools and algorithms to be implemented and tested.

In [15,16], Cuppens and Miège base their work on fusion on the IDMEF data model [12], implemented as Prolog predicates. As such, their model is simpler than M2D2, which means that less information is available for correlation purposes. The sole reliance on intrusion-detection alerts makes it impossible to assess the relevance of these alerts with respect to the monitored information system.

Cuppens and Miège suggest in [16] to correlate alerts by using attack scenarios specified in the Lambda language [17]. Such a language allows correlations between events belonging to a same attack scenario to be expressed, and could be used on top of M2D2.

The Tivoli RiskManager tool [13] is a correlation platform for security events provided by information systems. Originally developed for the correlation of intrusion-detection alerts, it now also correlates information from firewalls and antivirus systems. The data model used by Tivoli RiskManager was the source for the first version of IDMEF, but IDMEF has evolved since and the constraints of the Tivoli environment are such that it would be extremely difficult to implement the exact IDMEF model. Also, events from the three information sources (intrusion-detection systems, firewalls and antivirus systems) are not related together as is the case in M2D2. As such, the correlation algorithm used in RiskManager has to develop hypotheses concerning the relationships between events, creating relationships based on superficial characteristics of the events.

5 Conclusion

In this paper, we introduced M2D2, a formal information model designed for security information representation and correlation. The model includes the four types of information we think crucial to allow rich alert correlations: information system characteristics, vulnerabilities, security tools, and events and alerts. The model is a platform to provide answers to three important issues in intrusion-detection today, the sheer number of alerts provided, the quality of these alerts and the precision of the final diagnosis proposed to operators.

As far as possible, M2D2 reuses models proposed by others, for instance the Vigna's topological model. The first contribution of M2D2 is thus the integration of multiple interesting and relevant concepts into a unified framework. When needed, we extended these reused model to encompass the four types of useful information.

Moreover, M2D2 is formally defined, following Mc Hugh's suggestion that further intrusion detection significant progress will depend on the development of an underlying theoretical basis for the field. This formalism is the second contribution of M2D2, ensuring that processing of security information and in particular alert correlation is anchored on a rigorous model representing the information being processed. This formalism does lend itself to extensions, as we believe that progress on this path will lead us to incorporate additional information sources and components.

In order to illustrate how M2D2 could be used, we also presented three examples of correlation (actually of aggregation) methods. These examples are rigorously specified using the formal definition of M2D2. As opposed to already published correlation methods, these examples use more than the events generated by security tools; they make use of many concepts formalized in M2D2.

References

1. G. Jakobson and M. D. Weissman. Alarm correlation. *IEEE Network Magazine*, pages 52–60, 1993.
2. J. McHugh. Intrusion and intrusion detection. *International Journal of Information Security*, July 2001.
3. Icat vulnerabilities database. http://icat.nist.gov/icat.cfm.
4. G. Vigna and R. A. Kemmerer. Netstat: A network-based intrusion detection approach. In *Proceedings of the 14th Annual Computer Security Application Conference*, December 1998.
5. G. Vigna and R. A. Kemmerer. Netstat: A network-based intrusion detection system. *Journal of Computer Security*, February 1999.
6. R. P. Goldman, W. Heimerdinger, S. A. Harp, C. W. Geib, V. Thomas, and R. L. Carter. Information modeling for intrusion report aggregation. In *Proceedings of the DARPA Information Survivability Conference and Exposition*, June 2001.
7. G. Vigna. A topological characterization of tcp/ip security. Technical Report TR-96.156, Politecnico di Milano, 1996.
8. J.-R. Abrial. *The B Book: Assigning programs to meanings*. Cambridge University Press, 1996.

9. R. Shirey. Internet security glossary. RFC2828, 2000.
10. J. Arlat, J.P. Blanquart, A. Costes, Y. Crouzet, Y. Deswarte, J.C. Fabre, H. Guillermain, M. Kaaniche, K.Kanoun, J.C. Laprie, C. Mazet, D. Powell, C. Rabejac, and P. Thévenod. *Guide de la sureté de fonctionnement.* Cepadues editions, 1995.
11. D. E. Mann and S. M. Christey. Towards a common enumeration of vulnerabilities. In *Proceedings of the 2nd Workshop on Research with Security Vulnerability Databases*, January 1999.
12. Dave Curry and Hervé Debar. Intrusion detection message exchange format data model and extensible markup language (xml) document type definition. Internet Draft (work in progress), December 2001. http://search.ietf.org/internet-drafts/draft-ietf-idwg-idmef-xml-06.txt.
13. Hervé Debar and Andreas Wespi. Aggregation and correlation of intrusion-detection alerts. In Wenke Lee, Ludovic Mé, and Andreas Wespi, editors, *Proceedings of the 4th International Symposium on Recent Advances in Intrusion Detection (RAID 2001)*, number 2212 in Lecture Notes in Computer Science, pages 85–103, Davis, CA, USA, October 2001. Springer.
14. J. D. Howard and T. A. Longstaff. A common language for computer security incidents. CERT - SAND98-8667, http://www.cert.org/research/taxonomy_988667.pdf, 1998.
15. F. Cuppens. Managing alerts in multi-intrusion detection environment. In *Proceedings of the 17th Annual Computer Security Applications Conference (ACSAC'01)*, 2001.
16. F. Cuppens and A. Miege. Alert correlation in a cooperative intrusion detection framework. In *Proccedings of the IEEE Symposium on Security and Privacy*, 2002.
17. Frédéric Cuppens and Rodolphe Ortalo. Lambda: A language to model a database for detection of attacks. In H. Debar, L. Mé, and S. F. Wu, editors, *Proceedings of the Third International Workshop on the Recent Advances in Intrusion Detection (RAID'2000)*, number 1907 in LNCS, pages 197–216, October 2000.

Development of a Legal Framework for Intrusion Detection

Steven R. Johnston, CISSP

Communications Security Establishment
P.O. Box 9703
Terminal
Ottawa, Canada K1G 3Z4
Steven.Johnston@cse-cst.gc.ca

Abstract. To meet demands for increased interconnectivity, efficiency or competitiveness, organizations increasingly rely on technology. This trend creates significant opportunities to improve service delivery and to move into new areas of endeavour. But reliance on an inherently insecure infrastructure exposes organizations to a constantly evolving threat environment. Not only has the nature of the threat changed, so too has the scope of the protection problem. Protection of information systems is now seen as a component of national security. As organizational assets move online, so does the threat. Key sources of threat information are now online, including within the network communications themselves. This puts organizations in a position where they must monitor network communications in order to obtain intelligence, indications and warnings of intrusions and evidence to support criminal prosecution as appropriate. One method of performing this monitoring is through the use of intrusion detection systems (IDS). However, this may involve the monitoring of private communications, which introduces a number of legal (privacy and criminal law) concerns. While existing legislation adequately addresses interception by S&I and law enforcement agencies, they generally fail to address interception of network traffic by other public or private sector organizations. This paper seeks to identify and discuss some of the key legal issues affecting the development of a general legal framework for intrusion detection for network protection.

Keywords: Anti-terrorism law, criminal law, interception, intrusion detection, privacy, private communications, wiretap

1 Introduction

Organizations of all types, in both the public and private sectors, are increasingly dependent on information technology. This dependency results from demands for increased interconnectivity, efficiency or competitiveness – on the part of the organizations themselves and/or their clients. These organizations have increasingly

A. Wespi, G. Vigna, and L. Deri (Eds.): RAID 2002, LNCS 2516, pp. 138-157, 2002.

been turning to information system technology for a variety of purposes: e-commerce, e-government, and improvements in information access and sharing among others. Reliance on an inherently insecure infrastructure exposes organizations to a variety of new threats. They must now contend with new cyber-threats such as fast-spreading malicious code and criminal hacking.

The scope of the protection problem has also dramatically increased over the past few years. Protection of computer systems and networks has been an issue of concern for some time. In the last few years, however, the protection of these same systems and networks has come to be viewed as a component of national security. This is due in large part to the work of the President's Commission on Critical Infrastructure Protection (PCCIP), which broadened the definition of national security to include protection of critical infrastructures[1], one of which is the telecommunications sector. Arguably, the telecommunications sector is the most important of the sectors, as all of the others are dependent in some way or another (many of which may not be well understood) on telecommunications.

Traditionally, the emphasis of the protection effort has been on the use of cryptography, firewalls, anti-virus applications and so on. Although this is still very important, greater emphasis is being placed on detection of and response to anomalous events – in recognition of the fact that perfect protection is impossible to achieve. A key element of this strategy is the use of intrusion detection systems IDS), which examine network or host activity to detect indications of malicious activity. However, this may involve the interception of personal or private information, which introduces a number of legal (privacy and criminal law) concerns.

2 Aim

The ability to copy, save and/or log personal information transmitted over a network exists in technologies that are already in widespread use, such as firewalls, servers, and anti-virus software. Indeed, logging of personal information is a far easier proposition in the context of these network infrastructure and security administration tools. The implications of criminal and anti-terrorism legislation are thus not limited to IDS alone, however, this paper will focus on their application to intrusion detection systems.

It is the intent of this paper to examine key provisions of criminal and anti-terrorism legislation, and examine some of the implications for the use of IDS in protecting computers and networks. Reference will be made to legislation from Australia, Canada, the United Kingdom and the United States in order to demonstrate that these are issues of general interest. It is not the intent of this paper to substitute

---

[1] Critical infrastructures have been defined as " infrastructures which are so vital that their incapacitation or destruction would have a debilitating impact on defense or economic security." Anonymous, "Critical Foundations: Protecting America's Infrastructures", The President's Commission on Critical Infrastructure Protection, October 1997, page 19. Although the actual delineation of these infrastructures varies somewhat from country to country, it is generally agreed that critical infrastructures include energy, banking and finance, transportation, water and sewage, government services, emergency services and telecommunications.

for considered legal advice. The opinions expressed in this paper are strictly those of the author and do not reflect the position of the Communications Security Establishment (CSE) or of the Government of Canada (GoC).

3 Interception Requirements

The manner in which public and private sector organizations conduct business and offer services to their clients has changed dramatically over the past several years, as organizations move to e-business, e-government or any other form of online activity. An increasing percentage of organizational assets are moving online, stored in file and mail servers, data warehouses and storage area networks. These assets include financial, medical or personnel records, strategic plans, and trade secrets – all of which represent some of the organization's most valuable assets. Similarly, business communications are increasingly carried by networks of one form or another.

Individuals and organizations that pose a threat to these assets are also moving online. This is partly because this is "where the money is". It is also partly because the use of these technologies confers flexibility, speed and anonymity on their activities. Cyber-threats are difficult to anticipate (other than in a generic way), detect, verify and trace. As the threat is increasingly cyber-based, information about the threat can also be found online. Hacker websites, e-zines, Internet Relay Chat channels and e-mail can be important sources of threat information, as is the network traffic itself. A number of functions performed by public and private sector organizations depend upon access to and analysis of that information, including intelligence, evidence, and indications and warnings to name a few.

Intelligence is defined as information and knowledge about an adversary obtained through observation, investigation, analysis, or understanding. [2] The collection, analysis and reporting of timely, accurate intelligence on threats to national security is generally the responsibility of security and intelligence (S&I) agencies such as the National Security Agency (NSA) or the Central Intelligence Agency (CIA).[3] The intelligence function is no longer the exclusive domain of 'national security' agencies – private sector organizations need to generate competitive business intelligence. They may also receive intelligence from law enforcement and S&I agencies under the auspices of programs such as the FBI's Infragard)[4].

---

[2] From the Joint Doctrine Encyclopedia, dated 16 July 1997. Part of the US Department of Defense Joint Electronic Library.

[3] The CIA, for example, is mandated to "provide accurate, evidence-based, comprehensive and timely foreign intelligence related to national security". Extracted from "About the CIA: CIA Vision, Mission, and Values", posted to the CIA Website. See also: the Canadian Security Intelligence Service – "The Service shall collect and analyze and retain information and intelligence respecting activities that may on reasonable grounds be suspected or constituting threats to the security of Canada", taken from (Canada (CA)) An Act to establish the Canadian Security Intelligence Service (R.S. 1985, c.C-23), updated to 31 August 2001 (*CSIS Act*), Section 12.

[4] Infragard was developed by FBI Cleveland in 1996 to promote protection of critical information systems. It provides formal and informal channels for the exchange of information about infrastructure threats and vulnerabilities. Taken from the Infragard FAQ.

Intelligence includes identifying and monitoring groups or individuals deemed to pose a national security threat. Intelligence may be gathered in a number of ways, ranging from sophisticated technical means (e.g. satellites and signals intelligence efforts) to human intelligence sources (e.g. spies, defectors and informants). In order to fulfill their mandates, these agencies now undoubtedly include intercept and analysis (of both traffic and content) of open (i.e. public) information sources and network traffic as sources of intelligence about both conventional and cyber-based threat activity.[5]

Technological advances and the Internet provide expanded opportunities for criminal activity. As a result, law enforcement agencies face many new challenges, notably building the knowledge and skills necessary to effectively combat cyber crimes. Part of this knowledge may derive from an examination of open sources, including intercept and examination of network communications as a source of criminal intelligence. Interception activity is also undertaken to collect evidence of an (criminal) offense to support prosecution efforts.[6] The use of cyber-based evidence is becoming more important, and there is no reason to suppose that law enforcement agencies would not consider IDS logs as a potential source of cyber-based evidence.[7] While the collection of evidence to support criminal prosecution has traditionally been a law enforcement responsibility, S&I agencies are increasingly being tasked to support law enforcement in this regard, particularly in support of anti-terrorism efforts.[8]

Specific legislation exists to govern interception activities of S&I and law enforcement agencies. A number of conditions must be met before an interception authorization can be granted, and certain conditions must be also be met while conducting the interception. These conditions may be spelled out in the relevant

---

[5] European Parliament Report on the existence of a global system for the interception of private communications (ECHELON interception system) (2001/2098(INI)), report reference A5-0264/2001, dated 11 July 2001. The report refers to claims that ECHELON has 'the ability to intercept any telephone, fax, Internet or e-mail message sent by any individual and thus to inspect its contents' (Section 1.6, page 23) while acknowledging that limitations on interception and analysis make this kind of global surveillance 'impossible in practice' (Preamble, page 11, Item D).

[6] Gellman, B., Washington Post Staff Reporter, "Cyber-Attacks by Al Qaeda Feared", dated 27 June 2002. In this case, FBI review of network audit and monitoring logs revealed that al Qaeda operatives were spending time on sites dealing with supervisory control and data acquisition (SCADA) systems – the systems that control power, water, transport and communications grids. See also: Pruitt, S., IDG News Service, "FBI gets new Web searching powers", dated 31 May 2002.

[7] While the admissibility of electronic records has been the subject of past case law, it is not clear if IDS logs have been included. For issues associated with the use of IDS logs as evidence, see: Sommers, P., "Intrusion Detection Systems as Evidence", First International Workshop on the Recent Advances in Intrusion Detection, 14 – 16 September 1998, Louvain-le-Neuve, Belgium; and Stephenson, P., "The Application of Intrusion Detection Systems in a Forensic Environment" (extended abstract), Third International Workshop on the Recent Advances in Intrusion Detection, 2 – 4 October 2000, Toulouse, France.

[8] See e.g.: (UK) An Act to give the Security Service the function of acting in support of the prevention and detection of serious crime, and for connected purposes (1996 Chapter 35), 18 July 1996 (*Security Service Act 1996*), Section 1(1).

governing legislation (e.g. CSIS Act) or other legislation (e.g. criminal or anti-terrorism law).[9]

Other public and private sector organizations also have a requirement to intercept and examine network traffic. While these organizations may need to do this for intelligence or evidentiary purposes, most such interception would likely be conducted by law enforcement or S&I agencies. Instead, interception performed by these organizations is more commonly done in order to identify an attempted or actual intrusion into a protected system or network, and to initiate an incident response process (of course, this also applies to law enforcement and S&I agencies). While existing legal regimes adequately address interception by S&I and law enforcement agencies, they generally do not adequately address interception of network traffic by other organizations.

4 What Is a Private Communication?

What constitutes a private communication, particularly those over a network, for the purposes of criminal or anti-terrorism law? The only definition of private communications that could be found in the selected legislation is that in the *Criminal Code of Canada*. A private communication is defined as "any oral communication or any telecommunication, ... that is made under circumstances in which it is reasonable for the originator to expect that it will not be intercepted by any person other than the person intended by the originator to receive it."[10] Telecommunications is further defined as "the emission, transmission or reception of signs, signals, writing, images, sounds or intelligence of any nature by any wire, cable, radio, optical or other electromagnetic system, or by any similar technical system".[11] Network communications, although not explicitly mentioned anywhere in the legislation, would almost certainly fall within this definition. Note that it is the expectation of the

---

[9] See e.g. (AU) An Act to prohibit the interception of telecommunications except where authorized in special circumstances or for the purposes of tracing the location of callers in emergencies, and for related purposes, Act No. 114 of 1979 as amended (the Telecommunications (Interception) Act of 1979), Part III, Sections 9 – 11; (CA) *CSIS Act*, Section 21; (CA) An Act respecting the Criminal Law (R.S., C-34), updated to 31 August 2001 (*Criminal Code)*, Section 184.2(3); (UK) Regulation of Investigatory Powers Act 2000 (2000 Chapter 23), dated 28 July 2000 ((UK) RIPA 2000), Part I, Sections 6 – 11; (US) US Code, Title 18, Part I, Chapter 119, Section 2516 (Authorization for interception of wire, oral, or electronic communications (in criminal cases)); and (US) US Code, Title 50, Chapter 36, Subchapter I, Section 1804 (Applications for court orders (for Foreign Intelligence Surveillance Act interceptions)).

[10] (CA) *Criminal Code*, Section 183. Other legislation defines private communication services, but not private communication. See also: (UK) RIPA 2000, Part I, Section 2(1); (US) US Code, Title 26, Subtitle D, Chapter 33, Subchapter A, Section 4241 to 4243, Subchapter B, Section 4252(d).

[11] (CA) An Act respecting the interpretation of statutes and regulations, updated to 31 August 2001 (*Interpretation Act*), Section 35(1). The Interpretation Act provides the authoritative basis for the definition and interpretation of selected terms that appear in other Canadian legislation. See also: (AU) Telecommunications (Interception) Act 1979, Section 5(1); (US) US Code, Title 18, Part I, Chapter 119, Section 2510(14); and (UK) RIPA 2000, Part I, Section 2(1).

originator of the message that no one other than the intended recipient will intercept the message that matters to the determination of whether the communications is a 'private communication'.

It might be useful at this point to try to distinguish between personal information that is public and that which is private[12]. Personal information is defined as information about an identifiable individual that is recorded in any form including any identifying number, symbol or other particular assigned to the individual, or the address of the individual.[13] Personal information has been construed as being very broad and probably includes Internet Protocol (IP) addresses. Some personal information should be considered private (e.g. financial or medical information). Some personal information, on the other hand, clearly resides in the public domain (e.g. name, street address and phone number in a phone book) and would likely not be considered private, although there are exceptions (e.g. unlisted phone number). In a network context, the equivalent information would be user name and domain (e-mail) and the corresponding addressing information as published, for instance, in a public key certificate. Although personal, this would likely constitute public information.

Is the issue one of keeping the content of a particular communication private? There is little argument that the user-entered portion of a communication would be considered personal and private. If the originator does not want the contents of the message to be read by anyone other than the intended recipient, then he/she can take steps to protect the message content through the use of encryption. In fact, criminal law specifically refers to electronic or other treatment of radio-based communication for the purpose of preventing intelligible reception by any person other than the intended recipient.[14] There is no mention of similar treatment for other forms of telecommunication (i.e. network communications). However, should an individual consciously take steps to protect the communication against intercept, given the generally inclusive definition of telecommunications, it is the author's opinion that a reasonable expectation of privacy would be created for the content of network communications.[15]

The situation is not quite so clear when it comes to packet header information. Are headers an integral part of the communication (i.e. part of the content of a message)?

---

[12] Public is defined as "open to or shared by all the public", or "reveal previously unknown information." Private is defined as "confidential; not to be disclosed to others", or as that which is "kept or removed from public knowledge or observation". The Canadian Oxford Dictionary, Oxford University Press, 1998.

[13] (CA) An Act to extend the present laws of Canada that protect the privacy of individuals and that provide individuals with a right of access to personal information about themselves (R.S. 1985, c.P-21), updated to 31 August 2001 (*Privacy Act*), Section 3. The identification of the individual may also include one or more factors relating to his physical, physiological, mental, economic, cultural or social identity and includes any expression of opinion about the individual.

[14] (CA) *Criminal Code*, section 183. See also: (US) US Code, Title 18, Part I, Chapter 119, Section 2510(16).

[15] However, provisions for the mandatory disclosure of encryption keys are being enacted into law. See e.g.: (UK) RIPA 2000, Part III, Section 50. In light of this, some authors are advocating measures intended to circumvent the provisions of this Act. See: Brown, I. and Gladman, B., "The Regulation of Investigatory Powers Bill – Technically inept: ineffective against criminals while undermining the privacy, safety and security of honest citizens and businesses", undated.

An argument could be made in favour of this position as at least part of the header (e.g. the destination address in an e-mail, or the URL of a website) is generated in response to user input. Intuitively, one would consider this to be private information – individuals may not want anyone else to know with whom they are communicating. On the other hand, there is a certain amount of information in network communications that must be public. This, of course, is the routing and handling information contained in the packet header. If one considers conventional mail, the header information would be analogous to the addressing on the outside of an envelope. The majority of this information is generally available to the public (other than on the envelope itself, which is afforded some privacy by virtue of being handled within the postal system) and would therefore be considered public, not private, information.

Is the issue then one of keeping the fact that a particular communication has occurred private? Simply intercepting the fact of a communication (i.e. traffic data) can lead to the development of a detailed user profile (through traffic analysis) that could reveal personal, private information (e.g. web surfing habits) even if the data from which the profile is constructed is all considered to be public.[16] Traffic data is defined as any computer data relating to a communication by means of a computer system, generated by a computer system that formed a part in the chain of communication, indicating the communication's origin, destination, route, time, date, size, duration, or type of underlying service.[17] In a telephone context, traffic data is generally considered to be a list of telephone numbers dialed to or from a specific telephone line. It is straightforward to separate the dialed number from the content of the telephone conversation.

In a network context, there is a difference of opinion as to what constitutes the equivalent of the dialed number list. The range of possibilities extends from the header (or addressing) portion of network traffic to the address and subject lines of an e-mail and Web URLs.[18] The only thing that seems to be consistent across the differing views is that traffic data will contain IP addresses. Given the nature of the information that can be logged by an IDS, it would almost certainly fall within the

---

[16] This will increasingly be an issue now that data retention legislation is being passed. See e.g.: (UK) An Act to amend the Terrorism Act 2000; ...; to provide for the retention of communications data; ...; and for connected purposes (2001 Chapter 24), 14 December 2001 (*Anti-terrorism, Crime and Security Act 2001*), part 11; and Reuters, "Spain passes law to regulate Internet content", dated 27 June 2002. This article makes mention of the Law on the Information Society and Electronic Commerce (LSSI), which includes provisions for ISPs to keep details on users for over a year.

[17] (Council of Europe(COE)) Convention on Cybercrime (ETS 185), opened for signature at Budapest, 23 November 2001, Chapter I, Article 1(d). "Origin" refers to a telephone number, Internet Protocol (IP) address, or similar identification of a communications facility to which a service provider renders services. "Destination" refers to a comparable indication of a communications facility to which communications are transmitted (from Explanatory Memorandum to the Convention, article 30). See also: (US) US Code, Title 18, Part II, Chapter 206, Section 3127(3), as amended by the USA PATRIOT Act; and (UK) RIPA 2000, Part I, Section 2(9).

[18] Black, J., "Uncle Sam Needs Watching, Too", published in Business Week Online, 29 November 2001. See also: Weinstein, L. and Neumann, P.G., "PFIR Statement on Government Interception of Internet Data", published by People for Internet Responsibility (PFIR), dated 7 September 2000.

definitions of traffic data, which has potentially significant implications with respect to data retention. The difficulty comes in separating addressing information from content, given that both travel together in network packets. Even configuring IDS to log only the header portion of the packets does not adequately address this shortcoming – IDS must still 'intercept' the entire packet in order to scan it for indications of malicious traffic.

There seems to be, at least in certain legislation, a definite effort to distinguish between 'traffic' data and content. For example, the Council of Europe (COE) Convention on Cybercrime refers to interception and real-time access to traffic data.[19] United States Code (criminal law) contains provisions for the use of pen registers and trap and trace devices. These devices are capable of monitoring and identifying the specific phone numbers dialed from a particular telephone line - they do not capture or record the content of any such communication.[20] That certain legislation distinguishes between traffic data and content implies that traffic data is not considered to be private communications.[21]

If the originator wants to disguise the fact that they initiated a particular message (e.g. a web session), then he/she has the option of using pseudonymizing techniques. At the very least, the use of these techniques would increase the expectation on the part of the originator that his/her communication will be private, and therefore would be subject to the relevant provisions of criminal or anti-terrorism law. However, even this is not definitive as many of these techniques provide a mechanism for associating the pseudonymized communication with a particular individual in certain circumstances.[22] Even the use of anonymizing techniques might not be sufficient to address this issue – if the disclosure of encryption keys can be forced, could providers of anonymizing services be forced by law to retain sufficient records to re-associate anonymized traffic with the originator?

---

19 (COE) Convention on Cybercrime (ETS 185), Title 5, Article 20. Article 21 deals separately with the interception of content data.

20 (US) US Code, Title 18, Part II, Chapter 206, Sections 3127(3) and (4). Pen registers are devices that identify the numbers dialed or otherwise transmitted on the telephone line to which such device is attached. Trap and trace devices are devices which identify the originating number of an instrument or device from which a wire or electronic communication was transmitted. See also: (US) An Act to deter and punish terrorist acts in the United States and around the world, to enhance law enforcement investigatory tools, and for other purposes, dated 26 October 2001 (Uniting and Strengthening America by Providing Appropriate Tools Required to Intercept and Obstruct Terrorism (USA PATRIOT ACT) Act of 2001) (hereinafter USA PATRIOT Act). The USA PATRIOT Act amends US Code Title 18, Part II, Chapter 206, Section 3127(3) and (4) to refer to recording or decoding dialing, routing, addressing or signaling information, but not including the contents of such communication.

21 Lee, S.C. and Shields, C., "Tracing the Source of Network Attack: A Technical, Legal and Societal Problem", published in the proceedings of the 2001 IEEE Workshop on Information Assurance and Security, pages 239 – 246. This is at least the case in the U.S. In their paper, the authors state "legally, there is no expectation of privacy for packet headers" (page 245).

22 For a discussion of the use of pseudonymization techniques to enhance the expectation of privacy in network communications, see Johnston, Steven R., "The Impact of Recent Privacy and Data Protection Legislation on the Sharing of Intrusion Detection Information". In W. Lee, L. Me, A. Wespi (Eds.), Proceedings of Recent Advances in Intrusion Detection 2001 (RAID 2001), pgs. 150 – 171, Springer-Verlag, Berlin Heidelberg, 2001.

What expectation of privacy does an individual involved in malicious activity have – do they forfeit any expectation of privacy with respect to those activities? According to provisions in the USA PATRIOT Act, a computer trespasser is "a person who accesses a protected computer without authorization and thus has no reasonable expectation of privacy in any communication transmitted to, through or from the protected computer."[23] In any event, expectations of privacy are probably considerably different in each case. Whatever the expectations, IDS are not sufficiently discriminating to distinguish between malicious activity (which should be monitored and logged) and benign activity (the privacy of which should be respected).

If the above analysis is correct, then the header portions of communications over the Internet would probably not be considered private. This distinction may prove to be important when interpreting the legislation in particular, and in configuring IDS to conform to the law. What exactly are the relevant provisions of criminal and anti-terrorism? How do organizations use IDS in a lawful manner?

5 Criminal Law

Criminal law generally prohibits the intercept of private communications. For example, the *Criminal Code* states: "every one who, by means of any electro-magnetic, acoustic, mechanical or other device, willfully intercepts a private communication is guilty of an indictable offence".[24] Electro-magnetic, acoustic, mechanical or other device includes any device or apparatus that is used or is capable of being used to intercept a private communication.[25] Intercept includes "listen to, record or acquire a communication or acquire the substance, meaning or purport thereof".[26]

It is important to note that an IDS is a computer in its own right, with a processor, primary and secondary storage, and input/output elements. The input element is the interface to the network, typically a network interface card (NIC) designed to IEEE 802.x standards. This interface operates in promiscuous mode, and it 'captures' every well-formed link-level frame – the format for the transmission of IP datagrams across a network. Once they have been 'captured' by the NIC, a copy of the frames and their contents (i.e. the information they contain) are placed in the sensor's primary memory for analysis. Even if the IP datagrams are eventually discarded (because they are deemed to be legitimate traffic), the IDS is in fact copying all of the network traffic.

---

[23] (US) USA PATRIOT Act of 2001, dated 26 October 2001, section 217. This section amends US Code, Title 18, Part I, Chapter 119, Section 2510 by adding a new subsection 2510(21) containing the definition of a computer trespasser. None of the other legislation contains a statement of this nature.

[24] (CA) *Criminal Code*, section 184(1). Similar offences exist under: (AU) The Telecommunications (Interception) Act 1979), Section 7(1); (US) US Code Title 18, Part I, Chapter 119, Section 2511(1); and (UK) RIPA 2000, Part I, Sections 1(1) and 1(2).

[25] (CA) *Criminal Code*, section 183. See also: (AU) Telecommunications (Interception) Act of 1979, Section 5(1); (US) US Code, Title 18, Part I, Chapter 119, Section 2510(5); and (UK) RIPA 2000, Part V, Section 81(1)

[26] (CA) *Criminal Code*, section 183. See also: (AU) Telecommunications (Interception) Act 1979, Section 6; (US) US Code Title 18, Part I, Chapter 119, Section 2510(4); and (UK) RIPA 2000, Part I, Section 2(2).

For the purposes of criminal law, is it essential that packet contents be intercepted, or does it still constitute an intercept if only the packet header is captured? Does the intercept take place when the link-level frame is copied into memory, or does it only take place when the packet is logged because it has been deemed to be malicious? These questions will probably only be answerable in a court of law. In the author's view, however, given that IDS make a copy of all network traffic, regardless of what they actually log, the use of IDS will almost certainly be considered to fall within the definition of device, and to be considered an intercept within the meaning of criminal law.

5.1 Exemptions in Criminal Law

The above analysis led to the conclusion that the use of IDS would likely, by strict definition, constitute an offence against relevant provisions in criminal law. There are, however, exemptions in the criminal law that specify conditions under which intercept would not be an offence. In general, the exemptions are where consent to the interception, express or implied, has been given by the originator of the private communication or the intended recipient; where an authorization has been obtained; and where the interception is by a person engaged in providing a telephone, telegraph or other communication service to the public.

Consent

Consent provisions imply that one party consent to the interception (i.e. either the originator or the intended recipient) is sufficient to exempt the individual performing the interception from the effects of criminal law. Obtaining the consent of the recipient (e.g. the employee) may be straightforward by making consent part of the conditions of being granted a network account, or use banners indicating that the information system being used is subject to being monitored for security purposes, and that continued use of the system constitutes consent to such monitoring.[27] What happens if an individual sends a communication to the wrong address? Does the (unintended) recipient have a right to consent to its interception? The answer to this question is not at all clear.

Obtaining the consent of originators (i.e. persons outside the organization where the monitoring is taking place) may be more problematic. Two key concepts related to consent are knowledge and choice. Does the individual know that the intercept is taking place? Ensuring that the originator knows the interception is taking place may be problematic. Does the individual have a realistic choice – is there an alternative communications path that is not subject to monitoring (intercept)? In terms of network communications, the answer to the latter question is probably no – most organizations will try to ensure that all points of interconnection to the Internet are adequately protected against intrusions (including the use of IDS).

---

[27] However, mere employee consent to surveillance is no longer sufficient to justify unlimited surveillance activities. Surveillance is to be limited to that which a reasonable person would consider appropriate. See: Geist, M., "Shift to more workplace privacy protection", dated 28 June 2002, Globe and Mail newspaper (online).

There is also the issue of implied consent. For an individual external to the organization, does the continued use of the computer system really constitute consent to monitoring for security purposes? Does the posting of a legal or privacy policy to a website create a diminished expectation of privacy? Should consent to intercept be implied by the simple act of sending an e-mail? In the absence of definitive proof that the individual initiating the communication knew that the communication would be subject to intercept, it is likely that the courts would tend to be conservative and deem that the originator has a reasonable expectation that the communication would not be intercepted.[28]

The Supreme Court of Canada has held that one party consent to the monitoring of private communications, even though it does not contravene the provisions of the *Criminal Code*, violates the protection provided by Section 8 of the *Charter*, dealing with search and seizure.[29] This implies that organizations could not rely on this provision to provide adequate legal authority to conduct intrusion detection.

Authorization

In order for an authorization for the interception of private communications to be granted under criminal law, there must be reasonable grounds to believe that an offence against criminal law has been or will be committed; either the originator of the private communication or the person intended by the originator to receive it has consented to the interception; and there must be reasonable grounds to believe that information concerning the offence will be obtained through the interception sought. [30] Authorizations are required for each instance of interception, and requests for authorization must specify the particulars of the offence. When the authorization is granted, it will specify the identities of the persons, if known, whose private communications are to be intercepted, and the period for which the authorization is required.[31] Most authorizations can only be granted for a maximum of 60 days before they need to be renewed.

Criminal law provisions for interception of private communications are not generally suited to interception of private communications for network protection. Given the unpredictable nature of network intrusions, it would be difficult to provide particulars of the offence that will be committed. In most cases it will be impossible to accurately identify the responsible individuals, and for intrusion detection purposes, the authorization would need to be permanent.

---

[28] Rubinkam, M., "Court to Decide on Web Wiretapping", Los Angeles Times article, dated 19 February 2002. In this article, the author refers to a case heard by the Pennsylvania Superior Court. The court ruled that the accused "had consented to the recording by the very act of sending e-mail and instant messages". The court further stated that "any reasonably intelligent person, savvy enough to be using the Internet... would be aware that messages are received in a recorded format, by their very nature". While not authoritative, a court in Canada may find this case informative/instructive.

[29] Insert relevant case law reference.

[30] (CA) *Criminal Code*, Section 184.2(3). See also: (AU) Telecommunications (Interception) Act of 1979, Sections 9 – 11; (US) US Code, Title 18, Part I, Chapter 119, Section 2518(3); and (UK) RIPA 2000, Part I, Sections 5(2) and 5(3).

[31] See e.g.: (CA) *Criminal Code*, Section 184.2(4).

Telecommunications Service Providers

Criminal law provides exemptions relating to interception of private communications by the provider of a telephone, telegraph or other communication service[32] to the public. The terms telephone and telegraph are not explicitly defined in criminal law, however, the definition of telecommunications is sufficiently broad that telephone and telegraph communication services would likely cover network services.

Interception must be necessary for the provision of the service, or it must relate to service observing or random monitoring for the purposes of service quality control checks. Depending on the service being provided, an argument could easily be made that the use of IDS as an integral component of a layered security architecture is essential to ensure the quality of the service, especially confidentiality, integrity and availability. Other methods of detecting intrusions, such as detailed review of all entries in device logs, are potentially more invasive than the use of IDS, which act as a filter to reduce the volume of information examined by an analyst.

These conditions, which would seem to apply to the use of IDS, apply to a person who is providing a telephone, telegraph or other communication service to the public. A key reason for public and private sector organizations moving online is to be able to improve the service they provide to the public. However, it is not at all clear if this is sufficient for them to be considered a telecommunications service provider for the purposes of criminal law. This is an issue that will require further analysis.

Previous analysis led to the conclusion that the header portions of network traffic probably do not constitute private communications. However, the use of IDS for network traffic monitoring would probably still constitute an intercept under criminal law, as the initial copying of network traffic includes content, even if it is not subsequently logged. An examination of existing criminal law exemptions suggests their application to public or private sector organizations would be problematic at best. They were intended for the intercept of private communications for the purposes of collecting evidence of a criminal offence, not for network protection purposes.

6 Anti-terrorism Law

Prior to September 11, the interception of private communications was governed by criminal law as discussed above. The general conclusion was that the use of IDS may constitute an interception as defined in criminal law, and that the existing exemptions did not adequately address the interception of private communications for network protection purposes. In response to the attacks on the World Trade Center and the Pentagon, a number of countries either initiated or accelerated plans to introduce

---

32 See e.g.: (COE) Convention on Cybercrime (ETS 185), Section 1A, article 1(c), and Explanatory Report, article 26. "Service provider" means any public or private entity that provides to users of its service the ability to communicate by means of a computer system. The term "service provider" is deliberately broad, and may include a closed group or the facilities of a provider that are open to the public, whether free of charge or for a fee. The closed group can be, for example, the employees of a private enterprise to whom the service is offered by a corporate network.

comprehensive anti-terrorism legislation.[33] Not all of the bills were passed - Australia's Telecommunications Interception Amendment Bill 2002 (designed to give government agencies authority to read e-mail, SMS and voice messages without an interception warrant) was defeated in the Australian Senate.[34] Most of these bills were omnibus bills, meaning they introduced a variety of new measures, frequently by amending existing criminal law. Of these, the most controversial and of most relevance to intrusion detection are those dealing with interception of telecommunications and data retention.

6.1 Interception of Private Communications

Prior to the introduction of anti-terrorism legislation, the interception of private communications was conducted in accordance with strict rules. Specific conditions had to be met before an authorization could be granted, and strict conditions applied to the actual conduct of the interception. Interception was also generally limited to telephone communications. This tended to restrict the circumstances in which interception could be conducted. For instance, a wiretap order only applied to a single telephone number. There was some judicial discretion with respect to granting or denying requests for authorization. This combination provided a certain degree of assurance that the privacy of network communications would not be violated without cause.

In the wake of the September 11 attacks, there were calls to grant law enforcement agencies much broader powers to monitor private communications and access personal information. The bills generally responded to these calls by amending existing criminal law provisions governing interception of private communications. The nature of the information that can be captured has been broadened to include dialing, routing and addressing information, effectively enabling law enforcement agencies to monitor and intercept electronic mail, web surfing and other forms of electronic communications.[35] Law enforcement agencies are now permitted to obtain

---

[33] The US signed the Uniting and Strengthening America by Providing Appropriate Tools Required to Intercept and Obstruct Terrorism Act (USA PATRIOT Act) into force on 26 October 2001. Canada's Bill C-36 (An Act to amend the Criminal Code, the Official Secrets Act, the Canada Evidence Act, the Proceeds of Crime (Money Laundering) Act and other Acts, and to enact measures respecting the registration of charities, in order to combat terrorism (the Anti-Terrorism Act)) came into force 24 December 2001. The UK's Anti-terrorism, Crime and Security Act became law on 14 December 2001. See also: Hayes, Ben, "EU anti-terrorism action plan: legislative measures in justice and home affairs policy", Statewatch post 11.9.01 analyses: No. 6. See also EU Press Release "Action by the European Union following the attacks on 11 September", MEMO/01/327 dated 15 October 2001, available from RAPID - The Press and Communication Service of the European Commission.

[34] Fitzsimmons, C., "Email snooping bill knocked down", dated 28 June 2002, AustralianIT (online).

[35] (US) USA PATRIOT Act, Section 216(c)(2) and (3) amend the definitions of pen register and trap and trace device respectively to permit recording of dialing, routing and addressing information, although there are still restrictions on the recording of communications content. It is important to note that concerns surrounding interception of e-mail pre-date September 11 – witness the controversy surrounding the FBI's Carnivore (now DCS1000) program.

a single authorization to tap any communications that a suspect may use. Combined with claims that the level of judicial discretion has essentially been drastically reduced, this has prompted fears of the potential for massive invasions of privacy[36]. Whether these fears are justified or not, the amendments still do not provide a general authorization for interception of network communications by public and private sector organizations – they must still operate under the prior, more restrictive regime.

There appear, however, to be two exemptions to the above statement. The first comes in Canada's *Anti-Terrorism Act.* This Act includes a specific clause authorizing intercept of private communications for the purposes of protecting GoC computers and networks.[37] However, this clause only applies to a single government agency (CSE) and then only under strict conditions, similar to those in the *Criminal Code.*[38] An authorization made under this section may contain any conditions that the Minister considers advisable to protect the privacy of Canadians, including additional measures to restrict the use and retention of, the access to, and the form and manner of disclosure of, information derived from the private communications. There is at least an indication, therefore, that attempts will be made to respect the privacy of network communications.

It is important to remember, however, that this authorization can only be provided to CSE, leaving all other GoC departments and agencies without a specific legislative basis for the conduct of intrusion detection. Provision does exist within the act for persons who assist with the execution of the authorization to be covered by the authorization[39], but the implications of this for the GoC, from an operational perspective, have yet to be examined.

The second exception is a provision in the USA PATRIOT Act, which amends US Code to add a provision that "it shall not be unlawful under this chapter for a person acting under color of law to intercept the wire or electronic communications of a computer trespasser transmitted to, through, or from the protected computer, if the owner or operator of the protected computer authorizes the interception of the computer trespasser 's communications on the protected computer; and such interception does not acquire communications other than those transmitted to or from the computer trespasser.''[40]

At first glance, this would appear to provide, in the United States at least, owners of a protected computer[41] the legal right to intercept private communications.

---

[36] See: Anonymous, "How the USA PATRIOT Act Limits Judicial Oversight of Telephone and Internet Surveillance", dated 23 October 2001. See also: Anonymous, "Analysis of Provisions of the Proposed Anti-Terrorism Act of 2001 Affecting the Privacy of Communications and Personal Information", dated 24 September 2001.

[37] (CA) *Anti-Terrorism Act*, Part 5, clause 273.65(3). The Minister (of National Defence) may, for the sole purpose of protecting the computer systems or networks of the Government of Canada from mischief, unauthorized use or interference, ….„authorize the Communications Security Establishment in writing to intercept private communications in relation to an activity or class of activities specified in the authorization. This clause was developed specifically with the use of IDS in mind.

[38] (CA) *Anti-Terrorism Act*, Part 5, clause 273.65(4).

[39] (CA) *Anti-Terrorism Act*, Part 5, clause 273.67.

[40] (US) USA PATRIOT Act, article 217, amending US Code, Title 18, Chapter 119, Section 2511(2).

[41] (US) US Code, Title 18, Part I, Chapter 47, Section 1030(e)(2).

However, it may not be as simple as that. First, the definition of protected computer appears to be fairly narrow, being a computer that is exclusively for the use of a financial institution or the United States Government or which is used in interstate or foreign commerce or communication. In relation to the private sector, for instance, what computers would this cover – individual workstations, web or mail servers (which could arguably be for foreign communication)? The answer to this question is not at all clear.

Second, the interception must only acquire the communications of the computer trespasser. Whether the term 'acquire' refers to the logging of specific communications, or to the copying of communications performed by an IDS prior to scanning, IDS are not sufficiently discriminating to ensure that only communications from the computer trespasser will be 'acquired'. There does not appear to be any issue with the IDS sensor scanning network traffic looking for indications of malicious activity and generating alerts/alarms as a result of the scanning. The problems appear to arise when a human analyst must examine any flagged traffic to validate the alarm (i.e. is it a valid alarm, or a false positive?). Even if they only log traffic deemed to be malicious, the current state IDS technology almost guarantees that at least some of this traffic will be benign (i.e. false positives), thereby violating this condition. In the absence of judicial interpretation of this provision, it is unlikely that organizations could generally rely upon it for legal authority to intercept private communications.

6.2 Data Retention

The second controversial provision in most anti-terrorism law concerns data retention.[42] A distinction must be made between data preservation, where data is stored and retained in response to a specific request and data retention, where the data is stored as a routine practice. As a general rule, there seems to be very little controversy associated with data preservation. On the other hand, data retention is causing a great deal of concern, especially among ISPs. Prior to the introduction of the anti-terrorism legislation, traffic data was only retained as long as was necessary for billing purposes, and at the end of that period was either to be anonymized or destroyed. Combined with the increasing trend to billing flat rates for network access, this ensured that a minimum of private information was collected and stored.

The anti-terrorism legislation not only increased the duration for which this information had to be stored, it also considerably expanded the information to be retained[43], although it appears as if it is predominantly traffic data that is to be

---

[42] Provisions for data retention are also found elsewhere. See e.g.: Anonymous, "Proposal for a Directive of the European Parliament and of the Council concerning the processing of personal data and the protection of privacy in the electronics communications sector", document reference /*COM/2000/0385 final – COD 2000/0189*/ (replacing Directive 97/66/EC, adopted 15 December 1997). This proposal was accepted in a vote in the European Parliament 30 May 2002.

[43] Anonymous, "List of Minimum and Optional Data to be Retained by Service Providers and Telcos", Expert Meeting on Cybercrime: Data Retention, The Hague, 28 December 2001 (File No. 5121-20020411LR-Questionnaire). The list of information to be retained is quite extensive: user-id and password (remote login); sender and receiver (login@domain), identifying information of e-mail retrieved (e-mail); and hostname or IP address, nickname used during session (IRC).

retained. While this was aimed at e-mails and web traffic, there is no reason to suppose that data collected by IDS would not be covered by these provisions, as it likely falls within the definition of traffic data. While this poses a number of data storage and management problems, it also poses problems with respect to privacy of network communications. As mentioned, IDS are prone to generating false positives. Unless procedures are in place to examine all of the flagged traffic, and then to destroy that which is determined to be benign, examination of the logged data will almost certainly result in privacy violations. Even if the data is anonymized, it is likely that law enforcement agencies would require its association with a particular originator for investigative purposes (if provisions exist to force disclosure of encryption keys, then requiring this association is likely to follow).

7 Conclusions

This paper has looked at some of the issues that will need to be addressed in order to develop a general legal framework for use of IDS, including trying to define what is meant by a 'private communication'. Depending on the nature of the communication, the originator might want to hide the content of the communication (through cryptography), the fact that a communication has taken place (through anonymization or pseudonymization) or both. The use of encryption to ensure the privacy of the contents of a communication is no longer an assurance that the contents will in fact remain private – legislation exists to force the disclosure of encryption keys. Similarly, the use of pseudonymizing techniques is no guarantee that the fact of a communication will remain private – most schemes permit the re-association of the pseudonymized traffic with the originator. An attempt to legislate the creation and retention of similar records by providers of anonymizing services is not beyond possibility. If law enforcement agencies decide that IDS data is essential to their investigation, there is little reason to believe that they will not require this reassociation.

IDS log data probably falls within the definition of traffic data, and is, therefore, probably subject to the provisions of legislation that require the long-term retention of traffic data by service providers. If this is the case, this poses numerous challenges for the service providers, and creates an increased risk of privacy violations.

Based on the definitions found in criminal law and supporting legislation, the use of IDS by public and private sector organizations will likely constitute an intercept. In order not to be an offense, the use of IDS must be in accordance with one of the exemptions, namely where consent to the interception, express or implied, has been given by the originator of the private communication or the intended recipient; where an authorization has been obtained; or where the interception is by a person engaged in providing a telephone, telegraph or other communication service to the public.

These provisions do not adequately address the situation facing public and private sector organizations. Single party (i.e. recipient) consent, although lawful, has been held to violate constitutional law. Two party consent is considered the minimum standard – something that will not be achieved if one of the parties to the communication has malicious intent. While satisfying the conditions necessary to the granting of an authorization should generally be possible, authorizations issued under criminal law are for evidence collection, not network protection. It is unlikely that a global, open-ended authorization for network protection would (or could) ever be granted under criminal law or withstand legal challenge. Narrower authorizations issued under the *Anti-Terrorism Act* (Canada) or to law enforcement or S&I agencies to intercept private communications for network protection (i.e. conduct intrusion detection) would be more likely to withstand challenge.

In a limited sense, a case could be made for the application of the criminal law exemption for telecommunications services providers to public and private sector organizations. However, it is not at all clear whether or not these organizations would actually qualify as a telecommunications service provider under criminal law and so organizations could probably not rely on this exemption. Anti-terrorism legislation greatly complicates the task of protecting the privacy of network communications and the personal information that they contain. Not only can a broader range of information be intercepted under authorization, but a similarly broad range of information is to be retained by ISPs and telephone companies. In addition, there appears to have been a reduction in the level of judicial discretion in the granting or denying of interception requests.

It is, therefore, reasonable to suggest that the key to addressing these deficiencies is the creation of a new exemption under criminal law. This exemption would provide the necessary legal basis for the interception of private communications for the purpose of protecting public and private sector computer systems or networks from mischief, unauthorized use or interference.

There is still a great deal of work that needs to be done to develop a general legal framework for the conduct of intrusion detection within the public and private sectors. Once the legal framework has been developed, additional work will be required in order to develop appropriate policies, standards and procedures for the use of IDS, especially with respect to what can be collected, how that information is to be handled, stored or disposed of and who has access to the information and under what circumstances. The assistance of the legal community would be invaluable in this endeavour.

References

[1] Anonymous, "About the CIA", undated. URL: http://www.cia.gov/cia/information/info.html (25 June 2002)

[2] Anonymous, "Analysis of Provisions of the Proposed Anti-Terrorism Act of 2001 Affecting the Privacy of Communications and Personal Information", 24 September 2001, Electronic Privacy Information Center (EPIC). URL: http://www.epic.org/privacy/terrorism/ata_analysis.html

[3] Anonymous, "Critical Foundations: Protecting America's Infrastructures", The President's Commission on Critical Infrastructure Protection, October 1997, Critical Infrastructure Assurance Office. URL: http://www.ciao.gov/resource/pccip/PCCIP_Report.pdf

[4] Anonymous, "How the USA-PATRIOT Act Limits Judicial Oversight of Telephone and Internet Surveillance", 23 October 2001, American Civil Liberties Union. URL: http://www.aclu.org/congress/1102301g.html

[5] Anonymous, Infragard Frequently Asked Questions. URL: http://www.infragard.net/faq.htm

[6] Anonymous, "List of Minimum and Optional Data to be Retained by Service Providers and Telcos", Expert Meeting on Cybercrime: Data Retention, The Hague, 28 December 2001 (File No. 5121-20020411LR-Questionnaire). URL: http://www.statewatch.org/news/2002/may/europol.pdf

[7] Anonymous, "Report on the existence of a global system for the interception of private and commercial communications (ECHELON interception system) (2001.2098 (INI)), dated 11 July 2001, presented to the European Parliament. URL (Federation of American Scientists): http://www.fas.org/irp/program/process/rapport_echelon_en.pdf

[8] (Australia) "An Act to prohibit the interception of telecommunications except where authorized in special circumstances or for the purpose of tracing the location of callers in emergencies, and for related purposes", (the Telecommunications (Interception) Act 1979), Act No. 114 of 1979 as amended. This compilation was prepared on 7 January 2002 taking into account amendments up to Act No. 166 of 2001. URL: http://scaleplus.law.gov.au/html/pasteact/0/464/pdf/TeleInt79.pdf.

[9] Black, J., "Uncle Sam Needs Watching, Too", published in Business Week Online, 29 November 2001. URL: http://www.businessweek.com/bwdaily/dnflash/nov2001/nf20011129_3806.htm.

[10] Brown, I. And Gladman, B., "The Regulation of Investigatory Powers Bill – Technically inept: ineffective against criminals while undermining the privacy, safety and security of honest citizens and businesses", undated. URL: http://www.fipr.org/rip/RIPcountermeasures.htm

[11] (Canada) An Act to amend the Criminal Code, the Official Secrets Act, the Canada Evidence Act, the Proceeds of Crime (Money Laundering) Act, and other Acts, and to enact measures respecting the registration of charities, in order to combat terrorism, 24 December 2001 (*The Anti-Terrorism Act*). URL: www.parl.gc.ca/37/1/parlbus/chambus/house/bills/government/C-36/C-36_4/C-36_cover-E.html.

[12] (Canada) An Act respecting the Criminal Law (R.S., c.C-46), updated to 31 August 2001 (*The Criminal Code*). URL (Department of Justice Canada) http://laws.justice.gc.ca/en/C-46/index.html.

[13] (Canada) An Act respecting the interpretation of statutes and regulations, (R.S. 1985, c.I-21), updated to 31 August 2001 (*The Interpretation Act*). URL (Department of Justice Canada): http://laws.justice.gc.ca/en/I-21/index.html.

[14] (Canada) An Act to establish the Canadian Security Intelligence Service (R.S., C-23), updated to 31 August 2001 (the *Canadian Security Intelligence Service Act*). URL (Department of Justice Canada): http://laws.justice.gc.ca/en/C-23/index.html

[15] (Canada) An Act to extend the present laws of Canada that protect the privacy of individuals and that provide individuals with a right of access to personal information about themselves (R.S. 1985, c.P-21), updated to 31 August 2001 (the *Privacy Act*). URL (Department of Justice Canada): http://laws.justice.gc.ca/en/P-21/index.html

[16] Canadian Oxford Dictionary, Oxford University Press, 1998.

[17] (Council of Europe) Convention on Cybercrime (ETS 185), opened for signature at Budapest, 23 November 2001. URL: http://conventions.coe.int/Treaty/en/Treaties/Word/185.doc.

[18] (Council of Europe) Explanatory Memorandum to Convention on Cybercrime, dated 8 November 2001. URL: http://conventions.coe.int/Treaty/en/Reports/Html/185.htm

[19] Data Protection Working Party, "Opinion 7/2000 On the European Commission Proposal for a Directive of the European Parliament and of the Council concerning the processing of personal data and the protection of privacy in the electronic communications sector of 12 July 2000 COM (2000) 385", dated 2 November2000. URL: http://europa.eu.int/comm/internal_market/en/dataprot/wpdocs/wp36en.pdf.

[20] (European Parliament) Anonymous, "Proposal for a Directive of the European Parliament and of the Council concerning the processing of personal data and the protection of privacy in the electronics communications sector", document reference /*COM/2000/0385 final – COD 2000/ 0189*/. URL: http://europa.eu.int/eur-lex/en/com/pdf/2000/en_500PC0385.pdf

[21] EU Press Release "Action by the European Union following the attacks on 11 September", MEMO/01/327 dated 15 October 2001. URL (RAPID - The Press and Communication Service of the European Commission): http://europa.eu.int/rapid/start/welcome.htm.

[22] Geist, M., "Shift to more workplace privacy protection", dated 28 June 2002, Globe and Mail newspaper (online version). URL: http://www.theglobeandmail.com/servlet/ArticleNews/printarticle/gam/20020628/EBGEISY

[23] Gellman, B., Washington Post staff writer, "Cyber Attacks by Al Qaeda Feared", 27 June 2002, Washington Post (online version). URL: http://www.washingtonpost.com/wp-dyn/articles/A50765-2002Jun26.html

[24] Hayes, Ben, "EU anti-terrorism action plan: legislative measures in justice and home affairs policy", Statewatch post 11.9.01 analyses: No. 6: URL: http://www.statewatch.org/news/2001/oct/analy6.pdf.

[25] Johnston, Steven R., "The Impact of Recent Privacy and Data Protection Legislation on the Sharing of Intrusion Detection Information". In W. Lee, L. Me, A. Wespi (Eds.), Proceedings of Recent Advances in Intrusion Detection 2001 (RAID 2001), pgs. 150 – 171, Springer-Verlag, Berlin Heidelberg, 2001.

[26] Joint Doctrine Encyclopedia, dated 16 July 1997. US Department of Defense Joint Electronic Library. URL: http://www.dtic.mil/doctrine/joint_doctrine_encyclopedia.htm

[27] Lee, S.C. and Shields, C., "Tracing the Source of Network Attack: A Technical, Legal and Societal Problem", published in the proceedings of the 2001 IEEE Man, Systems and Cybernetics Information Assurance Workshop, pages 239 – 246. URL: http://www.ai.usma.edu/Workshop/2001/Authors/Submitted_Abstracts/paperW1C1(09).pdf.

[28] Pruitt, S., IDG News Service, "FBI gets new Web searching powers", dated 31 May 2002, Computerworld Magazine (online version). URL: http://www.computerworld.com/securitytopics/security/privacy/story/0,10801,71599,00.html

[29] Reuters, "Spain passes law to regulate Internet content", dated 27 June 2002. Posted to SiliconValley.com. URL: http://www.siliconvalley.com/mld/siliconvalley/news/editorial/3556967.htm

[30] Sommers, P., "Intrusion Detection Systems as Evidence", as presented at the First International Workshop on the Recent Advances in Intrusion Detection, 14 – 16 September 1998, Louvain-le-Neuve, Belgium. URL: http://www.raid-symposium.org/raid98/Prog_RAID98/Full_Papers/Sommer_text.pdf.

[31] Stephenson, P., "The Application of Intrusion Detection Systems in a Forensic Environment", extended abstract, as presented at the Third International Workshop on the Recent Advances in Intrusion Detection, 2 – 4 October 2000, Toulouse, France. URL: http://www.raid-symposium.org/raid2000/Materials/Abstracts/47/47.pdf.

[32] (United Kingdom) Regulation of Investigatory Powers Act 2000, Chapter 23, 28 July 2000. URL: http://www.legislation.hmso.gov.uk/acts/acts2000/20000023.htm

[33] (UK) An Act to give the Security Service the function of acting in support of the prevention and detection of serious crime, and for connected purpose (1996 Chapter 35), dated 18 July 1996 (the *Security Service Act 1996*). URL: http://www.legislation.hmso.gov.uk/acts/acts1996/1996035.htm

[34] (UK) An Act to amend the Terrorism Act 2000; to make further provision about terrorism and security; to provide for the freezing of assets; to make provision about immigration and asylum; to amend or extend the criminal law and powers for preventing crime and enforcing that law; to make provision about the control of pathogens and toxins; to provide for the retention of communications data; to provide for implementation of Title VI of the Treaty on Eurpean Union; and for connected purposes (2001 Chapter 24), 14 December 2001 (the *Anti-terrorism, Crime and Security Act 2001*). URL: http://www.legislation.hmso.gov.uk/acts/acts2001/10024—a.htm

[35] (United States) United States Code Collection, Legal Information Institute, Cornell Law School. URL: http://www4.law.cornell.edu/uscode/.

[36] (United States) An Act to deter and punish terrorist acts in the United States and around the world, to enhance law enforcement investigatory tools, and for other purposes, dated 26 October 2001, (Uniting and Strengthening America by Providing Appropriate Tools Required to Intercept and Obstruct Terrorism (USA PATRIOT Act of 2001). URL: http://www.epic.org/privacy/terrorism/hr3162.pdf.

[37] (United States) United States Supreme Court, Record of Opinion, "Katz v. United States, 389 US 347 (1967)", decided 18 December 1967. Summary of opinion available at FindLaw http://findlaw.com/US/389/347.html.

[38] Weinstein, L. and Neumann, P.G., "PFIR Statement on Government Interception of Internet Data", published by People for Internet Responsibility (PFIR), dated 7 September 2000, available at http://www.pfir.org/statements/interception.

Learning Unknown Attacks – A Start

James E. Just[1], James C. Reynolds[1], Larry A. Clough[1], Melissa Danforth[2], Karl N. Levitt[2], Ryan Maglich[1], and Jeff Rowe[2]

[1] Teknowledge Corporation, 3900 Jermantown Road, Suite 100
Fairfax, VA 22030 USA
{jjust, reynolds, lclough, rmaglich}@teknowledge.com
[2] University of California, 3061 Engineering II, One Shields Avenue
Davis, CA 95616 USA
{danforth, levitt, rowe}@cs.ucdavis.edu

Abstract. Since it is essentially impossible to write large-scale software without errors, any intrusion tolerant system must be able to tolerate rapid, repeated unknown attacks without exhausting its redundancy. Our system provides continued application services to critical users while under attack with a goal of less than 25% degradation of productivity. Initial experimental results are promising. It is not yet a general open solution. Specification-based behavior sensors (allowable actions, objects, and QoS) detect attacks. The system learns unknown attacks by relying on two characteristics of network-accessible software faults: attacks that exploit them must be repeatable (at least in a probabilistic sense) and, if known, attacks can be stopped at component boundaries. Random rejuvenation limits the scope of undetected errors. The current system learns and blocks single-stage unknown attacks against a protected web server by searching and testing service history logs in a Sandbox after a successful attack. We also have an initial class-based attack generalization technique that stops web-server buffer overflow attacks. We are working to extend both techniques.

1 Introduction

Designing secure systems to survive cyber-attacks is both hard and complex. Initially designers built stronger locks (e.g., put strong security mechanisms into and around the systems) to keep out the attackers. However, protection wasn't perfect and some attacks still got through. Then designers added alarms (e.g., intrusion detectors) to alert us to the attackers. Unfortunately, the detectors weren't perfect and some attacks still got through undetected. More recently, designers have shifted their focus to building systems that continue operating despite the attacks by leveraging the well-developed techniques of dependability and fault tolerance.

It is impossible to build an intrusion tolerant system that survives for any meaningful time without solving the problems of unknown attacks and finite failover resources. It is always useful to harden a system so the adversary's work factor to penetrate it is increased but there are limits to this approach. The time and effort required to identify a new vulnerability in your system and develop an exploit for it may be quite large. To a determined opponent, it's just time and money.

A. Wespi, G. Vigna, and L. Deri (Eds.): RAID 2002, LNCS 2516, pp. 158-176, 2002.

The real problem is that, once an attack is developed and put in place, the time required to execute it is very small and, in many cases, the time and effort required to create simple variants of the attack are quite small. If the threat environment for an intrusion tolerant system includes a well-resourced adversary (e.g., a state-sponsored cyber-terrorist group or organized crime), the system must be capable of dealing with many unknown attacks -- possibly repeated quickly (seconds or minutes). This attack scenario will rapidly exhaust any redundant components in the system and represents, in our opinion, the worst case design point.

Many fault tolerant mechanisms work because most faults are independent, low probability events and hence are easily masked. Common mode failure is a well-understood problem and one to be avoided often through the use of diversity. Attacks do not allow the first assumption of independence and the nature of the critical service to be protected may belie the use of diversity.

This project is the first attempt to build a prototype system that combines intrusion detection, responses that block attacks, failover to remove compromised elements, learning to create rules for blocking future occurrences of attack and generalization to block even significant variants.

1.1 Background

Commercial organizations, the Government and even the military have reduced their cost and, arguably, improved their reliability through the increased use of COTS software and hardware, even for critical applications. Unfortunately, they have also increased their vulnerability to well resourced adversaries who want to do serious damage to critical infrastructure, steal information, and disrupt services. Most researchers are saying that it is essentially impossible to build large scale software without faults and it is certainly impossible to prove such software contains no faults [1], [2], and [3]. Moreover, as two damaging recent attacks (Code Red 1 and Code Red 2, which exploited a known buffer overflow vulnerability in Microsoft's web server, Internet Information Server) have amply demonstrated, faults are being exploited long after patches are available to fix the problems. This is not to say that software security cannot be improved but it is important to begin examining other approaches to security.

The US Defense Advanced Research Project Agency (DARPA) began a program in 2000 to apply fault tolerance techniques to building intrusion tolerance systems. As part of this effort, a number of organizations, including Teknowledge Corporation and University of California (Davis), are developing intrusion tolerant clusters.

The specific goal of our project (Hierarchical Adaptive Control of QoS for Intrusion Tolerance or HACQIT) is to provide continued COTS or GOTS-based application services in the face of multiple hours of aggressive cyber-attacks by a well-resourced adversary. This focus on COTS/GOTS applications means we do not have access to source code so the protections must be added around or to the binaries. We recognize that our defense cannot be perfect so two implied goals include (1) significantly increasing the adversary work factor for successful attacks and (2) significantly increasing the ratio of the attacker's work factor to generate successful attacks to the defender's work factor for responding to successful attacks. We also recognize that our system is expensive in terms of processing and overhead so we have modularized the components so that the amount of protection can be varied according to the need and budgets available.

1.2 Organization

The HACQIT project, its architecture, and basic intrusion tolerant design approach have been described in other articles [4, 5, 6]. The next section will provide enough information on HACQIT to enable the reader to understand the context, uses, and limitations of the learning and generalization as it exists today. The remaining sections will summarize the problem, the learning and generalization approach, its current implementation, test results, and conclusions / next steps.

2 The HACQIT Context

2.1 The General Problem and System Model

Formal environment and attack assumptions have been made to specify the research problem as developing dependability in the face of network-based cyber attacks rather than dealing with denial of service attacks, insiders, Trojans and other lifecycle attacks. These assumptions include:

- Users and attackers interact with services via their computers and the LAN. There are no other hidden interactions.
- The LAN is reliable and cannot be flooded, i.e., denials of service (DoS) attacks against LAN bandwidth are beyond the scope of the research. The LAN is the only communication medium between users and services. DoS attacks directly against critical users or firewalls are also beyond the scope of the research.
- Critical users and the system administrators for the cluster are trusted. No hosts on the external LAN are trusted.
- The protected cluster hardware and software are free of Trojans and trapdoors at startup and have been patched against known vulnerabilities. Attackers do not have and have not had physical access to the cluster hardware or software. This prevents planting Trojan software/hardware and trapdoors through lifecycle attacks.
- Other unknown vulnerabilities exist throughout the system.

Figure 1 describes the "formal" system model of the problem and design environment that is being addressed by intrusion tolerant systems. The goal is to protect critical application(s) so that critical users can continue to access them while under attack.

2.2 HACQIT System Model

HACQIT is not designed to be a general-purpose server connected to the Internet. Anonymous users are not allowed. All connections to the system are through authenticated Virtual Private Networks. We assume that the configuration of the system has been done correctly, which includes patching of all known vulnerabilities.

An attacker can be any agent other than the trusted users or HACQIT system administrators. Attackers do not have physical access to HACQIT cluster. An attacker may take over a trusted user's machine and launch attacks against HACQIT.

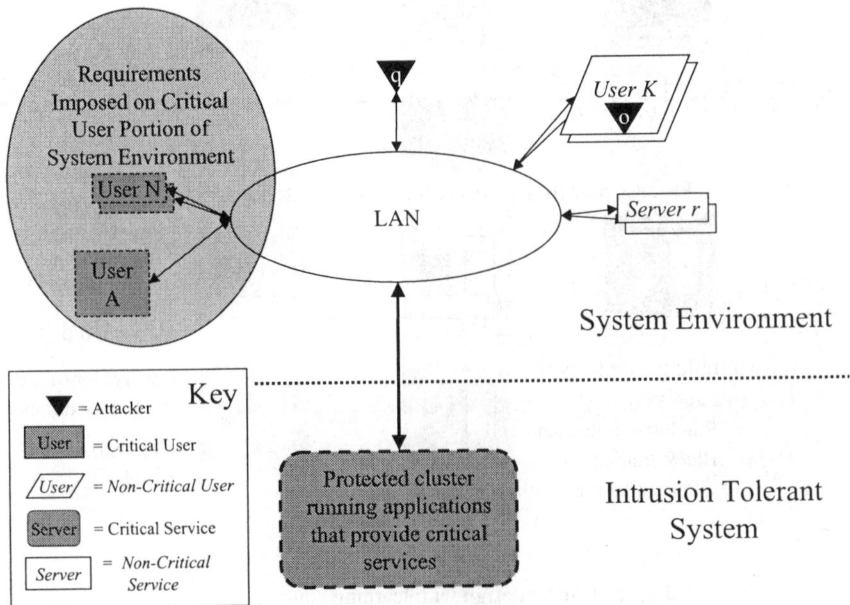

Fig. 1. Intrusion Tolerant System Environment

A failure occurs when observed behavior deviates from specified behavior. For HACQIT, we are concerned with software failures. Software failures are either repeatable or not. The causes of repeatable failures would include attacks (maliciously devised inputs) that exploit the some vulnerability (bug) in one of our software components. Non-repeatable failures may be caused by intermittent or transient faults. We cannot divine intent, so all inputs that cause repeated failures are treated the same. On the other hand, we recognize that the system may fail intermittently from certain inputs, in which case we allow retry.

To develop a system that meets these requirements, most designers would make the cluster very intrusion resistant, implement some type of specification-based monitoring of server and application behavior and use some set of fault tolerant mechanisms (e.g., redundancy and failover, process pairs, triple modular redundancy, n-version programming) for the servers to enable rapid failover and recovery. Our design employs these approaches and a few additional ones.

Our design is summarized in Figure 2. The HACQIT cluster consists of at least four computers: a gateway computer running a commercial firewall and additional infrastructure for failover and attack blocking; two or more servers of critical applications (one primary, one backup, and one of more on-line spares); and an Out-Of-Band (OOB) machine running the overall monitoring and control and fault diagnosis software. The machines in the cluster are connected by two separate LANs.

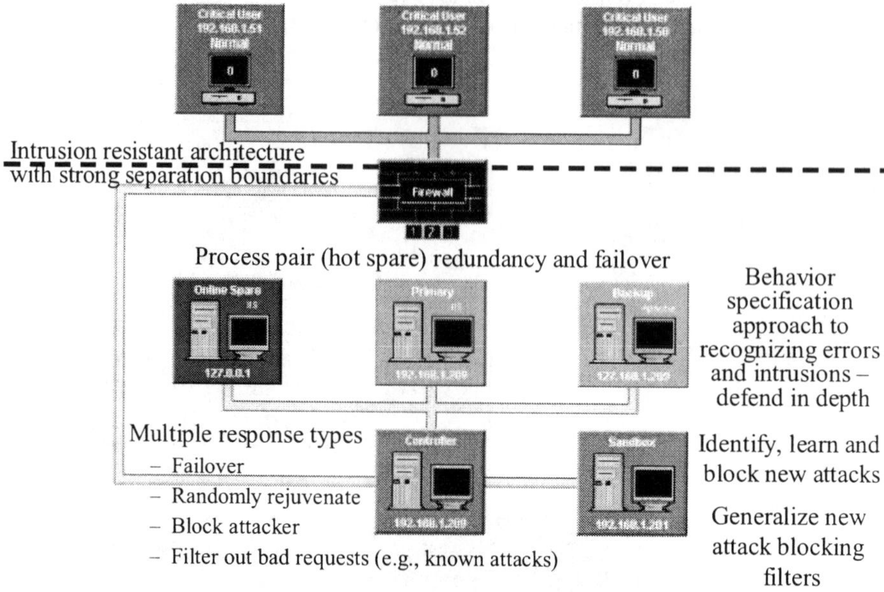

Fig. 2. Cluster design with learning components

HACQIT uses primary and backup servers running as a process pair, but they are unlike ordinary primary and backup servers for fault tolerance. Only the primary is connected to users. The virtual private network (VPN), firewall, gateway, and IP switch together ensure that users only talk to the critical application through the specified port on the primary server and vice versa. The primary and backup servers are not on the same LAN; they are isolated by the OOB computer, so no propagation of faults, for example by a automated worm or remote attacker, directly from the primary to the backup, is possible.

The potential for propagation from the primary to the Controller is limited by sharply constraining and monitoring the services and protocols by which the Controller communicates with the primary. When a failure is detected on the primary or backup server (possibly caused by an attack), it is taken off line. Continued service to the end user is provided by the remaining server of the process pair. A new process pair is formed with the on-line spare (if available), and both attack diagnosis and recovery of the failed server begins. Depending on policy, the Controller can also block future requests from the machine suspected of launching the attack.

The current critical application is a web-enabled message board that is duplicated on both the Microsoft IIS web server and the Apache web server machines. It contains dynamic data so HACQIT must maintain consistent state across the hosts and resynchronize data (checkpoint and restore) when failover and new process pair formation occurs. The spare server does not have current state when it is promoted into the process pair so a restore process is necessary to synchronize it.

2.3 HACQIT Software Architecture

The simplified software architecture is shown in Figure 3. The software implements a specification-based approach [7, 8] to monitoring and control for intrusion detection as well as defense in depth. It uses software wrappers [9], application and host monitors, response monitors, etc. to ensure that component behavior does not deviate from allowed. It does this in a protect-detect-respond cycle. Strong protections (and isolation) are melded with host and application-based fault, error, and failure detection mechanisms in a defense in depth design. Deviation from specified behavior may be indicative of an attack and thus, when such an alert is received, it triggers failover, integrity testing and other responses.

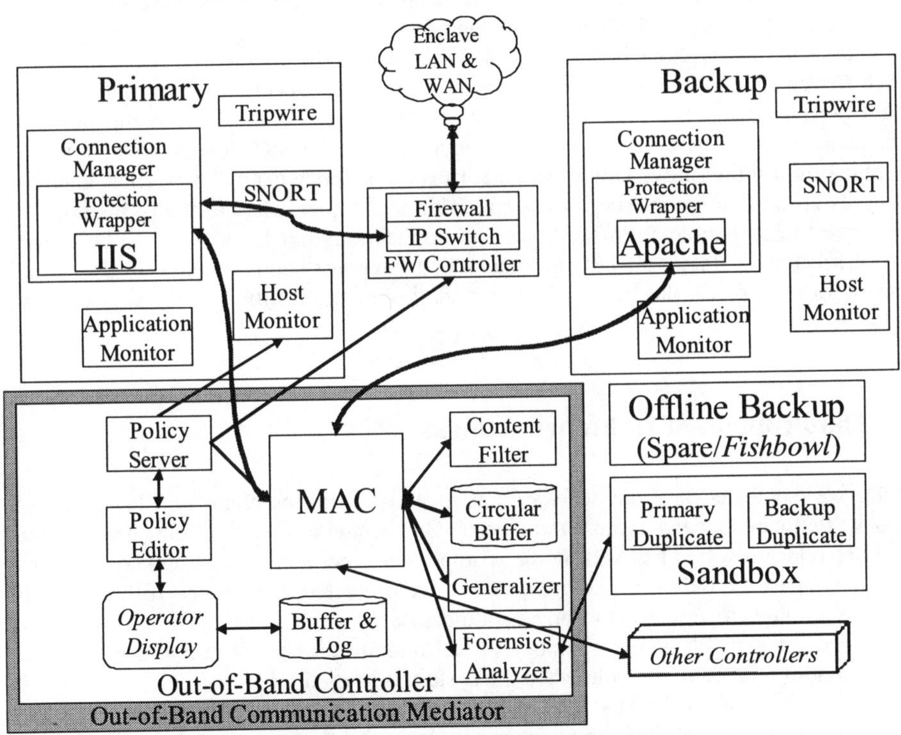

Fig. 3. Simplified software architecture

Wrappers are used to monitor and strictly control an application's ability to read and write files, execute or kill other processes, modify registry entries, change memory protection, etc. Any violations trigger an alert and failover response. The host monitor allows only specified processes to run and ensures that allowed processes do not consume more resources than allowed. It also monitors application liveness and generates a heartbeat for the Controller. The out-of-band communication mediator ensures that only authorized processes use the out-of-band communications network to attack the mediator-adapter-controller (MAC) or any other software on the controller machine. Integrity of control messages is enhanced with digital signatures. The MAC monitors responses from both servers and periodically orders a "self test" through the Host Monitor (HM). This entails issuing a known request to the servers and checking its response against the known response.

The Out-of-Band Communication Mediator shown in Figure 3 is implemented by host firewalls (specifically the firewall by Tiny Software Inc.) on each machine. They only allow access to the server from the OOB machine by authorized programs like the MAC through authorized ports and only allow access to/from the gateway machine through port 80. Similarly, they only allow access to the OOB machine from the server by authorized programs like the wrappers and the HM through authorized ports. To change this configuration, a user must enter a password at the server's keyboard. An attacker would have to gain system privileges and by-pass or shutdown the host firewall to have access to the OOB machine. Creating an unauthorized process to remove a file would trigger an "Unhealthy" state by the HM and cause a failover and clean-up of the "Unhealthy" machine.

One of our innovations is that the Controller also performs random rejuvenation of each server on an average interval that is set by policy. During random rejuvenation, the Controller takes the selected machine off-line, performs the appropriate failover, starts up a new process pair with the spare machine, shuts down and restarts all applications on the rejuvenating machine, runs an integrity check on its files, and sets it up as an on-line spare when it is done. This techniques was originally intended to preserve the performance of the hosts and applications but it turns out to be a very powerful to avert latent successful attacks that have not been detected. It also limits the required size of the history buffer and limits the ability of any attack to go undetected.

3 The Problem of Unknown Attacks

Since we cannot ensure that software has no vulnerabilities, there will be unknown attacks that can succeed against the cluster. Our behavior specifications are tight but bad behavior must be observed by the wrappers or other sensors. An unknown attack can sneak past the defenses and compromise the protected web server. If the attack is stealthy enough to not execute any unauthorized processes, write any unauthorized files or use too much cpu or memory, it can remain undetected and active until the next random rejuvenation cycle when the system is purged.

While the emphasis of HACQIT is on availability of critical services, we need to say a few words about what an attacker can do in the above circumstances and what the countermeasures would be. First it would be easy to corrupt the critical application data since the web enabled application has permission to write to that file. The solution to this is a secure storage system such as that developed under the Perpetually Available and Secure Information Storage (PASIS) [10, 11]. Such a storage system captures all changes to files in a host-independent and secure manner. This enables recovery from such incidents.

Another possibility would be for the attacker to simply monitor what was happening within the application (spy) and exfiltrate data. Since the most likely avenue of attack is by compromising a critical user machine, the attacker would effectively have access to the critical application and data anyway. This essentially becomes the insider problem. Exfiltration via other routes is difficult because of the firewall settings and isolation of the cluster.

For the purposes of critical application availability, the central concern is that an attacker has found an unknown attack that can be used to penetrate the cluster. Such an attack can be used to shutdown the vulnerable web server or the application behind it. The HACQIT goal is to maintain at least 75% availability in the face of on-going attacks. For the attacker to win, all that he/she must do is to find a small set of vulnerabilities in each of the diverse web servers or other critical applications. This essentially guarantees that the attacker will succeed in shutting down the cluster more than 25% of the time. As long as that vulnerability remains and the exploit succeeds, the attacker can just keep hitting the cluster with it and cause another failover. It does not matter how expensive these vulnerabilities are to find, once they are found and exploits developed for them, the time to launch successive attacks is minimal. The results of this will be devastating on the defenders.

Even if the IP address of the attacker is blocked or that user cut off in other ways, the attacker can always come back unless the cluster is cut off from users. Such an action amounts to a self-inflicted denial of service and is clearly unacceptable. Since the attacker has the ability to automate his attack, even physically capturing the attacker would not necessarily stop the attacks. Since it takes time to clean up a server after an attack before it can be put back into service with any confidence, unless the cluster has an indefinitely large number of backup servers for failover, it seems like a losing game for the defender. If the attacker has found a simple, inexpensive way to vary the attack signature, the problem becomes even more difficult for the defender

Can this problem be fixed? In principle there is no solution. But, as the reader will see, we are using classical machine learning methods (using observed instances of the attack to learn the most general description of an attack that has variants, followed by the most general blocking rules) combined with the use of sandbox to experiment offline with the observed instances to create other instances. Short of analyzing source (or object) code, that's the best we can do, and it is likely to be very effective. Our experiences with Code red and its variants can attest to this.

4 Solution Concept

Cyber attacks (network-based intrusions into a system) have several important differences from other natural or man-made faults: They are repeatable, they are not random (although certain types of attacks may depend on timing of events), and, if known, they can be filtered out at system or sub-system boundaries. These distinctions enabled us to develop a set of learning techniques to help deal with the unknown attack problem.

Given an observed failure on a cluster server, our goal is to identify an attack in the recorded cluster traffic. Repeatability of the attack against the critical application server is the key criteria of an attack, particularly given the difficulty of establishing malicious intent. We developed a set of components that learn an attack after it is first used, develop blocking filters against it, and generalize those filters to disallow simple variants of the attack that depend upon the same vulnerability. By preventing reuse of an unknown attack, would-be adversaries are forced to develop a large number of new attacks to defeat the cluster for any significant period. This raises the bar significantly on the amount of effort that an adversary must expend to achieve more than momentary success.

Clusters can communicate with one another so that the protective filters developed at one site can be propagated to clusters at other sites that have not yet experienced the same attack. This ability to do group learning is a very powerful feature of the design and implementation.

The information necessary for the forensics based learning system to work is provided by several key components including (1) logs of all network inputs to the cluster for, at least, the last N minutes, (2) logs of all system sensor readings and responses that indicate errors or failures in components, and (3) a "Sandbox" for testing attack patterns and filters. The Sandbox is an isolated duplicate of the critical application servers, i.e., the redundant process-pair software, sensors, and hardware. Note that it is most effective if the number of minutes of buffering (N) is equal to or slightly greater than the number of minutes between random rejuvenation. Search speed is obvious faster if N is a smaller number of minutes rather than larger.

Our approach to identifying, learning, and blocking unknown attacks begins when an error (i.e., a deviation from specified behavior)) is observed in the cluster, usually associated with the critical application. It proceeds in parallel with and independent from the failover process that guarantees continuity of service.

Since our goal is to prevent the success of future versions of this newly observed unknown attack, it is not necessary to understand the details of the attack after the initiating event that puts control into the attacker's code. What we want to do is to prevent the initiating event, which is often a buffer overflow, and we would like to do this as quickly as possible.

While it is useful to have a general process with guaranteed convergence to a solution, the practical aspects of the time required to test many hypotheses of attack sequences against a Sandboxed application are formidable. It can take several minutes to restart some applications after a failure and some applications cannot be run in multiple processes on the same computer. Our more practicable approach involves examining a variety of heuristics and specification / model-based protocol analyzers that can be used to shrink the search space of suspect connection requests to a very small number of candidates that must be verified in the Sandbox.

Given the observer error in the cluster, the essential functional steps in our learning and generalization "algorithm" are shown in Table 1. The first two steps rapidly produce an initial filter rule that blocks the previously unknown attack. The remaining steps then incrementally improve the rules by shortening and generalizing them if possible.

A few caveats are required before proceeding. Since we are dealing with Turing complete languages and machines, Rice's theorem implies that we cannot prove intrusion tolerance for the system. Nevertheless, within the assumptions imposed on the system model, we believe we can deliver very useful and usable results.

The fundamental metric in determining the success or failure of the HACQIT cluster is whether an attacker can generate an effective attack rate higher than the cluster's effective learning and generalization rate. Intrusion resistance and intrusion tolerance don't have to be perfect. They just have to be good enough to convince the attackers to try a different, less expensive approach.

There are also several responses that the cluster controller can take to thwart attacker or to make learning easier. For example, random rejuvenation can be used to force an attacker to start over again with a stealthy attack. It is also useful for limiting the size of the history file that must be analyzed after a successful attack. It is also possible to cut off the attacker or "excessive user" via blocking his IP address at cluster or enclave level firewall. Since all users come into the cluster over a VPN and spoofing is not possible, this is particularly effective if the address or user ID of the attacker can be learned from the captured attack sequence in the history log. All inputs can be stopped for short periods if the attacks are overwhelming the system.

Table 1. Steps in Learning and Generlization of Unknown Attacks

No.	Step Description
1.	Determine if observed error is repeatable based on connection history file since last rejuvenation. If repeatable, declare attack and continue. If not, return.
2.	Determine which connection request (or requests) from history file caused the observed error.
3.	Develop filter rule to block this connection request(s) pattern, test it, and send to content filter. Also block the associated user ID and IP address.
4.	Characterize the observed attack (i.e., classify it according to meaningful types).
5.	Shorten the blocking filter, if possible.
a.	Determine if the observed attack sequence has an initiating event
b.	If the initiating event is smaller than the observed attack sequence, shorten the blocking filter to block just the specific initiating event and test it.
6.	Based on characterization and observed attack specifics, generalize the blocking filter to protect against simple variants of the attack and test it.
7.	Return.

5 Analysis of Approach, Implementation, and Results

Each step in the learning and generalization process can be implemented differently for efficiency. For example, the most general approach to identifying an unknown attack once it has occurred is a search process (essentially "generate and test") using the Sandbox to establish truth [12]. Thus, Step 1 (determine if the component error was the result of an attack) could be implemented by simply rerunning the recorded service requests from the history log through the Sandbox.

We have found that it is faster to run a model-based service request checker against the log entries to eliminate "obviously correct" requests while retaining potentially suspicious requests. The actual selection rules are discussed later in this section. This approach speeds up both the "Was it an attack" testing and the subsequent "Which request(s) is the attack" testing by significantly reducing the search space. Since it takes significant time to restart a failed application or server, this is important. Generally, the more requests we can "test" analytically, rather than empirically in the Sandbox, the faster the search process.

While the last steps (shorten and generalize the filter rules) can be viewed as a search process, it is much more efficient to implement it using a knowledge-based matching approach. That is the reason for the classification step just before it. Since it is impossible to work with "unknown" attacks as a whole, we are working on meaningful taxonomies of unknown attacks to give us traction on the problem of reducing the search space. These sub-classes should be more amenable to informed search approaches and, even if we cannot identify all sub-classes of unknown attacks, we still obtain very useful search speed-up in the majority of practical cases. This is discussed further in the next two sections.

Using the component names shown in Figure 4, fault and intrusion tolerance are provided by the following control flow:

1. When the primary receives a request, it is forwarded to the MAC before it goes to the web server.
2. The MAC calls the Content Filter with the new request as the parameter. It also assigns a serial number and time stamp to the request.
3. The Content Filter, which contains a list of all bad requests received to date, checks the bad request list for a match with the new request.
4. If there is a match, the Content Filter returns false to the MAC (meaning "reject the request"); otherwise, it returns true ("accept request").
5. The MAC returns the answer (reject or accept) to the primary where the request is either forwarded to the web server or dropped as appropriate.
6. If the MAC received a true response from the Content Filter, it also forwards the request (with serial number and timestamp) to the backup server so the process pairs can be kept in sync.
7. Assuming that the request is legitimate, the MAC also waits for each server to process the request and then compares the return codes or times out if the response takes too long. If there is a mismatch or time out, the MAC response is policy driven. Under most circumstances, it will remove the suspect server from the process pair and begin normal failover and recovery steps.

The unknown attack identification and blocking functions are performed by learning components highlighted in Figure 4. These consist of the Forensics Analyzer, the Sandbox, the Circular Buffer and Event Logs, and the Content Filter and Generalizer.

The Content Filter examines each service request to determine if it matches a known bad request on its list of bad requests before it is passed to the primary or backup web servers. The Circular Buffer captures the last N-minutes of complete service requests from users to the critical servers while the event log captures sensor and event (such as failure) data.

If MAC detects error, it initiates failover and asks the Forensics Analyzer to start forensics. The MAC passes to it the history buffer, the state of the sensors and servers that caused the failover, and any likely candidates for the bad request. The Forensics Analyzer tests if any repeatable attacks are present in Circular Buffer by testing suspicious requests in the Sandbox. If it determines that an attack has occured, it then identifies specific attack request(s), and develops a new filter rule to block it. The Generalizer improves content filter rules to stop simple variants of some types of attacks. More details on this are provided below.

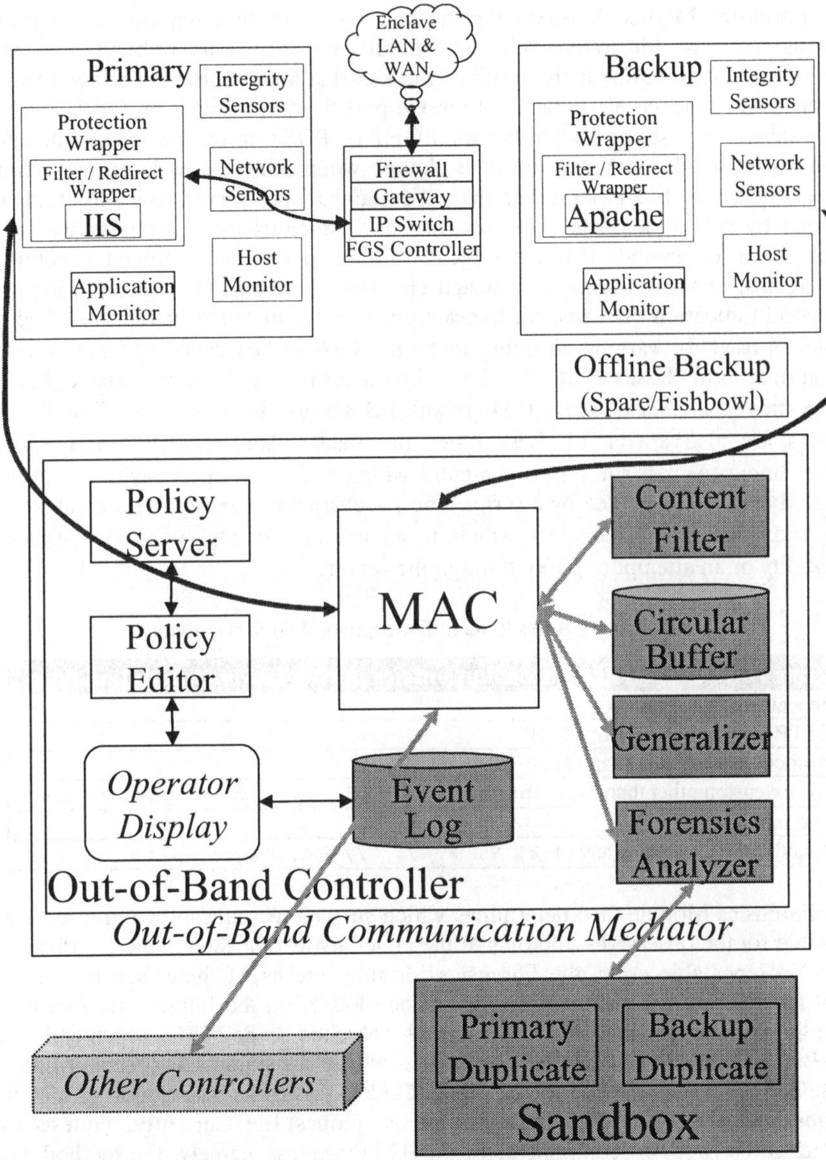

Fig. 4. Learning Components

There are also occasions when the MAC is able to determine relatively unequivocally that an intrusion has occurred. Examples include detecting unauthorized file writes or program executions. In many cases, the MAC itself can determine with high reliability which connection request is the likely attack. For example, if a particular request attempts to write an unauthorized file or start an unauthorized process, it is most likely an attack. In this case, the suspect request is forwarded to the Forensics Module as a prime candidate.

The Forensics Module looks in the circular buffer of past requests to identify suspicious requests. Illustrative rules for identifying suspicious web requests are shown in Table 2. Rule one is the result of the fact that many buffer overflow attacks use a repeated sequence of characters to move past the fixed length buffer. No valid HTTP transactions use methods other than GET or POST in our environment, thus rule two. This would obviously need to change when other methods are common. Attempts to access file types other than the standard set served are classified as suspicious by rule four. Rule five classifies as suspicious those requests that use unusually long commands that are typically found in remote command execution attacks against server-side scripts. Unusual characters found in the request string are also a good indication of a suspect transaction, and are included in rule six. The `%` character is used for various encoding methods, such as hex encoding, and is very common in several classes of attacks. The `+` character is interpreted as a space. Many directory traversal attacks against Microsoft IIS servers include them. The "`..`" characters are also a sign of these types of attacks. The `<`, `>`, and `<` characters indicate cross site scripting attacks which attempt to inject Javascript into a webpage dynamically created by a script. The `//` characters can represent a subset of a long sequence of `/` characters which is an attempt to exploit an old Apache vulnerability or an attempt to proxy through the server.

Table 2. Illustrative Rules to Identify Suspious Web Server Requests

Filtering rules to prioritize suspicious entries in web server transaction log
1) Repeated characters > 50
2) HTTP method not GET or POST
3) Protocol header other than HTTP/1.0 or HTTP/1.1
4) File extension other than `htm, html, txt, gif, jpg, jpeg, png,` or `css`
5) Command length > 20
6) Request string contains any of `%, ?, +, .., //, <, >, <, ;`

The Forensics Module then determines which suspicious request (or requests) was responsible for the observed symptoms of the attack by testing each in the Sandbox. If there is no repeatable error, the Forensics Module returns. If there is a repeatable error, it has determined what request should be blocked in the future. The forensics module then passes the known bad request to be blocked to the MAC, which calls the UpdateBadReq method of ContentFilterBridge (which implements the Content Filter) with the bad request as the parameter. UpdateBadReq adds the bad request to a static bad request list in memory and writes it to the bad request file. Currently, requests are truncated to the first two components of an HTTP request, namely, the method and URI.

Every time a request is received on the primary, it is forwarded to the MAC. The MAC calls the AllowRequest method of ContentFilterBridge with the new request as the parameter. The method checks the bad request list for an exact match with the new request. If there is a match, it returns false to the MAC, meaning block the request; otherwise, it returns true.

Thus far the learning is straight-forward and quite general. Unfortunately, the attack pattern that is being blocked is quite specific. If simple attack variants can be produced easily (e.g., by changing the padding characters in a buffer overflow attack or changing the order of commands in a cgi-script attack), then this specific learning approach is easily circumvented by an attacker. What is needed is a way to rapidly generalize the observed attack pattern so as to block all simple variants of an attack that are based on the same vulnerability initially exploited. This is a challenging area and is the subject of a continuing research effort.

As a proof-of-concept, we implemented generalization for a common but prevalent class of attacks: web server buffer overflows. Our initial approach was to enhance the AllowRequest method so that if an exact match is not found, it then analyzes the components of the requests (both new and bad) to determine if the new request is "similar" to a known bad request. If it is similar, AllowRequest returns false; otherwise, it returns true. In this way, learning is generalized from specific requests that have been identified as bad.

In principal, similarity is rule based and consists of two steps: classification and generalization. Classification categorizes bad requests into meaningful types such as buffer overflow or remote command execution and, as required, further into sub-types. Generalization develops a set of rules for determining similarity between an observed bad request and a new request based on the classification results. These rules can be implemented either as an active checking process or as comparison templates for use by another program.

For the proof-of-concept on web server buffer overflow attacks via http requests, we implemented one rule that acts as both a classifier and a generalizer. It is the following:

- If (1) the query length of the bad request is greater than (256+X) [this part of the rule classifies the request as a buffer overflow type[1]] and (2) the methods of the new request and the bad request are the same and (3) the file extensions of the new and bad requests are the same and (4) the query length of the new request is greater than or equal to the query length of the bad request, then return false (i.e., block the request).

Even with X=0 in this rule, many variants of Code Red I and II are blocked. The initial or padding characters in the query are irrelevant to how Code Red works; the length is critical; so whether "XXX..." or "NNN..." or "XNXN..." are in the query of the attack, the attack is blocked. In addition, the name of the file (minus the extension) is also irrelevant to how Code Red works, because it is the file extension that identifies the resource (Index Server) that is vulnerable to a buffer overflow, and it is the query that causes the buffer overflow, not the entire URI. (The URI contains the path identifying the resource and, optionally, the query.)

The reason for the first condition in the rule is to differentiate in a trivial way between bad requests that are buffer overflow attacks and bad requests that are some other type of attack, like remote command execution. Unfortunately, it introduces the possibility of false negatives, that is, a bad request that was a buffer overflow attack, but with the overflow occurring after less than 256 characters, would be ignored as an example to be generalized.

This rule has been constructed from extensive analysis of buffer overflows in general, buffer overflows in IIS and Apache web servers, and Code Red, in particular. Note that it only generalizes "learned" behavior. That is, if the HACQIT cluster has never been attacked by Code Red, it will not stop the first Code Red attack. It will

---

[1] X starts out equal to zero. Its role will be discussed later.

also not stop the first case of a variant of Code Red that uses the .IDQ extension[2]. This variant would first have to be "experienced", learned as a bad request, and then generalized by the above rule. Most importantly, the rule does not prevent use of a resource like Index Server; it prevents a wider variety of attacks that exploit an identified vulnerability in it from reducing availability of the web server.

Although this rule appears Microsoft-oriented, as the concept of file extensions does not exist under Unix, it would work against attacks exploiting vulnerabilities in other software, such as php and perl, because these resources also use file extensions. It might be possible to generalize this to file types under Unix. The key distinction to be made is, does the path in the URI identify a document to be returned to the client or does it identify an executing resource such as a search engine, a DBMS, etc.?

Finding the minimum length of padding characters for a buffer overflow attack is not difficult. We have implemented an enhanced version of the forensics and generalization modules that iteratively tests attack variants in the Sandbox with different padding character lengths. Specifically it successively tests padding character lengths between 256 and (Y-256) where Y is the length of the observed buffer overflow padding size. From this testing, it determines the value of X (which appears in the first condition of the generalization rule above) and passes it to the ContentFilterBridge for inclusion in the revised generalization rule. The observed padding size is currently determined by the number of characters before the first non-printing character (i.e., not ACSII character coder 32 through 126) in the query. While this is only an approximation that depends on certain assumptions being true, it proved to be a very useful approach for the proof-of-concept implementation. Our investigation with Code Red II shows the padding in the query that causes the buffer overflow is no more than one byte over the minimum required; that is, if you remove two characters from the query, a buffer overflow will not occur, and IIS will respond to the request correctly and continue to function according to specification.

It is worth comparing this automatic generalization with Snort's hand-coded rules for preventing Code Red attacks. Snort is widely used, open source, lightweight Intrusion Detection System. Immediately after the flurry of initial Code Red attacks, Snort aficionados began crafting rules to block these attacks. It took at least two days before rules were posted on the Snort site. These were not generalized and did not work against trivial variants. Some three months later, the rules block on ".ida" and ".idq" in the URI and "payload size" greater than 239 [13]. The use of the file extensions shows some generalization but the use of 239 as a limit on legitimate requests intended for Index Server in fact cause false positives because the payload can be much greater than 239 (at least 373) without causing the web server to fail.

Other improvements to generalization would use analysis based on HTTP headers and body content. These and other improvements are the central focus of the next phase of research.

One additional aspect of the design of the ContentFilterBridge software is worth discussing. It first calls AllowRequest with the bad request received from the MAC. If AllowRequest returns true, that means the bad request is not on the bad request list, so

---

[2] Index Server uses file types indicated by the extensions, ".IDA" and ".IDQ". These two extensions are used by IIS to identify the Index Server resource, which is then passed either the whole URI or the query component of the URI. The "path" component of the URI does not affect the behavior of the Index Server, except for the file extension identifying it as the resource target. Any file name other than "default" in "default.ida" works as well.

it is added. If AllowRequest returns false, this means it is on the bad request list, so it is not added to the list. This prevents duplication.

With the addition of generalization, not only will duplicates be prevented, but also trivial variants will not extend the bad request list to a performance-crippling length. As there are over 2^{1792} (or more than the number of atoms in the universe) variants of Code Red, this is an important and effective aspect of the design.

6 Next Steps

6.1 Software Improvements

In its initial implementation, the Forensics module truncates bad requests to the first two components of the HTTP request, namely, the method and URI. This makes sense in the case of the buffer overflows on web servers but it needs to be enhanced so there is a more robust way to identify the initiating event of an attack. In addition, there is much work to do to enhance the Forensics module's process for finding initial attack sequences efficiently, especially for multi-request attacks.

Similarly, the initial generalization rule base will be moved into a separate Generalization module that reflects the architecture. This module will attempt to generalize all requests or patterns returned by the forensics module to the content filter and insert specific new rules into the content filter. More broadly, we want the Generalizer to be able to task the Forensics Module to run Sandbox tests on any proposed set of filter rules and generalization parameters to what works, e.g., which contain the essential initiating event. In this way, we can refine the generalization while providing continued protection at the Content Filter level.

There is a great deal of work to be done in developing rules for generalizing attack patterns so that simple attack variants won't work. We would like to do this by focusing on meaningful attack classes. The literature contains many works on classifying various aspects of computer security including fault tolerance, replay attacks in cryptographic protocols, inference detection approaches, COTS equipment security risks, and computer attacks. Essentially all of these authors have emphasized that the utility of a taxonomy depends upon how well it accomplishes its purpose and that there is no such thing as a universal taxonomy.

Another module that we will likely need is one that allows us to simulate vulnerabilities in applications and generate resulting sensor reading. It is difficult to rely on real world attacks on our specific applications. There are simply not enough of them in circulation to give us the breadth of attack types that we need for the research.

6.2 Theory Improvements

As Krsul [14] states, "Making sense of apparent chaos by finding regularities is an essential characteristic of human beings." He laid out the essential characteristics of successful taxonomies: (1) They should have explanatory and predictive value. (2) Computer vulnerability taxonomies should classify the features or attributes of the vulnerabilities, not the vulnerabilities themselves. (3) Their classes should be mutually exclusive and collectively exhaustive. (4) Each level or division should have a fundamentum divisionis or basis for distinction so that an entity can be unequivocally placed in one category or the other. (5) The classification characteristics should be objective, deterministic, repeatable, and specific. Note that item (3) above is very difficult to achieve in practice outside the realm of mathematics and should be probably be replaced by extensibility as a goal.

Krsul developed a very extensive list of classes particularly focused on erroneous environmental assumptions. Unfortunately, his and most of the previous efforts (see review by Lough [15]) on developing taxonomies have focused on identifying and characterizing vulnerabilities in source code so that programmers could identify and eliminate them before the software was deployed. At one level this are fine in that they can give us insights into types of vulnerabilities. For example, the classic study by Landwehr et al. [16] lists the following type of inadvertent software vulnerabilities:

- Validation error (incomplete/inconsistent)
- Domain error (including object re-use, residuals, and exposed representation errors)
- Serialization/aliasing (including TOCTTOU errors)
- Identification/authentication errors
- Boundary condition violation (including resource exhaustion and violable constraint errors)
- Other exploitable logic errors

While these are important efforts and give us insights, we really need a taxonomy of remote access attacks, particularly one that characterizes the initiating events that can be exploited via network-based attacks on COTS or GOTS software.

Since our focus is on unknown network-based attacks, recent work by Richardson [17] is of interest. He developed a taxonomy for DoS attacks that identifies the following attack mechanisms:

1. Buffer overflows
2. IP fragmentation attacks
3. Other incorrect data attacks
4. Overwhelm with service requests
5. Overwhelm with data
6. Poor authentication or access control
    - Poor authentication scheme
    - IP spoofing
    - Data poisoning
    - Other miscellaneous protection shortcomings)

These categories will be informed by other studies of taxonomies [e.g., 18, 19]. The results will form the initial basis for our categorization of initiating events of unknown attacks. Priorities will be given to those attacks that are known not to have adequate protection measures built into the cluster currently and for which there are not easy fixes to the design that would prevent them. For example, IP fragmentation attacks against the primary can be prevented with a proxy on the firewall or gateway and IP spoofing is prevented by the VPN.

7 Conclusions and Recommendations

Our design for an intrusion tolerant server cluster uses a behavior specification-based approach to identify errors and failover to the hot spare. It then uses fault diagnosis to recognize the attack that caused failover (or violated QoS) and block it so repeated attacks won't defeat us again. We learn exact attacks by testing entries from complete log files in a “Sandbox” until we duplicate the observed failure. Single stage attacks can be recognized in seconds, automatically.

We have demonstrated that it is possible to generalize web server buffer overflow attack signatures after the initial identified attack so that simple variants that exploit the same vulnerability will be blocked also. We do this using a similarity measure for the class of attack. We have implemented rules that generalize a large subset of buffer overflow attacks aimed at web servers and have tested it using the Internet Information Server (IIS) by Microsoft, and believe that it will also work for Apache and other web servers also. For buffer overflow attacks, which have become the most common type of attack, we can also learn the minimum length of the request that causes the buffer overflow. This is important to minimize the probability of blocking legitimate transactions, i.e., the false positive rate.

We believe this knowledge-based learning is broadly applicable to many classes of remote access attacks and has significant uses outside of intrusion tolerance. We also believe that the generalization approach can be significantly extended to other classes of attack. The key, we believe, is generalizing an attack pattern to protect against all variants that exploit the same vulnerability rather than trying to generalize a specific attack to protect against all such attacks in the class. The ease of generalizing an attack pattern should be proportional to the ease of creating simple attack variants that work against the same vulnerability.

In summary, we have developed an approach to dynamic learning of unknown attacks that shows great promise. We have also implemented a proof of concept for generalization that works for a significant class of buffer overflow attacks against web servers on Microsoft NT/2000. Our results so far indicate that the generalization algorithms will be specific to particular types of attacks (such as buffer overflow), to particular protocols (such as http) and to particular application classes. More work is needed to determine whether they must be specific to particular applications but that is a likely outcome if the application class is not dominated by standard protocols.

We recommend that other researchers examine this knowledge-based approach to identifying unknown attacks. We hope they find it useful enough to apply it to other areas.

References

1. Schneier, B: Secrets and Lies: Digital Security in a Networked World. John Wiley & Sons, Inc., 2000 206, 210
2. Gray, J., Reuter, A.: Transaction Processing: Concepts and Techniques. Morgan Kaufmann Publishers, San Francisco, CA, 1993 107
3. Lampson, B.: Computer Security in the Real World. Invited essay at 16th Annual Computer Security Applications Conference, 11–15 December, New Orleans, LA, available at http://www.acsac.org/2000/papers/lampson.pdf

4. Just, J.E., et al.: Intelligent Control for Intrusion Tolerance of Critical Application Services. Supplement of the 2001 International Conference on Dependable Systems and Networks, 1–4 July 2001, Gothenburg, SW
5. Reynolds, J., et al.: Intrusion Tolerance for Mission-Critical Services. Proceedings of the 2001 IEEE Symposium on Security and Privacy, May 13–16, 2001, Oakland, CA
6. Reynolds, J., et al.: The Design and Implementation of an Intrusion Tolerant System. Proceedings of the 2002 International Conference on Dependable Systems and Networks, 23–26 June 2002, Washington, DC, pending
7. Ko, Calvin: Logic Induction of Valid Behavior Specifications for Intrusion Detection. IEEE Symposium on Security and Privacy 2000: 142–153
8. Ko, Calvin, Brutch, Paul, et al.: System Health and Intrusion Monitoring Using a Hierarchy of Constraints. Recent Advances in Intrusion Detection 2001: 190–204
9. Balzer, R., and Goldman, N.: Mediating Connectors. Proceedings of the 19th IEEE International Conference on Distributed Computing Systems, Austin, Texas, May 31-June 4, 1999, IEEE Computer Society Press 73-77
10. Strunk, J.D., et al.: Self-securing storage: Protecting data in compromised system. Operating Systems Design and Implementation, San Diego, CA, 23–25 October 2000, USENIX Association, 2000 165–180
11. Ganger, G.R., et al.: Survivable Storage Systems. DARPA Information Survivability Conference and Exposition (Anaheim, CA, 12-14 June 2001), pages 184–195 vol 2. IEEE, 2001
12. Russell, S., Norvig, P,: Artificial Intelligence: A Modern Approach. Prentice Hall, New York, 1995
13. Roesch, M.: Snort Users Manual, Snort Release: 1.8.3. November 6, 2001, available at http://www.snort.org/docs/writing_rules/
14. Krsul, I.V.: Software Vulnerability Analysis. PhD thesis, Purdue University, West Lafayette, IN, May, 1998, p. 17, available at https://www.cerias.purdue.edu/techreports-ssl/public/97-05.pdf
15. Lough, D.L.: A Taxonomy of Computer Attacks with Applications to Wireless Networks. PhD Thesis, Virginia Polytechnic and State University, Blackburg, VA, available at http://scholar.lib.vt.edu/theses/available/etd-04252001-234145/
16. Landwehr, C. E., Bull, A. R., McDermott, J. P., Choi, W. S.: A Taxonomy of Computer Program Security Flaws. ACM Computing Surveys, Volume 26, Number 3, September 1994
17. Richardson, T.W.: The Development of a Database Taxonomy of Vulnerabilities to Support the Study of Denial of Service Attacks. PhD thesis, Iowa State University, 2001
18. Aslam, T.: A Taxonomy of Security Faults in the Unix Operating System. Master's Thesis, Purdue University, Department of Computer Sciences, August 1995. http://citeseer.nj.nec.com/aslam95taxonomy.html
19. Du, W. and Mathur, A.: Categorization of Software Error that Led to Security Breaches. Technical Report 97-09, Purdue University, Department of Computer Science, 1997

Evaluation of the Diagnostic Capabilities of Commercial Intrusion Detection Systems

Hervé Debar and Benjamin Morin

France Télécom R&D, 42 rue des Coutures, F-14000 Caen
{herve.debar, benjamin.morin}@francetelecom.com

Abstract. This paper describes a testing environment for commercial intrusion-detection systems, shows results of an actual test run and presents a number of conclusions drawn from the tests. Our test environment currently focuses on IP denial-of-service attacks, Trojan horse traffic and HTTP traffic. The paper focuses on the point of view of an analyst receiving alerts sent by intrusion-detection systems and the quality of the diagnostic provided. While the analysis of test results does not solely targets this point of view, we feel that the diagnostic accuracy issue is extremely relevant for the actual success and usability of intrusion-detection technology. The tests show that the diagnostic proposed by commercial intrusion-detection systems sorely lack in precision and accuracy, lacking the capability to diagnose the multiple facets of the security issues occurring on the test network. In particular, while they are sometimes able to extract multiple pieces of information from a single malicious event, the alerts reported are not related to one another in any way, thus loosing significant background information for an analyst. The paper therefore proposes a solution for improving current intrusion-detection probes to enhance the diagnostic provided in the case of an alert, and qualifying alerts in relation to the intent of the attacker as perceived from the information acquired during analysis.

1 Introduction

There have been small number of publications on testing intrusion-detection systems, but we believe that important results have been left out of these publications. The emphasis of this work is quality evaluation. Our most important objective is the provision of a detailed and accurate diagnostic of the malicious activity occurring on our networks. Network operators with little security background operate the probes and handle daily alert traffic. Serious security breaches are left to trained analysts. These analysts are a scarce and valuable resource and must spend as little time as possible handling incidents. Therefore, alert information must be detailed and accurate to ensure that the analyst does not need to go back to raw data.

After a literature survey getting information from vendors and the community about multiple intrusion-detection products, we selected a small number of them for internal testing and comparative evaluation. The study was restricted

A. Wespi, G. Vigna, and L. Deri (Eds.): RAID 2002, LNCS 2516, pp. 177–198, 2002.

in scope in order to provide manageable results, and our interest focused on network-based intrusion-detection commercial products, with probe components available worldwide as a remotely manageable appliance. We deployed four commercial intrusion-detection systems on a test bed and carried out a comparative evaluation. Partial results from this evaluation are presented in the paper.

The remainder of the paper is organized as follows. Section 2 presents the goals and organisation of the tests. Section 3 presents our testing principles and our test bed. Section 4 presents the results obtained and the lessons that we learned from the tests. Section 5 proposes an enhanced model for an intrusion-detection system that would emphasize diagnostic accuracy.

2 Background on Testing Intrusion-Detection Systems

A number of papers related to testing intrusion-detection systems have been published in the literature. We ignore testing methodologies proposed by the developers of a given IDS method or tool, because we consider them biased towards enhancing the performances of their tools. During our literature survey, we found three independent testing methodologies close to our preoccupations and studied them.

2.1 The Lincoln Lab Experiments

One of the best-known testing experiments in the intrusion-detection community is the Lincoln Lab experiment [7], analysed by McHugh[8]. The purpose of this experiment is to fairly compare the various intrusion-detection technologies developed under DARPA funding. A network of workstations creates normal traffic and a set of attacks is inserted in the traffic. We have the following concerns with this test environment:

Focus on background traffic: As the test includes anomaly detection tools, realistic background data must be generated to ensure that these systems will be properly trained. None of the products tested includes anomaly detection features, and our need for background traffic is limited to performance evaluation. Our test bed includes the generation of profiled background traffic, geared towards maximizing the workload of the tested products.

Focus on research prototypes: The Lincoln Lab tests were commissioned by DARPA to evaluate DARPA-funded research work. No commercial IDS was ever taken into account. For example, intrusion-detection products provide configuration management features, that we want to evaluate, and these aspects are not available with the Lincoln Lab tests. Our test bed includes reporting on the installation, management and integration for each tool.

Focus on a broad set of attacks: The Lincoln Lab tests aim at exercising the largest possible set of attacks for the largest possible set of intrusion-detection systems. Our objective is to focus on network traffic close to firewalls; therefore the Lincoln Lab tests are too wide for our use. Also, attacks that are qualified as *local to root* are less relevant in telecommunication environment where monitoring is located on the wires. Our test bed focuses on specific types of applications that are representative of the traffic profiles seen on our networks.

Lack of reference point or baseline: The Lincoln Lab tests compare prototypes in a closed circle. There is no notion of a minimal set of requirements that a tested tool has to meet, only relative data comparing them with each other. Our test bed uses *Snort*[14] as a baseline.

The lack of a baseline that all tools would have to fulfil was felt as particularly lacking in the Lincoln Lab experiment.

2.2 The University and Research Work

The most representative work concerning university tests has been carried at UCDavis [12,11], with related and de facto similar work at IBM Zurich [4]. The UC Davis tests simulate the activity of normal users and the activity of attackers, using script written in the *Expect* language. *Expect* simulates the presence of a user by taking over the input and output streams of a TTY-based application, matching expected responses from the application with the displayed output, and feeding appropriate input as if the user typed it on its keyboard.

A similar approach was followed at IBM Zurich; in addition to user-recorded scripts the IBM approach introduced software testing scripts from the DejaGnu platform (also in *Expect*) to ensure that all aspects of an application were exercised, regardless of the fact that they were obscure features of an application or not. This testing methodology is closer to our own. In particular, the fact that tests are automated and reproducible is a very important property. However, the following points reduce the effectiveness of the test environment.

Heavy to manage: To achieve a significant level of background activity, the test bed must contain a significant number of machines, each of them piloting a number of users generating activity. This creates a complex environment to manage, and induces the risk of repetitiveness. Also, calibration to obtain data points at regularly spaced traffic rates is not taken into account.
Our test bed includes centralized distribution of software and management scripts that automate test run and result analysis.

Limited in attack testing: Requiring that attacks be scriptable using *Expect* makes it unfeasible to use a number of exploit scripts collected from "underground" sources. Also, verifying the actual execution of each attack, verify its effect, and correlate with the IDS system is a manual process.
Our test bed does not solve this issue; installing and configuring the vulnerable software, meeting preconditions, running the attack, verifying the results and restoring the environment for future tests is manual as well.

Applicability to commercial tools: The UC Davis test bed has been designed primarily for research prototypes, and no information is available as to how commercial intrusion-detection systems would be included in such testing.

2.3 Commercial Tests

Tests classified as commercial regroup the tests published by commercial test laboratories mandated by a particular company, and tests carried out by independent publications.

Several test labs have published test results related to intrusion-detection systems. In particular, Mier Communications has released at least two test reports, a comparative of BlackICE Sentry, RealSecure and NetProwler on one hand, and a test of Intrusion.com SecurenetPro Gigabit appliance. Appropriate queries on mailing list archives show that these test results have been the subject of many controversies. The general feeling is that it is appropriate for the testing laboratory to provide results that highlight the advantages of the product sponsored by the company paying for the tests. Even if direct test manipulation is not suspected, at least test configurations for the competing products is likely not to be handled by experts, thus resulting in unfair comparison.

Our feeling is that even this cannot be determined, as the description of the tests is very sparse and does not provide information that would allow an impartial judgement of the tests. Normal traffic generation, for example, is done using commercial products and the test report is considered complete with only the mention of the name of the traffic generator, without any information on the kind of traffic it generates or its configuration.

Journalists have also been involved in the testing of intrusion-detection systems. The most interesting article that we have found is by Mueller and Shipley [9]. It is the only comparative test of commercial intrusion-detection systems that we have found, where a number of intrusion-detection products are compared on equal footing and on live traffic, hence with a higher probability of either missing attacks or triggering false positives. The main drawback of this kind of test is reproducibility: when observing alerts it is quite difficult to reproduce the conditions in which these alerts are generated. Also, we believe that the tuning phase carried out during testing initiation is problematic; the testers describe a process in which alerts that they believe are false are turned off. Our approach has been the opposite, keeping a maximum number of signatures active and tabulating both appropriate and false alerts. Doing this enables us to demonstrate the current trade-off of signature-based (and probably misuse-detection only systems), that a large number of patterns catches more attacks but also generates more false alerts.

2.4 IDS Evasion

Another, related work, is the work on *evasion*, particularly pointed by Ptacek and Newsham [10], and enhanced by Vern Paxson and al.[6]. We consider this to be of extreme relevance to our work since network-based intrusion-detection systems periodically are shown vulnerable to evasion techniques.

However, we also believe that carrying out this kind of tests is extremely difficult, and that we would not be able to add much value to the state of the art in lower layers evasion (IP, TCP and UDP evasion). Therefore, our test bed focuses on evasion at the application protocol layer, because we believe that application protocol analysis is still an area of improvement for commercial products. There are in fact two distinct issues with application-layer protocols:

Misunderstanding of the protocol states or properties. Sometimes, vulnerabilities are only applicable to certain states of the application layer protocols. For example, *Sendmail* vulnerabilities usually apply only to the SMTP command mode. IDS products need to recognize these states and verify that state information is adequate before applying signatures. Keeping state information is processor-costly and infrequent states are sometimes ignored by the implementors to improve performance, creating evasion opportunities for attackers.

Misunderstanding of protocol encoding schemes. Sometimes, protocols encode data, hiding the information from the intrusion-detection system and inducing false positives. For example, unicode encoding on HTTP requests can result in false positives. Worse, implementers sometimes include hidden features or encoding schemes that deviate from the published protocol specifications, such as the infamous *%u unicode encoding* issue [2].

Our test bed currently uses *Whisker* [13] as a support for HTTP scanning and therefore has knowledge of HTTP evasion techniques. We are examining a transition to the *libwhisker* library, which provides additional evasive capabilities to be used in normal HTTP traffic as well. *Whisker* is an HTTP server scanner looking for vulnerable CGI scripts. Its two main properties are efficiency and stealth. While we do not care too much about efficiency, stealth is important for us because it makes the job of the tested intrusion-detection probes harder. The traffic is more difficult to analyse, and is also more likely to create false alerts.

3 The France Télécom R&D Intrusion-Detection Test Bed

From this background, we decided to develop our own test bed, reusing the interesting ideas and trying to improve where we felt there were weaknesses. Our test bed is segmented in five areas, described in Sect. 3.2. Each of these areas contains a set of tests that can be executed with different parameters. Our test bed repeats as many executions of each test set with as many parameter combinations as relevant for the expected results.

3.1 Objectives of the Test Bed

While designing the test bed, we set a number of objectives that this design must meet when performing the comparative tests.

Fairness. The test bed must ensure that all products receive the same input and have a chance to correctly detect the attack. Each appliance has two network interfaces to separate LANs, one for sniffing the malicious traffic and one for management purposes. During installation, all possible signatures are enabled to ensure that detection capabilities are at a maximum. This setting ensures that we receive as much information about the traffic as possible and enables us to review the proposed diagnostic.

Repeatability. Updates to both the detection software and the knowledge base are frequent. Since recent attacks are usually more dangerous, updates must be introduced as quickly as possible. The testing process will have to be repeated on a regular basis, to ensure that the tools deployed in the field still perform as expected, and to apply regression testing to discover whether "older" vulnerabilities are still being detected, whether the false alert rate has been improved, and whether performance is still comparable to the initial data.

Automation. Automation covers two different areas of the test bed, running the tests and exploiting the results. Automating the execution of the tests on the test bed and collecting the data has been an important goal of our design. Scripts keep the battery of tests running continuously for up to three days. Note, however, that a few tests described in Sect. 4.5 have not been automated. The main reason for this manual execution is the difficulty of checking all possible error conditions and verifying that the test actually succeeded. The tabulation and result extraction phase is also automated, reading data from syslog and computing the aggregated results from the several thousand alerts generated by one test run.

Baseline. Not only do we want to compare intrusion-detection products with each other, we also wish to establish if products are actually better than what the security community provides and maintain for free. The tested products are compared with *Snort*.

3.2 Description of the Test Protocol

The following five test sets have been implemented in the test bed:

IP Manipulation Tests. These tests are related to low level manipulations of the IP packet, such as *targa* or *winnuke*, which result in denial-of-service. The test bed runs and verifies seventeen different vulnerabilities.

Trojan horse traffic. These tests concentrate on the detection of management traffic, and as such do not carry out denial-of-service attempts against the server. We installed these four Trojans on the test bed, ensuring that all the components were indeed available and running. The Trojans run unmodified, and as such use default publicly known communication ports, default command sets and unencrypted traffic.

***Whisker* vulnerability scanning.** These tests use the freely available *Whisker* [13] CGI scanner, repeating the scan with the default database of vulnerabilities with multiple evasion parameters. We expect intrusion-detection vendors to use *Whisker* in their test suites, as we believe it is a tool actively in use for information gathering purposes, even though it is a bit old by Internet standards. Unfortunately, Sect. 4.3 shows that the diagnostic of *Whisker* scans is still lacking.

The same test database is used for 21 test runs, the first four with the `GET` method (`-M` command line flag), the next 9 with the `HEAD` method and the evasion parameter (`-I` command line flag) incremented from 1 to 9, and the next 7 with the `HEAD` method, and the evasion parameter set to `-I 1 -I n` with n varying from 2 to 9. Note that the `-M` command line switch is overridden for certain tests in the default database. Also, some tests are not carried out in some evasion modes, which results in a smaller number of requests reported by the Attacker.

Live cgi attacks. These tests carry out real attacks against our vulnerable HTTP server. These attacks have varying results, some of them giving a shell with HTTPD user privileges and the others displaying system files. At this stage, this set of tests has to be carried out manually to ensure that preconditions are met, that compromise effectively happens and to restore the server to its original state. Our file target is the `/etc/passwd` file, to ensure that all intrusion-detection products would have a chance to see abnormal activity. While other files are worthy of an attacker's attentions, the only common one that we have found registered across all intrusion-detection systems is the Unix™ password file.

***Whisker* signature evaluation.** These tests use the freely available *Whisker* cgi scanner and a specially crafted database taking into account the list of HTTP vulnerabilities that all products document, constructing creative requests to evaluate the extend and accuracy of each signature (see examples in Sect. 4.6). For each signature, the database contains a set contains URLs related to different attacker activity, such as scanning, normal activity (attempting to reproduce normal requests), outright malicious activity that would directly exploit the vulnerability, and abnormal activity that either extends the direct exploits with our own knowledge or derives from these direct exploits. The goal of this test is to verify that signatures listed in the product documentation do trigger and to approximate the trigger that sets the alert off. We expect this information to be very valuable for analysts assessing the alerts representing the diagnostic of the tested intrusion-detection systems, particularly if detailed alert description is not available.

Clearly, this does not cover all possible intrusive activity. Future extensions of the test bed include generating normal and abnormal DNS, mail (SMTP, POP and IMAP) RPC and file (NFS, Samba) traffic to broaden the coverage of the test. Our emphasis on HTTP is justified by its status today as an ubiquitous transport protocol, used not only for page serving, but also for additional traffic simply because it can traverse the firewall protecting most organizations today. Also, most content is served from web servers even if the server only acts as a mediation portal.

3.3 False Alerts

In this paper, a lot of results have to be interpreted in the light of what is a valid alert, and what is a false alert. At this stage, there is no straightforward answer.

During result analysis, we built a table of valid alerts and false alerts, according to the documentation associated with each alert, and our understanding of what the alert means. Since we practice black-box testing and do not have access to the signatures, we cannot understand why an alert appears and have to rely on our better judgment.

Let's take the example of the *Shopper Directory Traversal* vulnerability (bugtraq 1776, CVE-2000-0922).

An attack will take the following form:

```
$ lynx http://target/cgi-bin/shopper.cgi?newpage=../../../path/filename.ext
```

The combination of the `shopper.cgi` in the request and the `newpage` variable in the parameters is the mark of the attack. In addition, a directory traversal is required to break out of the default web directory, and a file of interest has to be specified.

An example of a false alert would be `shop`, because it is much too wide to work with. An approximate alert would be `shopper.cgi` without any addition, because it would not allow us to differentiate between a vulnerability assessment scan and a real attempt. What we expect of an alert in this case would be a single message pointing out `shopper.cgi`, `newpage`, the target file as a parameter, and any related evasion techniques.

4 Results Obtained during the Tests

This section presents the results obtained during a complete run of the tests. The intrusion-detection systems will be identified as IDS-A, IDS-B, IDS-C and IDS-D, representing four of the five commercial leaders in the field. We decided against explicitly naming the product because all of them exhibited significant (although not the same) shortcomings and we do not wish these results to be interpreted as an endorsement of these products. In fact, the shortcomings identified lead us to believe that none of them would be satisfactory for our demanding environment.

4.1 Results of the IP Manipulations Tests

Table 1 lists the attacks implemented and shows the results obtained. The first column identifies the attack, the next five the number of different alerts that were considered valid alerts for each IDS, and the final five the number of alerts that were considered false alerts for each attack. The summary counts the number of events flagged and the total number of different alerts for each tool in each category.

The tests carried out here are roughly two years old at the time of writing. We did observe a number of vulnerable machines around us and this observation lead us to believe in the relevance of this activity. Also, as a network operator we are concerned by the validity of traffic flowing through the core network. These signatures provide a measure of normality that is of interest.

Table 1. Results of the IP manipulation tests

Attack Name	Valid alerts					False alerts				
	Snort	IDS-A	IDS-B	IDS-C	IDS-D	Snort	IDS-A	IDS-B	IDS-C	IDS-D
papabroadcast	1	1	1	0	0	0	0	0	0	0
pinger	2	1	0	0	0	2	0	0	0	0
gewse	0	0	0	0	0	1	1	0	0	0
nestea (CAN-1999-0257)	1	1	1	0	0	1	0	0	0	0
newtear (CAN-1999-0104)	1	1	1	0	0	1	0	0	0	0
targa2-bonk (CAN-1999-0258)	0	0	0	1	0	1	0	0	0	0
targa2-jolt (CAN-1999-0345)	1	3	0	0	0	0	1	0	0	0
targa2-land (CVE-1999-0016)	0	0	1	1	0	1	1	0	0	0
targa2-syndrop (CAN-1999-0257)	1	1	1	0	0	0	0	0	0	0
targa2-winnuke (CVE-1999-0153)	0	1	1	1	1	0	0	0	0	0
targa2-1234	0	2	1	0	0	0	0	0	0	0
targa2-sayhousen	0	3	0	0	2	0	0	0	0	0
targa2-oshare (CAN-1999-0357)	0	0	0	0	0	0	0	0	0	0
kkill	0	0	1	0	0	0	2	1	0	0
octopus	0	0	0	0	0	0	0	0	0	0
overdrop (CAN-1999-0257)	0	0	0	1	0	0	0	0	0	0
synful	0	1	1	0	0	1	1	0	0	0
Number of events flagged	6	10	9	4	2	7	5	1	0	0
Number of alerts sets	7	15	9	4	3	8	6	1	0	0

First, none of the tested intrusion-detection probes detects the entire set of attacks. In fact, two of them, *gwese* and *octopus*, are not detected at all; this is not surprising since both of them are very specific, windows-95 denial-of-service attacks and as such are probably rare in the field. Note, however, that the best tool only detects ten out of seventeen trials; we believe that this number is quite low and should be improved.

Then, IDS-A has the best detection rate, but it is also the worst offender when looking at false alerts. This reveals an issue with signatures, that are probably not tuned enough to differentiate the real attack from symptoms that would also exist with other vulnerabilities, but are not significant simply by themselves. *Clearly, the trade-off between accuracy and coverage tilts towards accuracy for* IDS-B *and coverage for* IDS-A; hence the *better* intrusion-detection system depends on the particular needs of an organization. We would add that we are not happy by either side of the trade-off. Less accuracy means more expensive analyst time and a less precise diagnostic overall. Reduced coverage translates into missed attacks. Also, the Snort yardstick shows that IDS-C and IDS-D are not up to date with the possible attacks against an IP stack, letting the door open to potential denial-of-service attacks.

Note that table 1 displays the number of different alerts for each intrusion detection system, not the total number of alerts generated per attempt. This number of different alerts is important for an analyst because it gives him more information about the ongoing attack, and gives more meat to a correlation system. Using this measure shows that not only does IDS-A detect the largest number of attempts, it also is the one giving us the most information about the malicious activity going on. The conclusion reached here is similar to the one in the previous paragraph.

Finally, the table does not measure the number of alerts generated for each attack. Some attacks generate a large number of packets, each of them carrying the anomalous characteristic and triggering an alert. This characteristic has already been shown in [5], and we confirm this result; clearly this is a case where aggregation of consecutive, similar alerts is desirable and should be performed by the probe since it has all the elements to do so.

4.2 Results of the Trojan Horses Tests

Concerning the Trojan horse tests, all Trojans were detected by all intrusion-detection systems except IDS-D, which did not detect any of them and generated one false alert. Snort in addition generated two false alerts. Since default ports and keywords were in use, these results are exactly what's expected and the only valid test result is that functionality specified in the documentation matches our default expectations.

4.3 Results of the Whisker Vulnerability Scanning

The results are presented in Table 2. The first two columns indicate the total number of *Whisker* requests and the number of requests that actually deliver information outside of directory existence. Then, for each intrusion-detection system, the table gives the total number of alerts generated, the number of *Whisker* requests that triggered an alert, and the number of *Whisker* directory events that triggered an alert.

The table is incomplete, because of a number of circumstantial issues during testing. We were unable to configure IDS-D to send events to our syslog server; given that this test generates several thousands of events, it is impossible to manually reincorporate the results of IDS-D into the log file. It also proved impossible to use custom log files to carry out this task and we finally gave up, having the impression from screen observation that the box was performing in about the same way as the other ones. Also, during the final rounds of the tests our data collection network broke down and therefore the last three lines of the experiment should be discounted. Since such a test run takes more than 3 days, we were not able to keep the intrusion-detection systems reliably running for that amount of time, at a rate of about one alert per second each.

We propose the following comments for these results:

Missed events. The obvious remark from this table is that many scan events are not flagged as anomalous; the commercial intrusion detection systems flag between 10% and 20% of the scan events, and Snort goes up to 30%, but at the cost of many false alerts. This actually is a very reasonable trade-off, because this offers some resistance to alert flooding. The most verbose intrusion-detection probe across the board is Snort, which systematically generates 3 to 6 times more alerts than the other probes.

Missed summary of scan activity. Note that this test carries out a scan, not actual attacks. As such, there is no real attack traffic going on the network,

Table 2. Results of the *Whisker* vulnerability scanning

Whisker		Snort			IDS-B			IDS-A			IDS-C		
Total	Scans	Total	Event	Dir	Total	Event	Dir	Total	Event	Dir	Total	Event	Dir
548	347	145	144	7	75	63	5	33	33	4	131	77	5
548	347	145	145	7	30	28	2	33	33	4	131	77	5
548	347	145	145	7	30	28	2	33	33	4	130	77	5
548	347	145	145	7	30	28	2	33	33	4	132	77	5
548	347	145	145	7	75	65	5	33	33	4	197	67	3
548	347	92	92	5	75	65	5	47	33	5	1366	69	3
548	347	556	538	205	75	65	5	568	33	5	1176	70	3
548	347	872	539	322	75	65	5	33	33	4	319	60	5
548	347	554	542	203	75	65	5	33	33	4	117	69	3
548	347	144	144	6	75	65	5	33	33	4	121	70	4
241	71	26	26	6	1	1	0	2	2	0	17	9	1
522	325	76	60	6	73	65	5	17	17	4	118	66	3
95	68	792	95	27	9	8	1	7	7	1	220	17	2
548	347	100	92	5	75	65	5	47	33	5	1146	52	1
548	347	556	541	203	75	65	5	570	33	5	1171	69	3
548	347	874	541	319	75	65	5	33	33	4	326	57	4
548	347	553	541	203	75	65	5	33	33	4	116	69	3
548	347	144	144	7	75	65	5	33	33	4	120	70	4
241	71	26	26	6	1	1	0	2	2	0	17	9	1
522	325	76	60	6	73	65	5	17	17	4	111	66	3
39	35	344	38	4	4	3	0	2	2	0	0	0	0

only suspicious activity. The best response from an intrusion-detection probe facing this kind of traffic would be a single event, or the same event repeated at regular intervals with update information. What we see here is a chain of uncorrelated events (uncorrelated in the sense that the intrusion-detection probe does not link them, because an obvious correlation chain from IP address and monotonically increasing port number exists in the logs) that requires an operator manual analysis, not the complete diagnostic that would bring the expected value from this tool. Snort is the only tool to indicate that, with some evasion modes, it is facing a *Whisker* scan; this takes the form of additional alerts in evasion modes `-I 3, -I 4, -I 5` either alone or combined. IDS-A and IDS-C without explicitly mentioning *Whisker* at least give an indication of the evasion mode for some of them.

Missed identification of targets. A follow-up observation is that since this is merely a scan and not actual attacks, all of the alerts generated could be considered false alerts, from the point of view of an operator who would be looking for successful intrusions, or even intrusion attempts. A scan here attempts to assert that superficial traces of a vulnerability exist on the scanned web server. It does not attempt to use the potential vulnerability to break into the server. This unfortunately points out the fact that the intrusion-detection signatures implemented are attached to very superficial characteristics of the attacks and do not take into account symptoms indicating a really malicious attempt. The only anomalous symptom that the probes seem to look for is a request for the UNIX password file at the well-known location `/etc/passwd`. Our scan included attempts against other sensitive files (`.rhost`, `/etc/exports`, `/etc/shadow`) without the probes noticing them.

Ignorance of directory scanning. As shown in the *Dir* columns, most probes fail to detect directory scanning. This is a reasonable way to practice unless directory browsing is enabled on the web server.

Continuing issues with evasion modes. Attack detection is very much dependant of the kind of evasion mode that is activated. Evasion mode `-I 8` is particularly deadly, as it carries out most of the scanning activity with a very low detection probability. IDS-C resists better to the various evasion modes that the others, which indicates that the analysis engine has at least some notion of what an HTTP session should look like. The reason why this change from "/" to "\" is deadly for the probes is that many of them use the "/" character in the signatures to anchor the pattern; a signature pattern for the phf vulnerability could look like `''/phf?''` to indicate that the phf string is at the end of the request string and needs arguments. If the probe doesn't translate "\" to "/", the trigger is missed. Even though it also shows failures from the probes, mode `-I 7` is not as dangerous for the server, as it applies mostly to directory scanning; this gives an attacker some information about the directory structure of the server and as such some information about potential weaknesses, but it would still require a lot of work to actually break in the site.

Reaction of the various intrusion-detection probes to evasion varies greatly. Snort and IDS-C (and IDS-A to a lesser extend) become extremely verbose with certain modes. In the *long URL mode* (`-I 3, -I 4 or -I 5`), they also generate a number of false alerts because the random characters inserted match existing, too simple signatures. This test shows that even though the tested intrusion-detection probes have made progress with respect to interpreting the various subtleties of the HTTP protocol encoding (a similar test two years ago showed all probes failing detection even with the simplest encoding tricks), they could still be improved.

4.4 *Whisker* Scan and the Structure of Alerts

Table 3 focuses on the structure of the alerts generated by the intrusion-detection systems, in response to the same *Whisker* scan shown in Sect. 4.3. The total number of alerts is given in columns T. Then, alerts are classified into 4 groups: G for good alerts giving meaningful information about the event (although maybe not as exhaustive as what we would like), A for approximate, giving information that is correct but marginally relevant for the analysis of the actual event, B for bad alerts that are clearly false alerts, and E for alerts related to the evasion method used rather than the actual event. The definition of these groups is configured into the result analysis tool.

Concerning the E (Evasion) column, it means different things for each probe. For IDS-B, almost all alerts classified in that category indicate that the probe has caught a HEAD request. Such an alert is not systematically generated for every HEAD request. The most likely hypothesis at this stage is that there is a second factor that is required for the creation of the alert, yet unidentified. Some scan events not generating an alert in modes 1-3, but fairly close in patterns (e.g. PERL and PERL5) do give a *HEAD* alert. A second hypothesis is that IDS-B understands that there is an anomaly with the request. It then generates alerts with a slightly broader coverage in terms of pattern matching. Finally, the HEAD alert is sometimes doubled, e.g. two messages arrive in the syslog file for only one scan event. As such, it is unclear whether it is a bad alert or a useful one.

For IDS-A, evasion almost exclusively means that it has seen the infamous "../.." string in the URL. This pattern is inserted by whisker's mode 4 and is flagged by IDS-A as an attempt to exploit an IIS vulnerability, which is obviously an erroneous interpretation (even though the alert should not, in our opinion, be classified as bad because the request indeed presents the required characteristics). For IDS-C and Snort, it is a mixed bag of things, "../.." strings and also specific messages targeting *Whisker* modes such as the directory traversal of mode 4 and the splicing technique of mode 9.

Table 3. Structure of the alerts of the *Whisker* vulnerability scanning

	Snort					IDS-B					IDS-A					IDS-C				
L	T	G	A	B	E	T	G	A	B	E	T	G	A	B	E	T	G	A	B	E
0	145	140	5	0	0	75	1	3	0	71	33	24	9	0	0	131	60	16	1	0
1	145	140	5	0	0	30	12	15	0	3	33	24	9	0	0	131	59	17	1	0
2	145	140	5	0	0	30	12	15	0	3	33	24	9	0	0	130	59	17	1	0
3	145	140	5	0	0	30	12	15	0	3	33	24	9	0	0	132	59	17	1	1
4	145	140	5	0	0	75	1	3	0	71	33	24	9	0	0	197	53	13	1	51
5	92	88	4	0	0	75	1	3	0	71	47	24	9	0	14	1366	53	15	1	743
6	556	124	5	0	427	75	1	3	0	71	568	24	9	0	535	1176	53	15	2	553
7	872	105	3	8	756	75	1	3	0	71	33	24	9	0	0	319	40	16	4	112
8	554	138	5	1	410	75	1	3	0	71	33	24	9	0	0	117	53	15	1	0
9	144	139	5	0	0	75	1	3	0	71	33	24	9	0	0	121	50	19	1	0
10	26	24	2	0	0	1	1	0	0	0	2	2	0	0	0	17	7	2	0	0
11	76	50	4	0	22	73	1	1	0	71	17	14	3	0	0	118	54	11	1	3
12	792	0	0	0	792	9	1	3	0	5	7	7	0	0	0	220	10	7	0	102
13	100	88	4	0	8	75	1	3	0	71	47	24	9	0	14	1146	43	8	1	641
14	556	124	5	0	427	75	1	3	0	71	570	24	9	0	537	1171	53	15	1	551
15	874	96	5	4	769	75	1	3	0	71	33	24	9	0	0	326	41	14	2	121
16	553	138	5	0	410	75	1	3	0	71	33	24	9	0	0	116	53	15	1	0
17	144	139	5	0	0	75	1	3	0	71	33	24	9	0	0	120	50	19	1	0
18	26	24	2	0	0	1	1	0	0	0	2	2	0	0	0	17	7	2	0	0
19	76	50	4	0	22	73	1	1	0	71	17	14	3	0	0	111	54	11	1	0
20	344	0	0	0	344	4	1	3	0	0	2	2	0	0	0	0	0	0	0	0

Snort and IDS-C are the only intrusion-detection probes sending alerts that are obvious false alerts (note that the others do generate false alerts as well on other traffic profiles, but extensive testing for false alerts was beyond the purpose of the tests). For Snort, this is due to the overly simplistic signature database that matches on strings found in the *long URL* evasion mode (7 and 15). For IDS-C, the same *long URL* mode matches on pornographic signatures. IDS-C also matches ``perl'' on the ``perlshop.cgi'' scan event; this is considered a false alert because it matches on part of the file name only and should be corrected.

The classification of alerts in category *G*, *A* or *B* is not influenced significantly by the type of evasive encoding used (except for IDS-B). The total count drops a bit or stays steady, and only the *E* column increases. This means that filtering the evasive alerts would make the diagnostic acceptable for an operator, without loosing too much accuracy.

As a comparison between IDS-B and the other probes, only evasion modes 1 to 3 can be used (due to the *HEAD* phenomenon mentioned earlier). This shows that both IDS-B and IDS-A are almost equivalent in terms of performance. IDS-C has a slightly better score, with the cost of additional false alerts.

Although the ``perl'' signature from IDS-C is overly large, other signatures of the same probe are very precise and distinguish between locations. For example, it matches on /mlog.phtml, but not /cgi-bin/mlog.html, whereas Snort matches on both. This indicates that the signatures from IDS-C can probably be enhanced to better take into account the attack conditions, which is an important possibility to reduce the number of false positives.

4.5 Results of the Live cgi Attacks

The results of these tests are presented in Table 4. Note that when an attack is not caught through generation of one or several alerts, an actual compromise of our victim web server is not diagnosed and reported. Therefore, missed attacks are counted as a very bad point, especially since the attacks were carried out using parameters that are known to be embedded in signatures, such as /etc/passwd.

Our test includes 18 attacks. IDS-A and *Snort* are the two best probes on this test, generating at least one alert for 15 of the 18 attempts. This could look like a very nice result, but is actually not so. In cases where the tools give us only one alert they only catch the presence of the /etc/passwd string on the request; they do not identify the vulnerability itself. Quite clearly, the fact that a request targets the password file is important and must be flagged, even if the vulnerable CGI script is not known to the intrusion-detection system. However, the fact that only the password file is known as an anomalous symptom shows that the designers of intrusion-detection systems have little imagination in terms of attackers targets.

Also, this ranking changes when analysing the accuracy of the diagnostic proposed by the intrusion-detection system. IDS-C and IDS-D do a much better job of highlighting the extend of an attack. They both diagnose multiple aspects of the attack, such as the name of the vulnerable script used as the attack vector, directory traversal activity, request for /etc/passwd and indicate whether the

Table 4. Results of the *Whisker* vulnerability scanning

Attack	Appropriate alerts					Irrelevant alerts				
Attack name	Snort	IDS-A	IDS-B	IDS-C	IDS-D	Snort	IDS-A	IDS-B	IDS-C	IDS-D
accesscounter	0	0	0	0	0	0	0	0	0	0
aspseek-xpl (CAN-2001-0476)	0	0	0	0	1	0	0	0	2	0
bizdb (CVE-2000-0287)	1	1	0	0	1	0	0	0	0	0
clickrespond	0	1	0	0	0	1	0	0	0	0
clipper (CVE-2001-0593)	1	1	1	2	2	0	0	0	0	0
coldfusion	1	0	0	0	0	0	0	0	0	0
finger	1	1	1	1	1	0	0	0	0	0
handler (CVE-1999-0148)	1	2	0	3	1	0	0	0	0	0
htdig (CVE-1999-0978)	1	1	1	1	1	0	0	0	0	0
htgrep (CAN-2000-0832)	1	1	1	1	1	0	0	0	0	0
phf (CVE-1999-0067)	1	2	2	3	3	0	0	0	0	0
php-nuke (CVE-2000-0745)	1	1	0	0	0	0	0	0	0	0
php	1	2	2	2	2	0	0	0	0	0
search	1	1	1	1	2	0	0	0	0	0
search	2	1	1	2	2	0	0	0	0	0
viewsource (CVE-1999-0174)	1	2	2	2	3	0	0	0	0	0
webspirs (CAN-2001-0211)	1	1	1	3	2	0	0	0	0	0
whois (CAN-1999-0983)	1	1	0	2	0	0	0	0	0	0
Number of attacks found	15	15	10	12	13	1	0	0	1	0
Diagnostic accuracy	16	19	13	23	22					

attacker's actions have been successful. IDS-C in particular in one instance indicated us that the HTTP request for the password file was successful, and in another instance indicated that something that looked like a password file was being sent out of the protected network. These two elements show that extended diagnostic while not generally available is indeed possible and that this extended diagnostic eases the analyst's work.

Of course, this extended diagnostic still does not meet our expectations. In particular, it does require an analyst to manually correlate the three or four alerts related to the attack, as the intrusion-detection probe merely provides the alerts side by side without eliciting any relationship between them. In addition, each alert keeps the evaluation of its own severity, whereas the most relevant information would be an aggregated and appropriately weighted severity of all related alerts.

4.6 Results of the Signature Evaluation Tests

The results presented here are related to three vulnerabilities out of the 60 that the test bed currently checks. They are representative of all observed behaviours.

```
1.1: scan () @roots >> add.exe
1.2: scan () @roots >> add.exe?
1.3: scan () @roots >> add.exe?foo
1.4: scan () @roots >> add.exe?C:\inetpub\iissamples\default\samples.asp
```

Fig. 1. Signature evaluation for the `add.exe` vulnerability

Figure 1 shows an extract of the *Whisker* database file for the evaluation of the *add.exe* vulnerability. For this vulnerability, four requests are sent, although the first two could be considered equivalent because the HTTP standard does not differentiate between the absence of arguments to a CGI script and the empty argument. Therefore, the first two requests result in starting the add.exe script with an empty QUERY_STRING. The third request passes an argument to the CGI script in an attempt to simulate normal usage and the fourth one constitutes the effective attack.

For each of the requests the test bed receives the same ADD.EXE alert. This suggests that the pattern detecting the malicious activity for all of the intrusion detection systems is similar to (in perl-like regular expression syntax) `''/add\.exe''`. In each case, the attacker receives different information. In each case, the diagnostic provided by the intrusion-detection systems is the same. As such, we consider this diagnostic incomplete. A normal request to the script is flagged as anomalous by the intrusion-detection system with exactly the same severity as a scan or an intrusive attempt. Our conclusion is that for every alert generated, valuable analyst time must be spent on assessing the activity whereas the intrusion-detection system could automate the process.

Figure 2 shows an extract of the *Whisker* database file for the evaluation of the *Cart32* vulnerability. Again, the first two requests are related to scanning and the third one to normal activity (although normal activity is more complex than that). The two last lines are direct malicious attempts exploiting two different *Cart32* vulnerabilities.

```
2.1: scan () @roots >> c32web.exe
2.2: scan () @roots >> c32web.exe?
2.3: scan () @roots >> c32web.exe?foo
2.4: scan () @roots >> c32web.exe?TabName=Cart32%2B
&Action=Save+Cart32%2B+Tab &SaveTab=Cart32%2B &Client=foobar
&ClientPassword=e%21U%23_%25%28%5D%5D%26%25*%2B-a &Admin= &AdminPassword=
&TabToSave=Cart32%2B &PlusTabToSave =Run+External+Program &UseCMDLine=Yes
&CMDLine=cmd.exe+%2Fc+dir+%3E+c%3A%5Cfile.txt
2.5: scan () @roots >> cart32.exe/cart32clientlist
```

Fig. 2. Signature evaluation for the `Cart32` vulnerability

The tested intrusion-detection systems generate an alert only on the last request (line 2.5). Neither the scanning activity of the 2 first requests nor the attack attempt from the fourth generate any noise, although any analyst will consider that the one before last (line 2.4) looks extremely suspicious. This case illustrates that the signature for the Cart32 vulnerability is too restrictive, because additional vulnerabilities discovered more recently and affecting the same base software are not covered by the signature.

This does not contradict our analysis of the *Add.exe* vulnerability test (too simple signatures giving too many alerts). Our analysis proceeds from the same logic, namely that in both cases the diagnostic provided is extremely imprecise and requires manual analysis to verify the exact circumstances of the attack and assess the extend of the damage. Assessing the alert with the appropriate adjacent and related activity would allow broader signatures (hence less misses) with varying severity levels (hence facilitating the analyst's work). This analysis is expanded in Sect. 5.3.

Figure 3 shows an extract of the *Whisker* database file for the evaluation of the *shtml.exe* vulnerability. The real attack is on line 2.4, line 2.5 representing a variation of ours to check whether the signature implemented in the tested intrusion-detection systems includes the slash between the components or not.

```
3.1: scan () / >> _vti_bin/shtml.exe
3.2: scan () / >> _vti_bin/shtml.exe?
3.3: scan () / >> _vti_bin/shtml.exe?toto
3.4: scan () / >> _vti_bin/shtml.exe/prn
3.5: scan () / >> _vti_bin/shtml.exe?prn
```

Fig. 3. Signature evaluation for the `shtml` vulnerability

One of the tested commercial intrusion-detection systems crashed hard on the four requests not specifically targeting the vulnerability while correctly diagnosing the malicious attempt (e.g all requests except 3.4). This illustrates the difficulty of testing network-based intrusion-detection systems. Our suspicion is that vendors test products using a database of network frames representative of instances of the execution of a given vulnerability, without rebuilding the vulnerable environment and testing with live attacks. Therefore, attack variants that could be introduced accidentally because the operating system handles network traffic a bit differently, or intentional variants such as ours, are not tested and end up introducing faults in the software.

5 Proposed Model for an Intrusion-Detection System

Our test results clearly show that there is still a lot of work to be carried out within the intrusion-detection community to improve the systems that are being deployed today. We believe that the most serious issue highlighted by our tests is the *diagnostic* issue, namely the lack of details in the information transmitted to the analyst whereas useful information could be automatically extracted from the data source. From this, we take a look at the current model for intrusion-detection systems proposed by the intrusion-detection working group (IDWG) of the IETF and propose several improvements that would result in better diagnostic.

5.1 The IDWG Model of an Intrusion-Detection System

The model of an intrusion-detection system proposed by the IDWG in its requirements document [15] has several components. We refer the interested reader to the Internet Draft (or the upcoming RFC) for the complete description. We really are interested here in the analysis (boxes) and data (italics) components as shown in Fig. 4(a).

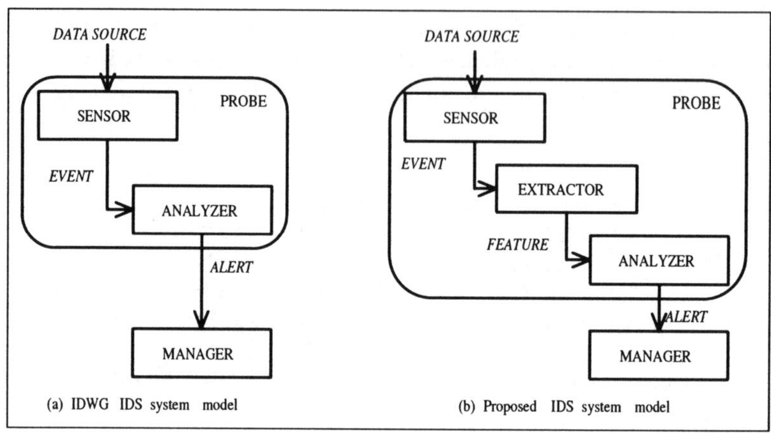

Fig. 4. The IDWG IDS model

Concerning the data components, the *data source* provides raw information tapped by the SENSOR. The SENSOR parses and formats information available in the *data source* and into *events* that are sent to the ANALYZER. The ANALYZER processes the *events* to realize the actual detection process and provide *alerts* indicating that malicious activity has been discovered. The *alerts* are sent to the MANAGER for management and further processing. Note that the model provides for multiple flows of information, for example for multiple SENSORS to send *events* to one ANALYZER. Also, a component could play both the role of an ANALYZER for dialogs with upstream components and the role of a MANAGER for downstream components.

Let's take the example of one of the tested network-based intrusion-detection systems. The *data source* is the network packets that are sent on the wire. The SENSOR taps the network and formats the information retrieved, for example applying the anti-evasion techniques presented in [10] and possibly further decomposing the information to retrieve protocol states. The ANALYZER then takes this information and decides according to its attack signatures to generate an alert. Note that this example collapses the SENSOR and ANALYSER components together into an opaque box performing data acquisition, formatting and analysis altogether.

5.2 Our Proposal for an Enhanced Model

Figure 4(b) presents our extension to the IDWG model. It is indeed a very simple extension, consisting of the insertion of a feature extraction mechanism (EXTRACTOR taking *events* and providing *features* to the ANALYZER. While this seems like a minor change, we do believe it is important particularly in light of the test results shown in Sect. 4.5 and 4.6. The intrusion-detection systems tested to not provide a complete analysis and limit themselves to the feature extraction stage. Each feature is extracted from the data source and sent to the management console independently of the others, regardless of the fact that the features may be linked and thus loosing the link information in transit.

In addition, the quest for performance seems to push intrusion-detection vendors into not performing the complete feature extraction stage. Matching multiple patterns has a cost; exiting early from the search loop allows vendors to claim higher processing rates and gigabit capacity, which is often needed by customers. However, customers often do not realize that trading performance for accuracy will deter them from actually using the intrusion-detection system. This quest for performance also impacts the SENSOR, limiting the reconstruction of network data to IP and TCP fragmentation rather than analysing the actual TCP session from establishment to teardown.

Let's take the example of an HTTP request for the password file using the *phf* vulnerability, against a vulnerable server. The request sent on the wire could look like `GET /cgi-bin/phf?/etc/passwd HTTP/1.0`. From this request, all tested intrusion-detection systems provide us with two alerts, PHF and PASSWORD, representative of the two salient features of the URL. However, the two alerts are juxtaposed, not linked, even though they are extracted from the payload of the same packet.

IDS-C and IDS-D in addition to these two provide HTTP-SUCCESS and/or PASSWORD-EXTERNAL, indicating that they are monitoring not only the incoming request but also the response from the server, and that they extract additional features from the response. This shows that monitoring sessions rather than packets is feasible and informative.

All the alerts are shown on the user interface without indication that they come from the same TCP session, or even from the same IP packet. The analyst must manually notice that IP, PORT and URL information is identical and that timing is identical or very close, and infer that they are related. Similarly for the response, PORT and IP information is reversed but the analyst can relate the inverted address and port information easily. The intrusion-detection probes can make this correlation between alerts with confidence, because they also definitively know if the features were in the same packet, or in the same TCP session due to matching sequence numbers. We see no valid reason for the probes not to carry out this simple task, and give us a single alert fully assessing the damage created by each malicious activity.

5.3 Damage Assessment and Alert Qualification

This notion of *assessment* also improves the notion of *severity* of an alert. Vendors determine the current severity level of the alerts generated by their products by taking into account the intrinsic damage caused by a successful exploit of a vulnerability, and sometimes the difficulty of exploiting it. If the vulnerability gives the attacker an administrative access, then the alert is severe, regardless of the parameters of the event (does the attacker perform an attack or a scan ?) and of its success or failure. We wish that the diagnostic would propose at least one of the following qualifiers:

Scan. The attacker tries to determine the existence of a vulnerability. The main characteristic of a *scan*-qualified alert is that no actual compromise can result from the attacker's actions even though he gains information about the environment.

Exploit. The attacker tries to exploit the vulnerability, possibly with well-known attack scripts, well-known targets or publicly available information. An *exploit*-qualified alert indicates a clearly identified attack that can result in actual compromise if the target is vulnerable.

Variant. The attacker tries to exploit the vulnerability by using the application outside of its specifications, but the target or exploit could not be definitely recognized. A *variant*-qualified alert indicates that there is a possibility for compromise or leakage of information.

In addition, the *success* of the malicious activity and its *relevance* to the monitored information system should be included in the alert. In some of our operations, we are simply not interested in unsuccessful malicious activity, because there is simply too much. We just want to know when they succeed, and then be able to react immediately and accurately. This would enable an evaluation of the severity of each alert based not only on the intrinsic characteristics of the vulnerability exploited but also on the actual exploit circumstances.

Also, attacks scripts are sometimes recognisable. This information should be provided as well. Finally, the age of the vulnerability is extremely relevant to assess its severity, as shown by this study of the Computer Emergency Response Team (CERT/CC) incidents database[1].

Note that these notions have been included in the message format of the IDWG working group [3] for some time, but have not yet found their way into products, although the relevance of attacks is starting to be evaluated in post-mortem commercial analysis products by crossing vulnerability assessment reports with intrusion-detection alerts.

5.4 Impact on Alert Correlation

Alert correlation is currently being pushed heavily in the research community as being the way to solve the diagnostic accuracy issue exhibited by many current commercial intrusion-detection systems. Indeed, we also are part of this trend and working towards alert correlation as a way to reduce the number of alerts that are being processed by operators and analysts. Adding the *feature* extraction in our model pushes some correlation functionalities down in the probes, thus reducing alert traffic and ensuring that the MANAGER is devoted to alert correlation from different probes.

We believe that the EXTRACTOR layer in the probe is in fact a fairly simple engine, extracting *features* from a single *event*. We do not envision a feature extraction mechanism that would search for a feature across two events. In such case, the *feature* should be split in two sub-*features*, each of them covering one event. The aggregation of the two sub-*features* is implemented by the ANALYZER. As such, the ANALYZER could provide a diagnostic based on a single *event* as well as multiple *events*.

A number of research projects have proposed mechanisms to use existing intrusion-detection alerts and correlate them to improve the quality of the diagnostic and lower the number of false positives. An example of this trend is the Tivoli RiskManager product [5]. Looking at it from the feature extraction viewpoint, a much better solution is to improve the probes themselves, for the following reasons:

- They have more data internally than in the alerts they provide, in particular for diagnosing that multiple features come from the same data source, and the same occurrence within the data source such as a single HTTP request or a single network packet.
- The exchange of information between the different components of an intrusion-detection system is minimized, enabling remote management on smaller network links or minimizing bandwidth required for in-band management.
- Communications with management consoles can be prioritised more easily, sending only what needs to be acted upon immediately and batching the alerts related to reporting and trending activities.

With more intelligent probes in place, work on correlation can concentrate on the most interesting aspects of crossing alerts provided by multiple probes while resting sure that each probe provides a trustworthy diagnostic.

6 Conclusion

In this paper, we have shown the design of a test bed for comparative evaluation of intrusion-detection systems. This test bed has been used to compare four commercial intrusion-detection systems with each other and with the open-source lightweight *Snort*. The results show that there is room for improvement in the tested probes. In particular, the probes must improve the alerts they send by including more relevant information that can be acquired from the data stream monitored.

Concerning future work, the test bed is under development to introduce additional applications and services and vary the traffic profile in order to make the task of the tested intrusion-detection systems more difficult. Also, we are developing a prototype probe to validate the EXTRACTOR concept and verify that the diagnostic is indeed improved; this prototype will be used as an additional yardstick along *Snort* in the test bed.

References

1. Browne, H. K., Arbaugh, W. A., Hugh, J. M., and Fithen, W. L. A trend analysis of exploitations. In *Proceedings of the 2001 IEEE Symposium on Security and Privacy* (Oakland, CA, May 2001).
2. Cert Coordination Center. Multiple intrusion detection systems may be circumvented via %u encoding. Cert-CC Vulnerability Note VU#548515, July 2001.
3. Curry, D., and Debar, H. Intrusion detection message exchange format data model and extensible markup language (xml) document type definition. Internet Draft (work in progress), December 2001.
4. Debar, H., Dacier, M., and Wespi, A. Reference Audit Information Generation for Intrusion Detection Systems. In *Proceedings of IFIPSEC'98* (Vienna, Austria and Budapest, Hungaria, August 31–September 4 1998), pp. 405–417.
5. Debar, H., and Wespi, A. Aggregation and correlation of intrusion-detection alerts. In *Proceedings of RAID 2001* (Davis, CA, USA, October 2001), pp. 85–103.
6. Handley, M., Kreibich, C., and Paxson, V. Network intrusion detection: Evasion, traffic normalization, and end-to-end protocol semantics. In *Proceedings of the 10th USENIX Security Symposium* (Washington, DC, August 13–17 2001).
7. Lippman, R., Haines, J. W., Fried, D. J., Korba, J., and Das, K. Analysis and results of the 1999 darpa off-line intrusion detection evaluation. In *Proceedings of RAID 2000* (October 2000), pp. 162–182.
8. McHugh, J. The 1998 lincoln laboratory ids evaluation, a critique. In *Proceedings of RAID 2000* (Toulouse, France, October 2000), pp. 145–161.
9. Mueller, P., and Shipley, G. To catch a thief. *Network Computing* (August 2001). http://www.nwc.com/1217/1217f1.html.
10. Ptacek, T. H., and Newsham, T. N. Insertion, evasion, and denial of service: Eluding network intrusion detection. Tech. rep., Secure Networks, January 1998.
11. Puketza, N. J., Chung, M., Olsson, R. A., and Mukherjee, B. A software platform for testing intrusion detection systems. *IEEE Software 14*, 5 (September–October 1997), 43–51.
12. Puketza, N. J., Zhang, K., Chung, M., Mukherjee, B., and Olsson, R. A. A methodology for testing intrusion detection systems. *IEEE Trans. Softw. Eng. 22*, 10 (October 1996), 719–729.
13. Rain Forest Puppy. A look at whisker's anti-ids tactics. http://www.wiretrip.net/rfp/pages/whitepapers/whiskerids.html, 1999.
14. Roesch, M. Snort - lightweight intrusion detection for networks. In *Proceedings of LISA'99* (Seattle, Washington, USA, November 7-12 1999).
15. Wood, M., and Erlinger, M. Intrusion detection message exchange requirements. Internet draft (work in progress), June 2002.

A Stochastic Model for Intrusions

Robert P. Goldman

Smart Information Flow Technologies (SIFT), LLC
2119 Oliver Avenue South
Minneapolis, MN 55405-2440 USA
rpgoldman@sift.info

Abstract. We describe a computer network attack model with two novel features: it uses a very flexible action representation, the *situation calculus* and *goal-directed* procedure invocation to simulate intelligent, reactive attackers. Using the situation calculus, our simulator can *project* the results actions with complex preconditions and context-dependent effects. We have extended the Golog situation calculus programming with *goal-directed* procedure invocation. With goal-directed invocation one can express attacker plans like "first attain root privilege on a host trusted by the target, and then exploit the trust relationship to escalate privilege on the target." Our simulated attackers choose among methods that can achieve goals, and react to failures appropriately, by persistence, choosing alternate means of goal achievement, and/or abandoning goals. We have designed a stochastic attack simulator and built enough of its components to simulate goal-directed attack on a network.

1 Introduction

In order to best develop the techniques of cyber defense, we must be able to explore them in a scientific way. Practically speaking, we must be able to test cyber defense components without having to bring on-line special network configurations and then conducting possibly very harmful actions on those networks. While it is not a substitute for actual experimentation (both in laboratories, honeypots and on-line), we must be able to simulate internet attacks based on analytic models, just the way the military must be able to simulate conflicts in the real world.

Our stochastic model design comprises multiple, modular, components that will allow security researchers and wargamers to repeatably exercise their sensors and defenses against goal-directed attackers. The core of our stochastic model is built upon an expressive representation of actions, the situation calculus [21, 23], permitting us to accurately model complex cyberworld processes. We have augmented the situation calculus, and its process-modeling extension, Golog [20, 13,23], with facilities for goal-directed procedure invocation. These permit convenient specification of cyber attackers' plans in terms of the goals they are to achieve, providing modularity in modeling and permitting us to experiment with a wide variety of attack techniques.

A. Wespi, G. Vigna, and L. Deri (Eds.): RAID 2002, LNCS 2516, pp. 199–218, 2002.

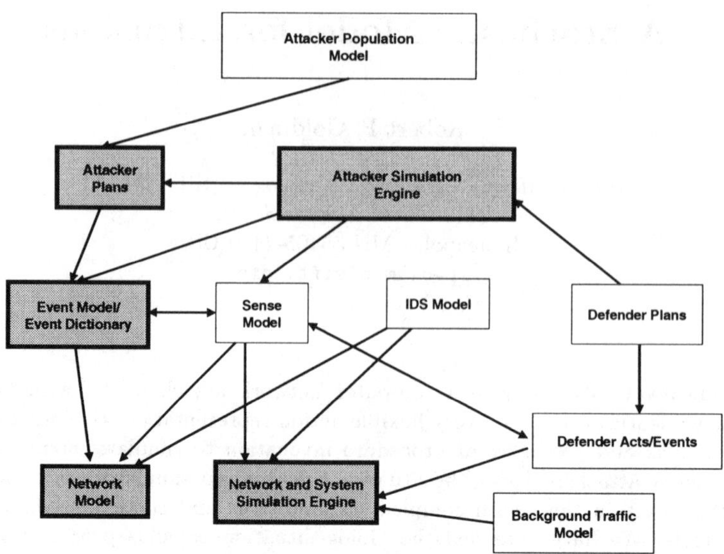

Fig. 1. Illustration of the stochastic model components. Note that the arrows are loosely intended to denote "depends-on" relationships. These are *not* indications of interfaces. The shaded, heavily-outlined components are the ones present in our current prototype.

In the next section, we present the architecture we propose for a full cyber warfare simulation. We have developed a proof-of-concept implementation of the components of that architecture required to simulate a goal-directed attacker. Discussion of that implementation and its theoretical background consume the rest of the paper. We present a full example of a simulated attack, and explain how it was generated by the simulator. Finally, we give proposed future directions and an account of related work.

2 Architecture

We have designed the architecture for a full network warfare model (see Figure 1). We have not implemented this entire architecture; we hope to do so in future work. What we have now is a subset of this architecture sufficient to simulate the actions of a single attacker. We have chosen to focus on attacker simulation because this provides the greatest theoretical and modeling challenges. The current proof-of-concept shows how a single attacker, with a library of plans and their component exploits, can attack a network.

Attacker population model. In order to determine how a cyber defense will react "in the wild," we need to be able to consider how it will react to a wide variety of possible attackers. One question of great concern at present is whether we can effectively defend against serious threats like organized crime and terrorists, without having all our resources absorbed in overreaction against nuisances. This raises the interesting question of whether attackers' intent can be inferred from their actions [10].

We propose to meet this need with a simple stochastic generator that is capable of randomly creating a set of attackers for a scenario. For particular scenarios, and for "wargaming," this could be replaced by having a user just state how many, and of what type, the attackers are, or some combination (randomly generate ankle-biters operating concurrently with a user-specified or even user-controlled team of more serious adversaries). Attackers would be drawn from a set of n types of attacker, $t_1 \ldots t_n$. These might include, for example: script-kiddie, cyber warrior, recreational hacker and cyber terrorist. To experiment with different threat conditions, the probability of appearance of each class could be varied.

Associated with each of these classes of attacker, we should have distributions that determine the attacker's objectives and knowledge. The attacker's objectives would be captured in terms of goals achievable by the plan library. It's not clear what an appropriate set of candidate goals would be, but some obvious choices include:

1. denial of service — for various services run at the site;
2. compromise integrity of service — for various services run at the site;[1]
3. compromise confidentiality of some data at the site;
4. compromise integrity of some data at the site;
5. web site defacement;
6. compromise of root access;
7. springboard for attack on other sites.

Given a finite set of possible goals, $\mathcal{G}$, then for any given attacker, a, who has a set of goals, $G(a)$ and goal, $g \in \mathcal{G}$, there is a set of probabilities as follows: $P(g \in G(a))$, which is a function of the type of the attacker, $t(a)$:

$$t(a_i) = t(a_j) \implies P(g \in G(a_i)) = P(g \in G(a_j))$$

As far as knowledge, each agent will know only a subset of the plans in our entire plan library. Much of the differences in knowledge will be in the specifics of the plan library. That is, overall how to do cracking attacks is pretty well-defined. But down below, when we get to particular exploits on particular services, skill levels get to be radically different. Further, each attacker will have only limited knowledge of the target systems.

Attacker simulation engine. Most of the work to date on the stochastic model has centered around this component and the attacker action model, and most of this paper will concern itself with those aspects of the work. We have developed an extension to the Golog logical representation of procedures. That extension adds

[1] Defacement of web site *could* be categorized as compromising the integrity of the service, but that's probably not appropriate, because it's significant as a special kind of attack.

goal-directed invocation facilities, loosely based on reactive planning languages such as PRS [11,12] and RAPS [9]. The attacker simulation engine is intimately intwined with the plan library and attacker actions.

Attacker plan library. The attacker plan library is made up not of specific exploits, but sub-goaling data structures that capture abstract strategies for achieving attack goals. For example, to attain access to a host, you can attain access to a trusted host and then exploit the trust relationship. These would correspond to the tops of attack trees [26], or the most abstract capabilities in JIGSAW [29]. We have developed a collection of simple attacker plans, primarily based on the "Frostbite Falls" scenario [15]. We discuss these plans in more detail in Section 4.1.

Note that attack plans are modeled separately from the attackers' individual actions and from background network traffic (the next two items on our list). There are three reasons to make this separation. First, attacker plans need little, if any, direct interaction with the world model. Second, factoring out the model of individual actions would allow us to experiment with attacker plan recognition [10] without involving the full weight of a simulation engine and a sensor model. Finally, we separate out the details of individual actions because those are the aspects of the model that change most rapidly. Avoiding mention of specific exploits makes it easier to maintain the plan library.

Why does the attacker model not need direct interaction with the world model? Any information consumed (sensing) would come from the model of the attackers' beliefs contained in the state of the attacker simulation engine. That set of beliefs would be populated as a side effect of executing actions. Any changes to the world model would be made through the execution of actions (for which see below).

Attacker actions (Event Model). The attacker actions are modeled using the situation calculus [23]. Attacker actions include not only exploits *per se*, but also conventional actions (e.g., writing a `.rhosts` file), that are used in attacks. The actions provide the bridge between the attacker's plans and the world model.

We model attacker actions in terms of their preconditions and effects. The situation calculus representation provides a crisp semantics for precondition-effects modeling of attacker actions [4,29]. The attackers' actions are invoked in the plans executed by the attacker simulation engine. Ideally these actions would also be executable in the real world, permitting us to replicate our simulation results in real networks. The model of attacker events is the most developed part of the current system. We have a small but substantial library of exploit actions.

Network simulation. At the moment, we have only a relatively static model that describes the current configuration of the network. The network is a passive field upon which the attackers' actions are played out. Eventually we would like to see the network simulation grow to a timed discrete-event model of a computer network. This model must be able to be affected by the invocation of attacker actions in the Event model. It must also provide some kind of time signal that can be used by other components (especially the attacker simulator so that time is handled correctly. Finally, it must be possible for simulated agents to sense this model and update their beliefs.

Ideally, the network simulation should be plug-compatible with the real world. I.e., we should be able to invoke the attack events (or at least a subset thereof) against the real world to replicate simulation results. We should also be able to have the attacker engine sense the real world and build beliefs the same way it would sense the network simulation.

Sensor Models. The proposed sensor model component comprises models of intrusion detectors and other components (e.g. system loggers, firewalls, etc.) that would provide information to a defender. We are particularly interested in developing this component in order to be able to experiment with IDS report aggregation without having to field large suites of IDSes. Such a model would let us do large numbers of experiments on sensor event fusion without having to handle actual traffic, just alerts.

3 Framework for the Single-Attacker Simulation

To successfully simulate a computer intruder, we must be able to model primitive actions and their composition. This includes being able to model the actions that are available to the attacker, and their effects on the world (in this case the computer network). The model must also provide a simulated attacker that will intelligently assemble primitive actions into a plan to achieve its objectives. Assembling plans together must include the ability to react to failed actions, either by trying again, choosing new methods that will achieve the same end, or abandoning failed goals.

We have assembled a theoretical framework that meets these needs, and developed a proof-of-concept implementation. We use the *situation calculus* [21,23] as the framework within which to model attacker actions. The situation calculus provides an expressive framework for encoding actions, including those whose effects are complex functions of the state in which they are executed. Further, the situation calculus includes solutions to the frame and ramification problems, deep problems about reasoning about actions. Golog (ALGOL IN LOGIC) [20] and its variants, Congolog [13] and Indigolog [14] provide ways to express complex procedures in terms that are consistent with the situation calculus. They thus let us describe complex actions, composed of primitive actions, conditionals, etc., in a way that preserves the crisp semantics of the situation calculus. We have augmented the Indigolog framework with *goal-directed invocation*, the ability to invoke procedures based on desired effect, rather than by name. This adds a crucial kind of abstraction to our plan models. We have added stochastic goal persistence and abandonment, so that our agents will react appropriately to failures in their plans. Finally, we have modified an existing Indigolog interpreter [19] to provide the simulation engine for our work.

3.1 The Situation Calculus

As the most mature representation for actions and dynamic change, the situation calculus provides the best framework for us to experiment with descriptions of cyber attacker's actions. The situation calculus is a formal representation for dynamic world models developed by artificial intelligence researchers. The situation calculus is a dialect of first order logic, with certain special features for representing dynamic change.[2] Situation calculus researchers have provided solutions for many knotty problems in representing dynamic change, including management the notorious frame, ramification, and qualification problems.

Recent work on attack modeling by Templeton and Levitt [29] and Cuppens and Ortalo [4] has argued for a "requires-provides" or "precondition-postcondition" model of attack actions. We were inspired by this work, and came to choose the situation calculus framework because it provided a semantics for the pre- and postconditions not yet provided by other work.

Fluents and Predicates. The situation calculus is a dialect of typed first order logic. For the purposes of this paper, we assume familiarity with the syntax and semantics of first order logic. The most important distinguishing feature of the situation calculus is the addition of a distinct type, the *situation.* A situation is a snapshot of the world state, together with the history that led to that state. The world is described in terms of static predicates and *fluents.* Static predicates are those that don't change, no matter what actions are taken. Fluents, on the other hand, are predicates that can vary over time, and thus must take situations as arguments. So, for example, *mortal(socrates)* is a formula with a static predicate. On the other hand, *loggedin(b0ri5, host123, s)* tells us that the cracker "b0ri5" is logged into host123 in situation *s*; *loggedin* is a fluent.

Doing actions. In order to reason about the effects of actions, we need to be able to refer to the situation that results from the execution of an action. This is done using the *do* function. $do(a, s)$ is a function that denotes the situation that results from doing action *a* in situation *s*. In its simplest form, the situation calculus assumes that actions are only executed sequentially, and that all actions are deterministic. There are a number of extensions that permit stochastic actions, concurrent actions (possibly by multiple agents) and reasoning about actions of varying duration.

Projecting the results of actions. In order to reason about dynamically-evolving situations, the situation calculus requires axioms of three sorts. The first are those that state the conditions under which it is possible to execute an action. For example,

$$Poss(login(user, host), s) \equiv atconsole(user, host, s)$$

[2] Actually, in part of the formulation of the situation calculus, some second-order quantification is resorted to. However, anyone *using* the situation calculus, will find him/herself on familiar ground.

gives a simple model for when it is possible for a user to log into a host.[3] The second are *successor state axioms* for the fluents. For example,

$$
\begin{aligned}
Poss(a,s) \supset [\ & loggedin(user, host, (a,s)) \equiv \\
& \{\ a = login(user, host) \vee \\
& \quad (loggedin(user, host, s) \wedge a \neq logout(user, host))\ \} \\
]&
\end{aligned}
$$

That is, if a new situation results from the execution of action a in situation s, then *user* is logged into *host* in the resulting state if and only if either (1) the action a is one of logging into *host* or (2) *user* is logged into *host* in the previous state, and the new action is not one of *user* logging out of *host*. While successor state axioms can be difficult to formulate manually, compilation techniques are available for deriving such axioms from sets of easier to manipulate action descriptions and some closure assumptions [17,22,27]. Likewise, when the action descriptions are of limited form, one may simply reason with them directly, without taking the intermediate step of explicitly formulating the successor state axioms. This is the technique adopted by the Indigolog interpreter [19] that we have used.

3.2 Golog and Its Variants

While the situation calculus allows us to reason about the execution of primitive actions, and even sequences of such actions, it is not sufficient to express scripts, or programs, involving looping, conditional execution, etc. The languages Golog [20], and its extension Congolog (Concurrent Golog) [13], were developed to meet this need. Congolog adds control constructs such as branching, conditional execution, etc., and concurrency constructs, to the situation calculus. Using Congolog, one can express a concurrent, branching program, whose atomic actions are described in the situation calculus. This makes it possible to describe a reactive program's interaction with a dynamically-evolving environment. We use these facilities to develop a script-based model of cyber attackers.

Golog allows actions to be combined into programs using the following constructs in Table 1: test, sequence, and nondeterministic choice of action, nondeterministic choice of action arguments and nondeterministic iteration. These constructs should be relatively familiar to those familiar with dynamic logic. Golog also permits procedures (really macros) to be defined (**proc**) and used.

For example, an attacker might want to use a login procedure like the following:

proc *login(host)*
  begin
    if *console_access(host)*
      then
        $(\pi uid)?(known\_uid(uid, host));\ (\pi sess)login(host, uid, sess)$
    fi
  end

[3] Here, as elsewhere, free variables are implicitly universally quantified.

Table 1. Golog control constructs.

Construct	Notation
Simple actions	a
Test actions	$?\phi$
Sequence	$\delta_1;\ \delta_2$
Nondeterministic choice	$\delta_1 \mid \delta_2$
Nondeterministic choice of argument	$(\pi x)\delta(x)$
Nondeterministic iteration	δ^*

Table 2. Congolog control constructs.

Construct	Notation
Conditional	**if** ϕ **then** δ_1 **else** δ_2
Looping	**while** ϕ **do** δ_1
Concurrent execution	$(\delta_1 \parallel \delta_2)$
Prioritized concurrency	$(\delta_1 \gg \delta_2)$
Iterated concurrency	$\delta_1^{\parallel}$
Interrupt	$\phi \rightarrow \delta$

The attacker wishes to use this to login to some argument *Host*. First, we check to determine whether the attacker has console access. If not, the procedure simply ends. If the attacker does have console access, then s/he chooses a known uid. Note the use of the nondeterministic choice of argument operator to bind the *uid* argument. Then the attacker completes the procedure by executing the login primitive action.[4] In this case, the environment will, effectively, bind the *sess* (session) argument.

Congolog adds a number of constructs to the above: **if-then-else**, **while** loops, concurrent execution, prioritized concurrency, iterated concurrency and interrupt. See Table 2. Concurrent execution and monitoring will be critical for our modeling of a goal-directed attacker's behavior. For example, a goal-directed attacker determined to shut down a target host might begin activating a number of DDoS servers she "owns," until such a time as she determines that the target host no longer responds to a ping message. We will discuss this further below. Likewise, in order to execute an "ip spoofing" attack, an attacker will concurrently attempt a denial of service attack on the host to be spoofed, while sending forged packets to the target host. This might be encoded as follows:

<u>**proc**</u> $ip\_spoof(host)$
  <u>**begin**</u>
    $(\pi t)?(trusted(host, t));\ DoS(t) \parallel spoof\_to(host, t)$
  <u>**end**</u>

[4] This is somewhat confusingly given the same name as the procedure. Names are disambiguated by arity.

3.3 Our Golog Extensions

The use of the Congolog framework provides us with two key components to our attacker simulation. The first is a representational scheme for actions with complex pre- and post-conditions, and the second is a representation for complex attack scripts or plans. However, in its "off the shelf" form, Congolog does not meet all of our needs. First, the semantics of Golog and Congolog rely on "angelic" nondeterminism. To overcome this problem, we have been working with an interpreter for Indigolog. Second, the Golog procedure mechanism is not sufficient to express goal-directed procedure invocation. Recall that we want our simulated attackers to assemble and modify their own attack procedures, based on their goals and the context, and using the actions and procedures available to them. We have developed a framework that adds goal-directed procedure invocation to Indigolog, along the lines provided by PRS [11,12].

The first concern in designing Golog was to come up with a language for AI agents that would provide clear semantics for plans, and that would allow agents to search for the right plan for a situation. In this case, an appeal to angelic nondeterminism may be appropriate. An agent that is looking for the right way to achieve its goal can use search to explore the possible deterministic sequences of actions that correspond to a given nondeterministic program. The nondeterministic program then provides a convenient shorthand to describe the problem of choosing the right course of action.

However, in situations where the world is not under the complete control of an agent, angelic nondeterminism is not an appropriate construct. For example, there may not be enough information at the point of nondeterministic decision for the agent to avoid painting itself into a corner. Consider a cracker that wants to break into host h by finding a vulnerable host h' that is trusted by h, and achieving root privilege on h'. There is no way for our cracker to know that she can actually achieve root privilege on h' before she tries (although there may be positive and negative indications), and if she tries and fails, she cannot simply backtrack to the state of the world before she has attempted the crack. Many times an alert security analyst will be able to mitigate or stave off a network attack by if she notices evidence of failed attacks. Certainly it would be harder to catch crackers, if they could miraculously undo all the effects of their failed attempts!

There have been a number of efforts to bring together Golog constructs with the needs of robotics control applications, which present these same problems [14, 16]. We have based our work on the Legolog interpreter [19], based on the Indigolog (INcremental Deterministic Golog) dialect of Golog. This Indigolog interpreter commits to particular strategies for interleaving concurrent actions, choosing between nondeterministic alternatives, etc. The interpreter's functioning is very easy to understand, because it is written in Prolog and directly implements the semantics of Indigolog as described in [14]. However, we will see that some enhancements were necessary to permit goal-driven execution.

In order to build flexible plans for our attackers, we need to be able to use goal-directed invocation or subgoaling. That is, we need to be able to specify parts of the attacker's plan by specifying *what* it is to achieve, instead of how it to do the job. For example, we should be able to specify that a plan requires the attacker to achieve root privilege on a specific Unix host, without specifying *how* the attacker is to do this. There are two reasons we would like to have this feature. The first is a simple matter of software engineering. We would like to be able to write procedures with straightforward interfaces, and the goal of the procedure is a good way to specify the its interface. The goal provides a stronger, more standard way of encoding the interface than conventional name and parameter methods, because unlike those methods, the goal expression has its own semantics. Further, using the goal as interface specification allows us to permit multiple, alternative, methods for the same goal. Finally, if procedures are characterized by their goals, our agents have an operational method for determining whether or not a procedure invocation has succeeded. In turn, this permits them to react appropriately to success and failure.

While we cannot develop a psychologically accurate model of a cyber attacker, we do want to develop a model that is capable of reasoned attack on a defended network. To do so, we have been broadly guided by the "beliefs-desires-intentions" model [3]. As the name suggests, this model proposes that agents have beliefs and desires about the world. Based on those beliefs and desires, the agents adopt goals, and intentions (plans) to pursue those goals. We have also been guided by the design of reactive programming languages like PRS [11,12] and RAPS [9].

The facility we need is the ability for our agents to pursue goals. Our agents should be able to choose methods to achieve their goals. They should be able to adopt sub-goals as part of a plan to achieve another goal. For example, our agents must be able to decide to take over one host as a stepping stone to a true target. Agents must persist in their goals, and choose alternative methods when one method fails. For example, if an agent is unable to use a specific exploit successfully against a host, that agent might try a number of alternatives: she might try the exploit again (some exploits, like race conditions, are not reliable); she might try a different exploit or she might try to attack a different host. Finally, sub-goals should not last too long. For example, if an agent is performing a denial-of-service attack as part of an ip-spoofing attack, the denial-of-service should not persist past the end of the ip-spoofing attack.

To meet these objectives, we significantly expanded the Indigolog language, adding goal-directed procedures, called KAs (after the similar PRS construct). A KA is like a normal Indigolog procedure, but has additional special features and components. KAs are associated with particular goals, the purpose for which they are to be executed. One may also specify the *context* that limits the conditions under which a KA may be executed. Figure 2 shows a KA for achieving root privilege on some host. Note that it is only applicable when the agent is already logged into that host.

Within a KA definition, one may invoke a special procedure, $achieve\_goal(G, S)$, to achieve a subgoal. We give pseudo-code for $achieve\_goal(G, S)$ in Figure 3. $achieve\_goal(G, S)$ works as follows: first, check to see if the goal already holds. If so, simply return success. If the goal does not hold, the agent must do something to make it hold. Check to see

```
KA user_to_root(Host)
  begin
    π sess
    ?(logged_into(Host, sess));
    achieve_goal(root_privileged(sess))
  to achieve
  root_privileged_on(Host)
  when
  logged_into(Host)
  end
```

Fig. 2. Example KA for escalating to root privilege from a local login session. Choose an existing login session on that host, and then try to escalate the privilege of that session to root privilege.

if there are any methods available. If not, then return failure. If there are methods, choose one and invoke it. If the chosen method has succeeded, we're done. Otherwise, decide whether or not to persist in achieving the goal, and either try again, or give up. In our model, the probability of persistence is a simple constant parameter, but it would be trivial to extend the model to make the persistence probability sensitive to the goal in question.

In general, a single KA may have multiple subgoals. For greater convenience, the KA will monitor all of its subgoals for failure. If a subgoal fails irrecoverably, then the parent KA will fail also (possibly triggering replanning). We do this because subgoals are meaningful only in the context of an over-arching plan. For example, one might have a plan to travel by buying a train ticket and then boarding the corresponding train. If the subgoal of acquiring a ticket fails completely, then the entire plan fails, and the agent will try some other means of achieving his goal, instead of boarding the train with no ticket.[5]

3.4 Future Enhancements

There are a number of issues in attacker modeling that we did not handle in this first version of our simulator. The first is that we allow only a single attacker. This is not as great a limitation as it sounds, since we *do* allow multiple sites of attack. So our current framework is adequate to model multiple, coordinated attackers. However, for future applications, multiple attackers should be added. Such models are necessary to allow researchers to experiment with techniques for attributing attacks to multiple different attackers and with disentangling the concurrent activities of truly threatening attackers from the background activities of "ankle-biters."

We also skirted the issue of the *duration* of actions. The concurrency semantics we use is strictly a matter of interleaving and no attempt is made to distinguish between actions that consume different amounts of time. There is

[5] Assuming it is impossible to purchase a ticket on the train.

```
proc achieve_goal(G, S)
  begin
    if G
      then S := true;
    elsif all_methods(G) = ∅
          then S := false;
          else
               π ms, m
               ms := all_methods(G);
               m := choice(ms);
               call m;
               if G
                 then S := true;
               elsif stochastic_persist(G)
                     then achieve_goal(G, S),
                      else S := false;
               fi
    fi
  end
```

Fig. 3. Pseudo-code definition of *achieve_goal*. G is an "in" parameter, the goal expression. S is an "out" parameter, a Boolean value indicating success or failure in achieving the goal.

already a treatment of actions with variable durations for the situation calculus, together with a corresponding Golog interpreter [23, Chapter 7]. Therefore, adding temporally-extended actions will not add any theoretical difficulty. However, it will add substantial complication to the modeling and knowledge-engineering process for the simulator. We simply do not have good estimates of the durations of the various actions and processes, and these durations bear complex relations to the state of the environment (e.g., network and host load). Nevertheless, if we are to incorporate features like (authorized) background traffic models, we will have to address this issue.

The current system has only deterministic actions. For example, the exploits will always work if and only if they are executed against a vulnerable target. Note that even though the actions are deterministic, *the simulated attacker is not.* The attacker's choice of plans, exploits and targets, and its reaction to failures are all stochastic phenomena, lending a great deal of variance to even simple scenarios.

Even so, limiting ourselves to deterministic actions is obviously a substantial oversimplification. Therefore we must incorporate actions that succeed only with a certain probability. The formal framework for stochastic actions already exists [2]. As with action durations, the primary difficulty is the challenge of knowledge acquisition.

Most of the systems for working with the situation calculus and golog assume a finite domain of entities. The finite domain is necessary for the practical inference methods to succeed (cf. [23, Chapter 9], [8]). This limitation is not at all appropriate to modeling computer intrusion situations. Modeling interactions with software entities require us to be able to model the construction and (to a lesser extent) destruction of entities. For example, if we wish to talk about login sessions, we must be able to talk about their creation by the act of logging in to a host. One solution would be to create *ab initio* a fixed pool of *potential* sessions, and talk about them being actually there or not [18]. This solution poses two problems. First, it commits us to a fixed pool, and our simulator will fail if we guess wrong and overflow the preallocated pool. Second, it means that we will often find ourselves quantifying over unnecessarily large domains. For the moment, we have programmed ourselves out of this problem by restricting the cases where open domain entities can appear and by making sure we do not make inappropriate queries about those entities. This is an unsatisfactory state of affairs, however, and calls for some further technical work. There has been some preliminary work in the area of "softbots" [6], on exploiting local closed world knowledge [7]. This is related, but limited to trying to provide intelligent agents enough knowledge about action effects to predict whether they can act successfully, not to handle projection for simulation.

The current system has a patchy treatment of agents' knowledge and beliefs. We have put in special-purpose fluents to track the attacker's knowledge of key facts like the identity of hosts and user passwords. However, we have not extended this to a full knowledge model for the attackers. As with stochastic actions and actions with variable durations, there is already some available theory for such a model [25], [23, Chapter 11]. However, the current solutions are not acceptable for our purposes. The problem is that the existing solutions assume that agents will differ only in their knowledge of the *state* of the world. They must agree on the *physics* of the world — particularly on how all the actions will affect the state of the world. This is simply not appropriate for simulating computer attackers. Typical attackers — especially the "script kiddies" — will shape their attacks to the particular tools that they have. Furthermore, one can often determine useful things about an attacker based on ill-chosen actions. For example, one sometimes sees a script kiddie gain access to a Unix host, and attempt to execute Windows commands there. This is an area for further research.

A final area for work is less theoretically interesting, but perhaps the most important. This is the creation of better tools to work with our attacker simulation tool. There are two most critical needs. The first is for some compile-time and run-time validation of model components. The indigolog interpreter is written in Prolog. This is desirable for many reasons, most notably because it permits a direct implementation of the semantics of the language. On the other hand, it means that we have a very crude compilation environment with very poor detection of ill-formed actions or tests. The system badly needs some form of type-checking or other verification. Currently all of these errors must be detected by the programmer using the debugger. This debugger should also be improved. Currently, one must use the Prolog debugger, which plunges the programmer into the details of the interpreter's implementation. It would be better to build a special-purpose debugger for the simulator application, that would focus on the golog semantics and suppress the details of their Prolog implementation.

Finally, we have made a number of extensions to the syntax of the language to make it easier to write. However, further work could certainly be done in this area.

4 Modeling the Cyber Attack Domain

In this section, we discuss how we model cyber attack domains using our tool. After that, we present a specific example, the "Frostbite Falls" scenario. We then describe a simulated attack, generated by our simulator. This attack will show how the simulated attacker can use multiple alternative plans and exploits, to achieve its end. We will see that the attacker adapts to the outcomes, both success and failure, of earlier actions.

Our modeling of cyber attack domains has been somewhat ad hoc, but our modeling has a clear rationale, and our modeling decisions should be understood in the light of this rationale. The first component to this rationale is that cyber attack modeling is being done in order to support more cost-effective intrusion research. The second component is that we are interested in studying the way intelligent systems can combine reports from multiple sensors into an overall situation assessment.

The purpose of the cyber attack modeling tool is to allow more cost-effective intrusion research, with repeatable tests. We are particularly interested here in extended attack scenarios: not just the isolated deployment of a given exploit. There are three obstacles we'd like to overcome. The first is the cost of testing an intrusion scenario. Doing so involves building a test network, and then restoring the network's state after any destructive modifications by the attacker. The second obstacle is the requirement to have expert humans involved in performing the intrusions we would like to study. The final obstacle is the sheer time cost of exploring multiple variations of a single attack plan. It's common for red teams to develop attack trees that contain many ways of attacking a particular network. But it is very difficult to explore these at all thoroughly because of the time needed and the requirement for direct human involvement.

We are not interested in the detailed modeling of individual events. Instead, we are interested in the way a number of different sensors will report on the same events. We are particularly interested in how those reports can be fused together in ways that exploit background knowledge.

Our purposes in this project, then, dictate a relatively abstract level of modeling. For example, we have avoided modeling the details of network traffic and network protocols. We have also avoided modeling the details of the file systems of the computers that are attacked. The primary consideration in this abstraction has been to develop a simulation that can be run far more quickly than a true version of the attacks. Compare this to network simulations intended to assess the performance of various network protocols, which typically run at only some fraction of the protocol itself. Further, since we are interested in what attacks will look like through the eyes of sensors, rather than experimenting with sensor designs, we avoid modeling the exact phenomena that will cause sensors

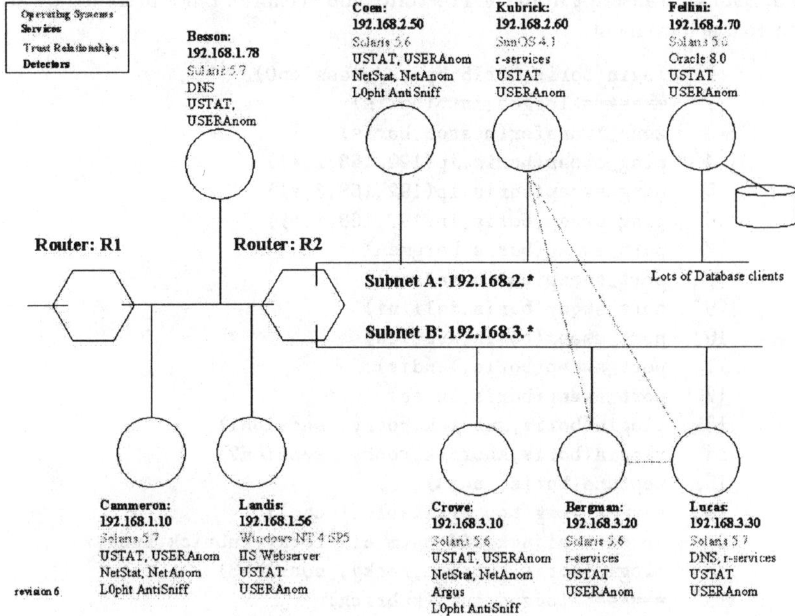

Fig. 4. Frostbite Falls network topology.

to trigger. Of necessity, this will cause some inaccuracies, but we feel the price we pay is worthwhile, at least for this application.

4.1 The Frostbite Falls Scenario

The Frostbite Falls scenario concerns the attack made by a cracker, whom we'll call b0r15, on a network that contains an Oracle database (see Figure 4). b0r15 wants to gain access to the Oracle database (on the host `fellini`) order to corrupt its contents. However, initially b0r15 only knows the IP address of Frostbite Falls' DNS server, he does not know that `fellini` is the target. Further, b0r15 does not know any exploit that will directly allow him access to `fellini`; he will have to gain access indirectly, through another host on the defended network. This means that the attack will go through all of the classic cycles of reconnaissance, initial foothold, and exploitation and consolidation.

4.2 Attacking Frostbite Falls

The easiest way to use the cyber attack modeling tool is to experiment to with our Frostbite Falls scenario model and the attached KAs and primitive actions. Table 3 shows our hacker, b0r15 trying to gain access to the Frostbite Falls Oracle database in order to insert phony orders.

The transcript in Table 3 was generated by a top-level plan that has three steps:

Table 3. Sample transcript from the Frostbite Falls scenario. Lines preceded by arrows indicate goal achievement.

```
login(boris,b0ri5,bpass,_session0)
=======>logged_into(boris)
zone_transfer(besson,boris)
ping_sweep(boris,ip(192,168,2,*))
ping_sweep(boris,ip(192,168,3,*))
ping_sweep(boris,ip(192,168,1,*))
port_sweep(boris,bergman)
port_sweep(boris,besson)
port_sweep(boris,fellini)
port_sweep(boris,kubrick)
port_sweep(boris,landis)
port_sweep(boris,lucas)
rlogin(boris,kubrick,rocky,_session1)
rlogin(boris,kubrick,rocky,_session2)
neptune(boris,lucas)
=======>neg(tcp_available(lucas))
session_hijack_add_perm_all(rocky,kubrick,lucas)
rlogin(boris,kubrick,rocky,_session3)
=======>logged_into(kubrick)
ftp(dtappgather)
=======>available(dtappgather)
dtappgather(_session3)
dtappgather(_session3)
email(sadmindex)
=======>available(sadmindex)
sadmindex(_session3)
=======>root_privileged(_session3)
=======>root_privileged_on(kubrick)
magic_transfer(sniffer)
=======>available(sniffer)
install_sniffer(_session3,kubrick)
=======>access(oracle,fellini)

yes
```

1. prepare to attack (this is primarily concerned with getting b0r15 logged into a workstation he owns so that he can attack);
2. conduct reconnaissance;
3. get access to the oracle database, wherever it is.

The Golog code was as follows:

begin
$(\pi sess) start\_work(sess);$
$b0ri5\_recon;$
$(\pi oh)?(known\_service(oh, oracle);\ achieve\_goal(access(oracle, oh)));$
end

In Table 3 we see b0r15 first login to his own workstation (lines 1-2). Then he conducts his reconnaissance (3-12). He first tries a zone transfer from the DNS server of the network, then does an IP sweep. Finally, he does portsweeps to see what services are being run on the individual hosts. b0r15 is not stealthy!

b0r15 has a number of possible means of attack on the Oracle host. In this transcript he has chosen to attack by attempting to sniff Oracle passwords out of network traffic. This plan requires him first to achieve root privilege on some other host on the network. With root privilege, he can install the sniffer and then log into the oracle host.

b0r15 chooses `kubrick` as his stepping-stone on the way to attacking `fellini`. Recall that his plan calls for him to gain root access to `kubrick` as a means to install a sniffer. To gain root access, b0r15 chooses to gain local user level access to `kubrick` and then escalate his privilege (he could also have tried a more direct remote-to-root attack). b0r15 first tries a simple rlogin to `kubrick`, on the off-chance that the rlogin services have been left unsecured (13). Note that b0r15 tries this twice (13-14); the simulator provides for the possibility of persistence.

After two tries, b0r15 gives up on trying to get through an unguarded rlogin, and adopts session hijacking as an alternative. `kubrick` trusts `lucas` for the purposes of rlogin. b0r15 chooses the "neptune" denial of service attack from the set of alternatives, and brings down `lucas`'s TCP stack (15-16), allowing him to impersonate a legitimate user, and set up an rhosts file on `kubrick` (17).

Now that he has user-level access, b0r15 moves on to the goal of being `root_privileged_on kubrick`. He first tries the `dtappgather` exploit, mistakenly, since `kubrick` is not vulnerable to it (20-23). Note that in order to try this exploit he first transfers the the exploit code to `kubrick` via ftp (20). b0r15 gives up on `dtappgather` after trying it twice, and seizes on the `sadmindex` exploit. This time, instead of ftping the exploit, he decides to use email (24) to achieve the goal of `available(sadmindex)` (25).

With root-privilege secured, b0r15 transfers a sniffer onto `kubrick` (29-30). He does this using "`magic_transfer` (29), an action which is meant to stand in for any kind of covert channel the defenders have not yet seen. Thus, to the defenders this channel is, effectively "magic." Note that the addition of "magic" actions like this allow us to experiment with situations where we confront attackers with exploits not previously known to us. With the sniffer installed, b0r15 can get the password to Oracle user accounts, so he has achieved his goal of `access(oracle,fellini)`, and is done (32).

5 Related Work

We have already mentioned some of the most directly relevant work on attack modeling. Both Cuppens and Ortalo [4] and Templeton and Levitt [29] have developed modeling langauges based on actions with preconditions and postconditions. One of our contributions is to marry precondition/postcondition intrusion modeling with the situation calculus, which provides a rich and sound semantics.

At least two groups have used model-checking techniques and attack models to assess network vulnerabilities [24,28]. This work is similar to ours in projecting the effects of action sequences. There are two important differences. First, the syntax and semantics provided by the model checkers' temporal logics are less expressively powerful and modular than the situation calculus. Second, these researchers are not interested in *naturalistic* modeling of computer attackers. Their interest lies in identifying network vulnerabilities, for which it is sufficient to consider the worst case attack, computed by the model-checkers' exploration of the attack space. We are interested in modeling the full situation, for which we must consider not only the most competent attacker, but the full range of phenomena. These two objectives are complementary, rather than conflicting, and it would be interesting to see whether the different efforts could benefit from each others' modeling efforts.

6 Conclusions

We have described a comprehensive approach to computer network attack simulation. Computer security researchers need such simulations in order to carry out large-scale, repeatable experiments in computer security. The situation calculus and the Golog situation calculus programming language, suitably extended, can provide a theoretical and practical basis for such simulations. We have provided a number of Golog extensions, most notably goal-directed procedure invocation, to better model cyber attackers. Our prototype simulator can simulate a single attacker, who is able to synthesize full network attacks from a library of plans and primitive actions, reacting to successes and failures it encounters.

Acknowledgements. Thanks to Maurice Pagnucco for much assistance working with Indigolog and to Maurice, Hector Levesque, and the University of Toronto for providing the Indigolog interpreter. Thanks to Keith Golden for helpful comments based on his experience with softbot planning. Thanks to the Argus/Scyllarus team, and Dick Kemmerer and Giovanni Vigna for the Frostbite Falls scenario. This material is based upon work supported by DARPA/ITO and the Air Force Research Laboratory under Contract No. F30602-99-C-0017. The work described here was done while the author was employed at Honeywell Laboratories.

References

[1] American Association for Artificial Intelligence, *Proceedings of the Seventeenth National Conference on Artificial Intelligence*, Menlo Park, CA, July 2000. AAAI Press/MIT Press.

[2] C. Boutilier, R. Reiter, M. Soutchanski, and S. Thrun, "Decision-theoretic, High-level Agent Programming in the Situation Calculus,", in *Proceedings of the Seventeenth National Conference on Artificial Intelligence* [1], pp. 355–362.

[3] M. E. Bratman, "What is Intention?," in *Intentions in Communication*, P. Cohen, J. Morgan, and M. Pollack, editors, chapter 2, pp. 15–31, MIT Press, Cambridge, MA, 1990.
[4] F. Cuppens and R. Ortalo, "LAMBDA: A Language to Model a Database for Detection of Attacks," in *RAID*, H. Debar, L. Mé, and S. F. Wu, editors, volume 1907 of *Lecture Notes in Computer Science*, pp. 197–216. Springer, 2000.
[5] DARPA and the IEEE Computer Society, *DARPA Information Survivability Conference and Exposition(DISCEX-2001)*, 2001.
[6] O. Etzioni, "Intelligence without Robots: A Reply to Brooks," *AI Magazine*, vol. 14, no. 4, pp. 7–13, 1993.
[7] O. Etzioni, K. Golden, and D. Weld, "Tractable Closed World Reasoning with Updates," in *Principles of Knowledge Representation and Reasoning:Proceedings of the Fourth International Conference*, J. Doyle, E. Sandewall, and P. Torasso, editors, pp. 178–189. Morgan Kaufmann Publishers, Inc., 1994.
[8] A. Finzi, F. Pirri, and R. Reiter, "Open World Planning in the Situation Calculus,", in *Proceedings of the Seventeenth National Conference on Artificial Intelligence* [1], pp. 754–760.
[9] R. J. Firby, "An Investigation in Reactive Planning in Complex Domains," in *Proceedings of the Sixth National Conference on Artificial Intelligence*, pp. 196–201. AAAI, Morgan Kaufmann Publishers, Inc., 1987.
[10] C. W. Geib and R. P. Goldman, "Plan recognition in intrusion detection systems,", in *DARPA Information Survivability Conference and Exposition(DISCEX-2001)* [5], pp. 46–55.
[11] M. Georgeff and A. Lansky, "Procedural Knowledge," *Proceedings of the IEEE, Special Issue on Knowledge Representation*, vol. 74, pp. 1383–1398, October 1986.
[12] M. P. Georgeff and F. F. Ingrand, "Real-Time Reasoning: The Monitoring and Control of Spacecraft Systems," in *Proceedings of the Sixth Conference on Artificial Intelligence Application*, pp. 198–204, 1990.
[13] G. D. Giacomo, Y. Lesperance, and H. Levesque, "ConGolog, A concurrent programming language based on the situation calculus," *Artificial Intelligence*, vol. 121, no. 1-2, pp. 109–169, 2000.
[14] G. D. Giacomo, H. J. Levesque, and S. Sardiña, "Incremental execution of guarded theories," *ACM Transactions on Computational Logic*, vol. 2, no. 4, pp. 495–525, October 2001.
[15] R. P. Goldman, W. Heimerdinger, S. A. Harp, C. W. Geib, V. Thomas, and R. L. Carter, "Information Modeling for Intrusion Report Aggregation,", in *DARPA Information Survivability Conference and Exposition(DISCEX-2001)* [5], pp. 329–342.
[16] H. Grosskreutz and G. Lakemeyer, "On-Line Execution of `cc-Golog` Plans," in *Proceedings of the 17th International Joint Conference on Artificial Intelligence*, pp. 12–18, Los Altos, CA, August 2001, Morgan Kaufmann Publishers, Inc.
[17] A. R. Haas, "The case for domain-specific frame axioms," in *The Frame Problem in Artificial Intelligence: Proceedings of the 1987 Workshop*. Morgan Kaufmann, 1987.
[18] Y. Lesperance, August 2001. Personal communication.
[19] H. J. Levesque and M. Pagnucco, "Legolog: Inexpensive Experiments in Cognitive Robotics," in *Proceedings of the Second International Cognitive Robotics Workshop*, Berlin, Germany, August 2000.
[20] H. J. Levesque, R. Reiter, Y. Lesperance, F. Lin, and R. B. Scherl, "GOLOG: A Logic Programming Language for Dynamic Domains," *Journal of Logic Programming*, vol. 31, no. 1-3, pp. 59–83, 1997.

[21] J. McCarthy and P. J. Hayes, "Some philosophical problems from the standpoint of artificial intelligence," in *Machine Intelligence*, B. Meltzer and D. Michie, editors, volume 4, Edinburgh University Press, Edinburgh, 1969.
[22] R. Reiter, "The Frame Problem in the Situation Calculus: A Simple Solution (Sometimes) and a Completeness Result for Goal Regression," in *Artificial Intelligence and Mathematical Theory of Computation: Papers in Honor of John McCarthy, Vladimir Lifschitz (Ed.), Academic Press*, 1991.
[23] R. Reiter, *Knowledge in Action*, MIT Press, Cambridge, MA, 2001.
[24] R. W. Ritchey and P. Ammann, "Using model checking to analyze network vulnerabilities," in *Proceedings 2000 IEEE Computer Society Symposium on Security and Privacy*, pp. 156–165, May 2000.
[25] R. B. Scherl and H. J. Levesque, "The Frame Problem and Knowledge-producing Actions," in *Proceedings of the Eleventh National Conference on Artificial Intelligence*, pp. 689–695, Menlo Park, CA, 1993, AAAI Press/MIT Press.
[26] B. Schneier, *Secrets & Lies*, John Wiley & Sons, 2000.
[27] L. Schubert, "Monotonic Solution of the Frame Problem in the situation calculus," in *Knowledge Representation and Defeasible Reasoning*, J. H.E. Kyburg, editor, pp. 23–67, Kluwer Academic Publishers, 1990.
[28] O. Sheyner, J. Haines, S. Jha, R. Lippmann, and J. M. Wing, "Automated generation and analysis of attack graphs," in *2002 IEEE Symposium on Security and Privacy (SSP '02)*, pp. 273–284, Washington - Brussels - Tokyo, May 2002, IEEE.
[29] S. J. Templeton and K. Levitt, "A Requires/Provides Model for Computer Attacks," in *Proceedings of the New Security Paradigms Workshop*, sep 2000.

Attacks against Computer Network: Formal Grammar-Based Framework and Simulation Tool

Vladimir Gorodetski and Igor Kotenko

St. Petersburg Institute for Informatics and Automation
39, 14th Liniya, St. Petersburg, Russia
gor@mail.iias.spb.su, ivkote@iias.spb.su

Abstract. The paper presents an approach and formal framework for modeling attacks against computer network and its software implementation on the basis of a multi-agent architecture. The model of an attack is considered as a complex process of contest of adversary entities those are malefactor or team of malefactors, on the one hand, and network security system implementing a security policy, on the other hand. The paper focuses on the conceptual justification of the chosen approach, specification of the basic components composing attack model, formal frameworks for specification of the above components and their interaction in simulation procedure. The peculiarities of the developed approach are the followings: (1) malefactor's intention-centric attack modeling; (2) multi-level attack specification; (3) ontology-based distributed attack model structuring; (4) attributed stochastic $LL(2)$ context-free grammar for formal specification of attack scenarios and its components ("simple attacks"); (5) using operation of formal grammar substitution for specification of multi-level structure of attacks; (6) state machine-based formal grammar framework implementation; (7) on-line generation of the malefactor's activity resulting from the reaction of the attacked network security system.

1 Introduction

Attacks against computer network form one of the many other dimensions of cyber terrorism and therefore detection of such attacks and prevention of their harmful effects have recently become the task of great concern. It is undoubtedly that substantial increase of Intrusion Detection System (IDS) efficiency could be achieved in case of using knowledge resulting from generalization and formalization of the accumulated experience with regard to computer attacks [28]. A lot of such data is hitherto accumulated and systematized in the form of taxonomies and attack languages ([2], [30], [38], etc.). Nevertheless, there are no serious attempts to generalize the accumulated data in order to develop a *mathematical model* of a wide spectrum of attacks and use this model for *attack simulation.* Perhaps this is due to the extreme complexity of the network attack and computer networks from modeling perspective: "*there is no widely accepted information physics that would allow making an accurate model*", and the network attacks are "*so complex that we cannot describe them with any reasonable degree of accuracy*" [5].

A. Wespi, G. Vigna, and L. Deri (Eds.): RAID 2002, LNCS 2516, pp. 219–238, 2002.

Attack formal model could be a powerful source of knowledge needed for IDSs development. It could provide for deeper study and understanding of the essence and peculiarities of attacks (intentions of malefactors, attack objects, structures and strategies of attacks, etc.). This model would play an important role in IDS learning both known and of unknown attacks if it were used as a generator of training and testing samples of attacks. Finally, formal model of attacks and attack simulation tool could be used as a testbed for security policy validation, i.e. *for testing, comparing and evaluating of IDS components and IDS on the whole.*

Development of such a model, its formal specification and implementation issues are the subjects of research in this paper. The rest of the paper is structured as follows. *Section 2* outlines conceptual aspects and general strategy of attack modeling and describes definitions of basic notions composing an attack specification. *Section 3* describes the developed ontology of the problem domain "Computer network attacks" which is considered further as a basis for consistent attack specification. *Section 4* gives an outline of the proposed formal grammar framework for specification of attacks and exemplifies such specifications for several classes of attacks. *Section 5* describes the model of the "counterparty" of attacks, i.e. model of the attacked computer network. *Section 6* presents architecture of the Attack simulation tool and its implementation issues. *Section 7* gives an overview of related works. *Section 8* summarizes the main results of the paper.

2 Attack Modeling Strategy

The computer network attacks concern to the class of complex systems possessing such features as large scale, multi-connectivity of elements, diversity of their connections, variability of structure, multiplicity of executed functions and structural redundancy. An *attack model* is understood as a formal object having a likeness in basic properties with regard to real-life attacks, serving for investigations by means of fixing known and obtaining new information about attacks. A *formal model* of attacks is a collection of mathematical dependences specifying attacks and allowing to study them formally and via simulation.

The *peculiarities of planning and execution of attacks*, influencing on choice of a formal model of attacks, are as follows:

1. *Any attack is target- and intention-centered,* i.e. it is directed against a particular object (network, computer, service, directory, file, etc.) and, as a rule, has a quite definite intention. *Intention* is understood as a goal or sub-goal a malefactor intends to achieve. We speak about malefactor's "intentions" according to the terminology used for mental concepts. Formally specified intention we call a "goal". *Examples of intentions*: reconnaissance (e.g. learning of network structure, identification of OS, hosts and/or services, etc.); penetration into the system; access to files of some directory; denial of service, etc. *Examples of targets*: IP-addresses of trusted hosts; password file; files of a particular directory; some resources of a particular host, etc. It should be noticed, that in some cases

intention cannot be determined in advance. It can be accepted by malefactor in progress of attack development as a decision made on the basis of the obtained information and successfulness or ineffectiveness of particular malefactor's actions fulfilled earlier.

2. *An attack intention can be represented in terms of partially ordered set of lower-level intentions.* A set of malefactor's intentions partially ordered in time is called an *attack scenario.* Intentions constituting attack scenario can be represented at different generalization levels. At the lowest level, each such intention is realized by a malefactor as a sequence of actions (network packets, commands of OS, etc.). Any malefactor's intention can be realized in multiple ways. Malefactor can vary the scenario implementing the same intention and the same attack object.

3. *Attack modeling corresponds to an adversary domain.* Attack development depends on the result of each particular step of attack, i.e. it depends on response of the attacked network. In turn, a network response depends on security policy implemented. The current attack "state" is determined by initial malefactor's information about the attacked network (or host), by information collected at preceding attack steps, and also by the successfulness or ineffectiveness of the preceding steps.

Thus, any attack development depends on many random factors and, first of all, depends on attacked network response. Therefore, even if a general malefactor's intention is determined, *the attack development scenario cannot be definitely specified beforehand.* An attack development depends on many uncertainties: (1) uncertainty in choice of the attack intention and attack object; (2) uncertainty caused by the information content with regard to the attacked network which a malefactor possesses at the beginning of attack and in progress of its development; (3) uncertainty of choice of attack scenario implementing the already selected intention; (4) uncertainty of the attacked computer network response. Let us describe conceptually the *scheme of attack generation (simulation).*

Selection of the attack intention and attack object is a subjective act. Let the list $\boldsymbol{X} = \{X_1, X_2, \ldots, X_N\}$ of possible attack intentions and the list $\boldsymbol{Y} = \{Y_1, Y_2, \ldots, Y_M\}$ of attack objects be given. To select some attack intention and an attack object, it is necessary to set some formal mechanism of choice, for example, randomization mechanism. Let an intention $X \in \boldsymbol{X}$ and an attack object $Y \in \boldsymbol{Y}$ be selected.

The next component of attack modeling is a mechanism for generation of the attack given upper-level intention X and attack object Y in the terms of hierarchy of lower-level malefactor's intentions and respective sequences of actions. Let us demonstrate the basic idea of such mechanism by example. Let us suppose that the malefactor's intention X consists in getting access to files of some directory of a host. If malefactor does not possess some basic information about computer network or host then he/she has to start from reconnaissance R, which corresponds to the first intention at the level that is lower with regard to the intention X of the top level. The reconnaissance R can be fulfilled, for example, by four different sub-attacks $\{A, B, C, D\}$. Only one of them can be selected on current step of the attack development as a sub-goal (intention) of the second level. We admit, that the malefactor has selected sub-goal C. Another malefactor in the same situation could make other decision. Therefore, it is quite reasonable to specify the above selection as a randomized step. Thus, generation of an attack in terms of lower level intentions given upper-level intention X and attack object Y can be formalized on the basis of randomization of selection among $\{A, B, C, D\}$.

Let the selected sub-goal C be to be realized as a sequence of "commands", first of which be the command a_1. The term "command" is used here in the generalized sense. Main difference between "command" and "intention" consists in the following. The command is a concrete action; it is not a mental concept, which represents a certain abstraction in malefactor's mind. It can be a sequence of IP-packages, a command of operating system, etc. An intention is a component of the plan of actions; it is an "abstraction" represented formally at respective level of detail. In other words, malefactor "thinks and plans" in terms of intentions but acts in terms of commands.

A set of sequences of commands, by which the malefactor tries to realize his /her intention, can be selected ambiguously. Therefore it is necessary to set a non-deterministic mechanism for generation of sequences of commands. It is obvious, that it can be randomization mechanism, however, probably, not so simple, as the random- number generator with a discrete distribution. Let a_1 be the first generated command. This command is dispatched to the attacked computer network (host). The hierarchy <attack intention X, attack target Y> $\rightarrow$ <lower level intentions> $\rightarrow$ <actions>, corresponding to the considered example scheme of the initial phase of the attack generation is shown in tree-like form in Fig. 1.

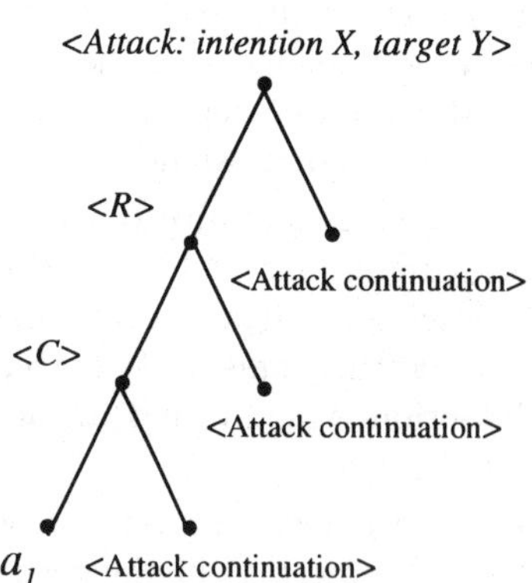

Fig. 1. Attack scheme

The formally determined process of choice can be represented as follows: *<Attack: intention X, target Y>* $\rightarrow$ *<R> <Attack continuation, detailing X>*, *<R>* $\rightarrow$ *<C> <Attack continuation, detailing R>*, *<C>* $\rightarrow a_1$ *<Attack continuation, detailing C>*. The response of the attacked system to each command can be characterized as "success" if the command is executed like the malefactor wanted, or "failure", if the attacked system reacts to the command in the way that is undesirable for the malefactor. The following commands are depend on response of the attacked object to the command a_1.

If the chosen intention C is failed then the attack modeling and simulation process can be stopped ("the attack is over"), or the attack can be continued starting with reselection of the choice associated with specialization of intention R — predecessor of the failed intention C in the tree (see Fig. 1) in terms of the rest of the set $\{A, B, C, D\}$, i.e. in terms of one of the lower level intentions

$\{A, B, D\}$. In the last case, the choice of a new alternative for the intention R specialization is made with the respective recalculation of the probability distribution given over the truncated set of lower-level intentions.

The next step of attack generation is similar to the previous one. If in the following steps no one of intentions A, B and D does not result in success then the attack can be either finished or continued with the subsequent modification of the attack object. It is worth to notice, that in both above cases the probability distribution given over the set of the potentially admissible next step selections of intention alternatives should be recalculated.

If the attack with intention C is successful, then the attack can be stopped (if the goal is reached), or can be continued. This choice is also non-deterministic and can be simulated by a probabilistic mechanism and so on and so forth.

To an arbitrary step n of the attack generation (simulation) its state can be specified by a sequence of the following sort: $A(n)$ = *<Attack prehistory>* *<Current state>* *<Attack continuation>*, where *<Attack prehistory>* is a sequence of the symbols corresponding to the preceding steps, in which each symbol is marked with a flag from a set {"success", "failure"}. This sequence can include symbols of intentions of different levels of detail, and symbols of actions. It is supposed, that the attack can be simulated at various levels of detail of the description; *<Current state>* is a partially unfolded sequence of the current attack step symbols; *<Attack continuation>* is still unknown part of the sequence $A(n)$, which generation is expected. In addition, current state of the attack development can also contain information collected at preceding steps.

It should be clear that it is impossible to enumerate and to specify all sequences $A(n)$, i.e. to specify completely in declarative form the total set of attacks and variants of their development mapped to total set variants of the attacked network responses. Therefore, the only way to specify attacks if it exists at all, is procedural way, which suppose to model attack by a generation algorithm. This way is used in this research.

While describing the developed model, let us start with terminology that actually corresponds to the basic notions that will be structured and formalized below as the domain ontology. In the developed formal model of attacks, the basic notions of the domain correspond to malefactor's intentions and all other notions are structured according to the structure of intentions. This is a reason why the developed approach is referred to as "*intention-centric approach*".

The following *basic classes of high-lever malefactor's* intentions and their identifiers are used in the developed formal model:

1. R — Reconnaissance aiming at getting information about the network (host). The followings are the particular cases of intentions of this class: IH — Identification of the running Hosts; IS — Identification of the host Services; IO — Identification of the host Operating system; CI — Collection of additional Information about the network; RE — shared Resource Enumeration; UE — Users and groups Enumeration; ABE — Applications and Banners Enumeration.

2. *I* — Implantation and threat realization. The followings are its lower level variants of its specialization: *GAR* — Getting Access to Resources of the host; *EP* — Escalating Privilege with regard to the host resources; *GAD* — Gaining Additional Data needed for further threat realization; *TR* — Threat Realization, *TR* can be detailed at the lower level in the following terms: *CVR* — Confidentiality destruction (Confidentiality Violation Realization), for example, through getting access to file reading, *IVR* — Integrity Destruction (Integrity Violation Realization) realizing through attacks against integrity of the host resources, *AVR* — Denial of Service (Availability Violation Realization); *CT* — Covering Tracks to avoid detection of malefactors' presence, *CBD* — Creating Back Doors.

An attack task specification (or a top-level attack goal) can be specified by the following quad: *<Network (host) address, Malefactor's intention, Known data, Attack object>*[1]. The task specification has to determine the class of scenarios that lead to the intended result. *Known data* specifies the information about attacked computer network (host) known for a malefactor. *Attack object* corresponds to the optional variable in attack goal specification and are specified in the following ways:

1. "_" — the attack object is not specified for the malefactor's intention "Reconnaissance" (*R*);
2. If the intention corresponds to the attacks like *CVR* or *IVR* then the attack object is specified as follows: [*Account*,] [*Process* {*<Process name>*/*<Process mask>*},] [*File* {*<file name>*/*<file mask>*},] [*Data in transit* {*<file (data) name>*/*<file (data) mask>*}], where *Account* is object's account, *Process* is running process(es), *File* is file(s) that is the attack target(s) to get, *Data in transit* is data transmitting, where the variables in [ ] are optional, the repeatable variables are placed in {}, and symbol "/" is interpreted as "OR";
3. "*All*" — all resources of the host (network);
4. "*Anyone*" — at least one of the resources of the host (network).

3 "Computer Network Attacks" Ontology

It is well known that the development of a model of an information system must start with the development of the domain ontology. The ontology is the set of notions structured in terms of relationships existing over them. The ontology has to be abstracted from specifics of the implementation issues. A peculiarity of the particular domain is reflected in data structures and algorithms interpreting ontology notions and relationships. Although at present a lot of work is being performed in order to develop ontologies [33], but there are no such works in the network attacks domain.

The developed ontology comprises a hierarchy of notions specifying activities of malefactors directed to implementation of attacks of various classes in different levels of detail. In this ontology, the hierarchy of nodes representing notions

[1] In the software tool this quad is used for specification of simulation task by user.

splits into two subsets according to the *macro-* and *micro-levels* of the domain specifications. All nodes of the ontology of attacks at the macro- and micro-levels of specification are divided into the *intermediate* (detailable) and *terminal* (non-detailable).

The notions of the ontology of an upper level can be interconnected with the corresponding notions of the lower level through one through three kinds of *relationships*: (1) "*Part of*" that is decomposition relationship (*"Whole"–"Part"*); (2) "*Kind of*" that is specialization relationship (*"Notion"–"Particular kind of notion"*); and (3) "*Seq of*" that is relationship specifying sequence of operation (*"Whole operation"–"Sub-operation"*). High-level notions corresponding to the intentions form the upper levels of the ontology. They are interconnected by the "*Part of*" relationship. Attack actions realizing malefactor's intentions are interconnected with the intentions by "*Kind of*" or "*Seq of*" relationship. The developed ontology includes the detailed description of the network attack domain in which the notions of the bottom level ("*terminals*") can be specified in terms of network packets, OS calls, and audit data.

Let us look at a high-level fragment of the developed ontology (Fig. 2). At the upper-level of the *macro-specification of attacks*, the notion of "Network Attack" (designated by A) is in the "*Part of*" relationship to the "Reconnaissance" (R) and "Implantation and threat realization" (I). In turn, the notion R is in the "*Part of*" relationship to the notions IH, IS, IO, CI, RE, UE, and ABE. The notion I is in the "*Part of*" relationship to the notions GAR, EP, GAD, TR, CT, and CBD. In the next (lower) level of the hierarchy of the problem domain ontology, for example, the notion IH is in the "*Kind of*" relationship to the notions "Network Ping Sweeps" (DC) and "Port Scanning" ($SPIH$). At that, the notion "Network Ping Sweeps" (DC) is the lowest ("terminal") notion of

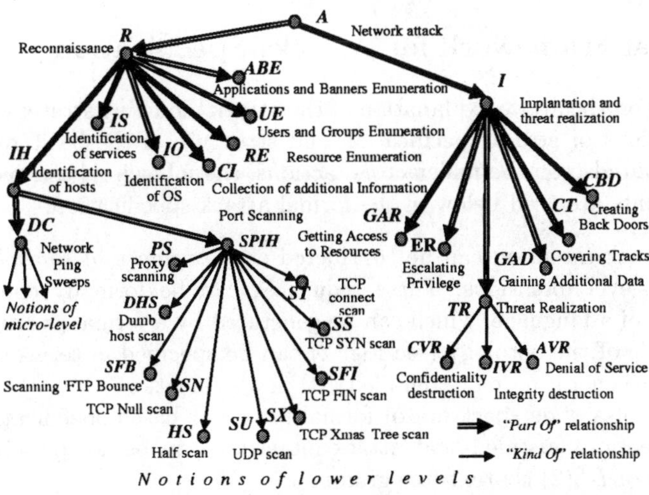

Fig. 2. Macro-level fragment of the domain ontology "Computer network attacks"

the macro-level of attack specification, and the notion "Port Scanning" ($SPIH$) is detailed through the use of the "*Kind of*" relationship by a set of "terminal" notions of the macro-level of attack specification.

The "terminal" notions of the macro-level are further detailed at the *micro-level of attack specification*, and on this level they belong to the set of top-level notions detailed through the use of the three relationships introduced above. Thus, for example, the notion "Network Ping Sweeps" (DC) is in the "*Kind of*" relationship with the notions "Network Ping Sweeps with *ping*" (PI), "Network Ping Sweeps with *Ping Sweep*" (PSW), etc., which, in turn, correspond to the names of utilities that perform "Network Ping Sweeps".

In turn, each of these notions, e.g. "Network Ping Sweeps with *Ping Sweep*" (PSW), is in the "*Seq of*" relationship to the "ICMP ECHO REQUEST" (IER) notions. The "ICMP ECHO REQUEST" (IER) notions correspond to network packets that are directed at the host (or the network) — the target of the attack.

In micro specifications of the attacks ontology, besides the three relations described ("*Part of*", "*Kind of*", "*Seq of*"), the relationship "*Example of*" is also used. It serves to establish the "*type of object — specific sample of object*" relationship. In Fig. 3, this type of relationship is used to establish the connection between the echo-request of the protocol ICMP ("ICMP ECHO REQUEST") and its specific implementation specified, for example, as a message **<time> <src_addr> > <dest_addr>: icmp: echo request**, where **<time>** — time stamp, **<src_addr>** — source IP address, **<src_port>** — source port, **<dest_addr>** — destination IP address.

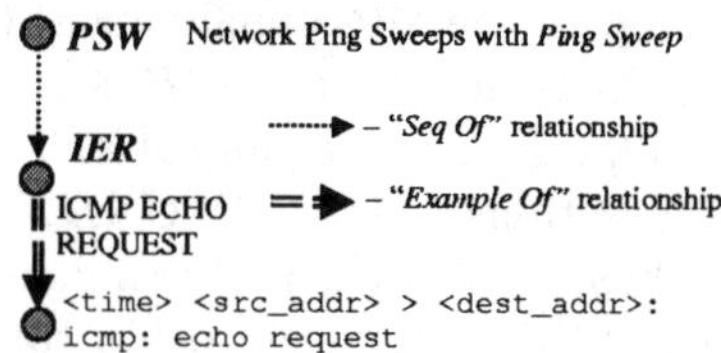

Fig. 3. Micro-specifications of the domain ontology "Computer network attacks"

4 Formal Framework for Attacks Specification

Being based on the above explanation of the attack modeling strategy, definition of basic notions of attack specification and structure of the basic malefactors' intentions and also on the malefactors' actions, the following basic assumptions and statements are used below in the formal attack specification:

1. Each attack intention can be considered as a *sequence of symbols* in terms of lower-level intentions. These sequences can be formally considered as "words" of a language, which can be generated by a formal grammar. Thus, each node of the ontology (see Fig. 2) can be specified in terms of a formal grammar generating more detailed attack specification;
2. Analysis of a wide spectrum of formal grammar-based specifications of attack intentions justified that attack intentions can be adequately specified in terms of $LL(2)$ context-free grammar;
3. Specification of uncertainties inherent to the attack development can be done in probabilistic terms through attributes and functions given over them.

Thus, in conjunction with the above conclusions the resulting framework for attack specification can be restricted to a stochastic attribute grammar;

4. Each node (grammar) of the ontology (like shown in Fig. 2) is interconnected with the upper level node (grammar) and this interconnection can be specified through "grammar substitution" operation [15] in which a terminal symbol of the parent node is considered as the axiom of the grammar corresponding to its child node.
5. Each malefactor's action has to be followed by an attacked network response.

The rest of this section presents the above summary in formal terms.

Thus, mathematical model of attack intentions is determined in terms of a set of *formal grammars* specifying particular intentions interconnected through "*substitution*" operations: $M_A =< \{G_i\}, \{Su\} >$, where $\{G_i\}$ — the formal grammars, $\{Su\}$ — the "substitution" operations.

Every formal grammar is specified by quintuple $G =< V_N, V_T, S, P, A >$, where G is the grammar name, V_N is the set of non-terminal symbols (that are associated with the upper and the intermediate levels of an attack scenario), V_T is the set of its terminal symbols (that designate the steps of a lower-level attack scenario), $S \in V_N$ is the grammar axiom (an initial symbol of an attack scenario), P is the set of productions that specify the specialization operations for the intention through the substitution of the symbols of an upper-level node by the symbols of the lower-level nodes, and A is the set of attributes and algorithms of their computation.

Attribute component of each grammar serves for several purposes. The first of them is to specify *randomized choice of a production* at the current inference step if several productions have the equal left part non-terminals coinciding with the "active" non-terminal in the current sequence under inference. These probabilities are recalculated on-line subject to the prehistory of attack development and previous results of attack. So, in order to specify a stochastic grammar, each production is supplemented with a specification of the probability of the rule being chosen in the inference process.

Also the attribute component is used to check *conditions determining the admissibility of using a production* at the current step of inference. These conditions depend on attack task specification, attacked computer network (host) response and also on the malefactor's previous actions. These conditions may depend on compatibility of malefactor's actions and attacked network or host properties, e.g., OS type and version, running services, security parameters, etc.

These are the examples of host parameters, which may form production conditions: (1) OS types — *Unix, Linux, Win* (all Windows OS), *9x* (*95, 98, Me*), *NT* (*NT, 2000*), *SunOS, Solaris,* etc.; (2) running applications — e.g., *PWS* — an initial version of Microsoft's Personal Web Server is running; (3) protection parameter — *CFP* (shared files and printers), *NS* (Null Sessions), *PA* (Password is Absent), *RR* (Remote Registry), etc.; (4) additional parameters — *AS* (Access to Segment of LAN), *THD* (Trusted Host Data), etc.

If it is necessary to specify several parameters, operations "OR" (signified by ",") and (or) "AND" (".") are used. Relationships of ownership and membership are also taken into account, e.g. *SunOS* $\in$ *Unix*; {*95, 98, Me*} $\subset$ *9x*; {*95, 98, Me, NT, 2000*} $\subset$ *Win*, *9x* $\in$ *Win*, etc.

Thus, in general case, the grammar production is recorded as follows: $[(U)]X \rightarrow \alpha$ *(Prob)*, where U — the condition for upholding the rule, [ ] — an optional element, X — non-terminal symbol, α — a string of terminal and non-terminal symbols, *Prob* — the initial probability of the rule.

Let us explain by example the operation of *grammar substitution* and its role in the formal model of attacks. Let $a \in V_T(G_i)$ be a terminal symbol of the grammar G_i in the sequence of symbols generated by the grammar G_i, a is a node of the ontology mapped to the grammar $G(a)$. Symbol a denotes the name of a particular intention or attack action and $G(a)$ is the grammar generating variants of the a implementation. Let also X be the axiom of the grammar $G(a)$. Then, *operation* $Su(a)$ *of substitution* $G(a)$ in place of symbol a is specified in the form $Su(a) : \{a \rightarrow G(a)\}$. *Semantics of this operation* is that in place of symbol a in already generated sequence any "word" generated by grammar $G(a)$ can be placed. In fact, this operation corresponds to a step towards the more detailed specification of an attack scenario.

When the micro specifications are used for modeling of attacks, it is necessary to use the ontology nodes of the lowest (terminal) level and substitute specific values for the variables that determine the attack task specification.

For example, let us suppose a ping attack is being implemented using "Network Ping Sweeps with *Ping Sweep*" (*PSW*). *PSW* is in the "*Seq of*" relationship to the "ICMP ECHO REQUEST" (*IER*) network packets that are directed at the target host (network). In micro specifications of attacks the *IER* node is in the "*Example of*" relationship to its specific implementation defined as the following message: <time> <src_addr> > <dest_addr>: icmp: echo request, where <time> — time stamp, <src_addr> — source IP address, <src_port> — source port, <dest_addr> — destination IP address.

The grammar that specifies *PSW* may look like this: $V_N = \{PSW,\ PSW1\}$, $V_T = \{IER\}$, $S = \{PSW\}$, $P = \{PSW \rightarrow IER\ PSW1$ (1), $PSW1 \rightarrow IER\ PSW1$ (0.2), $PSW1 \rightarrow IER$ (0.8)$\}$.

Let us suppose a ping attack with "*Ping Sweep*" is being implemented from host 244.146.4.20 on the hosts of the network 198.24.15.0 in the time interval [0:43:10.094644, 00:43:16.036735]. Let us suppose that the string "*IER IER*" was created as a result of using the *PSW* grammar. Then, based on the "*Example of*" relationship, the symbols of this string should generate two messages:

<time1> <src_addr> > <dest_addr>: icmp: echo request,
<time2> <src_addr> > <dest_addr>: icmp: echo request.

After the parameterization <time1> = 00:43:10.094644, <src_addr>= 244.146.4.20, <dest_addr> = 198.24.15.255, <time2>= 00:43:16.036735, these messages should look like these:
00:43:10.094644 244.146.4.20>198.24.15.255:icmp:echo request and
00:43:16.036735 244.146.4.20>198.24.15.255:icmp:echo request,
which correspond to the *icmp*-packets sent to the network hosts 198.24.15.0

(since the X.X.X.255 address is specified in the *icmp*-packets, the packets are sent to all the hosts of the specified networks).

The development of the family of grammars $\{Gi\}$ is conducted in the following order: (1) First, for each basic malefactor's intention, its own family of enclosed attributed stochastic context-free grammars is constructed; (2) Second, these families of grammars are transformed into the generalized grammars that correspond to each non-terminal node of ontology for all of the intentions.

It is assumed that if a value of the production condition is not determined at the moment of production selection all available productions may be used at the respective step of attack simulation. Also it is supposed that the terminal actions generated by productions are associated with the probabilities of successful realization of those actions (attacks) and the host response.

Let us consider, for example, the *grammars* for the intention "Users and groups Enumeration" (UE):

Level "Network Attack": V_N = {A, A1}, V_T = {R}, S = {A}, P = {A → A1 (1), A1 → R (0.7), A1 → R A1 (0.3)};

Level "Reconnaissance": V_N = {R, R1}, V_T = {UE}, S = {R}, P = {R → R1 (1), R1 → UE (0.7), R1 → UE R1 (0.3)};

Level "Users and groups Enumeration"
V_N = {UE, UE1, UE2, UE3, UE4}, S = {UE},
V_T = {DNNT, EUE, PIUD, IAUS, SNMPE, FUE, UTFTP},
$P_{for\ Windows\ 9x,Me,NT,2000}$ = {(Win) UE → UE1(1), (NS)UE1 → UE2(0.65),
UE1 → SNMPE(0.25), UE1 → SNMPE UE1 (0.05), (NS)UE1 → UE2 UE1 (0.05), (&)UE2 → CNS UE3(1), UE3 → DNNT (0.2), UE3 → DNNT UE4 (0.05), UE3 → IAUS(0.35), UE3 → EUE(0.2), UE3 → PIUD (0.2), UE4 → DNNT UE4(0.1), UE4 → DNNT(0.9)},
$P_{for\ Unix/Linux}$ = {(Unix, Linux) UE → UE1(1), UE1 → FUE(0.3), UE1 → SNMPE(0.2), UE1 → UTFTP(0.1), UE1 → FUE UE1(0.1), UE1 → SNMPE UE1(0.1), UE1 → UTFTP UE1(0.2)};

Level "Identifying Accounts with user2sid/sid2user":
V_N = {IAUS, IAUS1, IAUS2}, V_T = {ISU, IAS}, S = {IAUS},
P = {(NT) IAUS → IAUS1 (1), (&) IAUS1 → ISU IAS (0.8), IAUS1 → IAUS1 IAUS2 (0.2), (&) IAUS2 → ISU IAS (1)}.

In this set of grammars the following denotations are used: A — Network Attack; R — Reconnaissance; UE — Users and groups Enumeration; DNNT — Dumping the NetBIOS Name Table with *nbtstat* and *nbtscan*; EUE — Enumerating Users with *enum*; PIUD — Providing Information about Users with *DumpSec* (*DumpACL*); IAUS — Identifying Accounts with *user2sid/sid2user*; SNMPE — SNMP Enumeration with *snmputil* or *IP Network Browser*; FUE — Finger Users Enumeration; UTFTP — Use of Trivial File Transfer Protocol for Unix enumerating by stealing */etc/passwd* and (or) */etc/hosts.equiv* and (or) *~/.rhosts*; ISU — Identifying SID with *user2sid*; IAS — Identifying Account with *sid2user* using user's RID; A1, R1, UE1, UE2, UE3, UE4, IAUS1, IAUS2 — auxiliary symbols.

Algorithmic interpretation of the attack generation specified as formal generalized grammars is implemented by a family of state machines. The basic elements of each state machine are states, transition arcs, and explanatory texts for each transition. States of each state machine are divided into three types: first (initial), intermediate, and final (marker of this state is *End*). The initial and intermediate states are the following: non-terminal, those that initiate the operation of the corresponding nested state machines; terminal, those that interact with the host model; auxiliary states. Transition arcs are identified with the productions of grammars. The model of each state machine is set by specifying the following elements: diagram; main parameters; parameters of transitions that determine the stochastic model of the state machine for different relevant intentions; executable scripts; transition conditions.

5 Formal Model of the Attacked Computer Network

The attack development depends on the malefactor's "skill", information regarding network characteristics, which he/she possesses, some other malefactor's attributes [39], security policy of the attacked network, etc. An attack is developing as interactive process, in which the network is reacting on the malefactor's action. Computer network plays the role of the environment for attacker, and therefore its model must be a part of the attack simulation tool.

The peculiarity of any attack is that the malefactor's strategy depends on the results of the intermediate actions. This is the reason why it is not possible to generate the complete sequence of malefactor's actions from the very beginning. The malefactor's action has to be generated on-line in parallel with the getting reaction from the attacked network. The proposed context-free grammar syntax provides the model with this capability. At each particular step of inference, it generates no more than single terminal symbol that can be interpreted by the computer network model as a malefactor's action. The network returns the value of the result (success or failure). The model of attacker receives it and generates the next terminal symbol according to the attack model and depending on the returned result of the previous phase of the attack.

Model of the attacked computer network is represented as quadruple $MA =< M_{CN}, \{M_{Hi}\}, M_P, M_{HR} >$, where M_{CN} is the model of the computer network structure; $\{M_{Hi}\}$ are the models of the host resources; M_P is the model of computation of the attack success probabilities; M_{HR} is the model of the host reaction in response of attack. Let us determine the model M_{CN} of a computer network structure CN as follows: $M_{CN} =< A, P, N, C >$, where A is the network address; P is a family of protocols used (e.g., TCP/IP, FDDI, ATM, IPX, etc.); N is a set $\{CN_i\}$ of sub-networks and/or a set $\{H_i\}$ of hosts of the network CN; C is a set of connections between the sub-networks (hosts) established as a connectivity matrix. If N establishes a set of sub-networks $\{CN_i\}$, then each sub-network CN_i can in turn be specified by the model M_{CNi} (if its structure needs to be developed in detail and if information is available about this structure). Each host H_i is determined as a pair $M_{Hi} =< A, T >$, where A is the host address, T is the host type (e.g., firewall, router, host, etc.).

Models $\{M_{Hi}\}$ of the network host resources serve for representing the host parameters that are important for attack simulation. Let us determine the model of the network host resources as follows: M_{Hi} = <*A*, *M*, *T*, *N*, *D*, *P*, *S*, *DP*, *ASP*, *RA*, *SP*, *SR*, *TH*, etc.>, where *A* — IP-address, *M* — mask of the network address, *T* — type and version of OS, *N* — users' identifiers (IDs), *D* — domain names, *P* — host access passwords, *S* — users' security identifiers (SID), *DP* — domain parameters (domain, names of hosts in the domain, domain controller, related domains), *ASP* — active TCP and UDP ports and services of the hosts, *RA* — running applications, *SP* — security parameters, *SR* — shared resources, *TH* — trusted hosts.

Success or failure of any attack action (corresponding to terminal level of the attack ontology) is determined by means of the model M_P of computation of the attack success probabilities. This model is specified as follows: $M_P = \{R_j^{SPr}\}$, where R_j^{SPr} is a special rule that determines the action success probability depending on the basic parameters of the host (attack target). The rule R_j^{SPr} includes IF and THEN parts. The IF part contains action name and precondition (values of attributes constraining the attack applicability). The THEN part contains value of success probability (SPr). Examples of interpretations of the probability computation rules are as follows:

``If action is 'FF' (Connection on FTP and examination of bin-files in the directory /bin/ls) and OS Type is 'Unix, Linux' and Service is 'FTP' then SP is 0.7";

``If action is 'FCA' (Free Common Access) and OS type is 'Windows 9x' and Security parameter is 'CFP' (shared files and printers) then SP is 0.7".

The result of each attack action is determined according to the model M_{HR} of the host reaction. This model is determined as a set of rules of the host reaction: $M_{HR} = \{R_j^{HR}\}, R_j^{HR}$: *Input* → *Output* [& *Post-Condition*]; where *Input* — the malefactor's activity, *Output* — the host reaction, *Post-Condition* — a change of the host state, & — logical operation "AND", [ ] — optional part of the rule. The Input format: `<Attack name>: <Input message> : <Attack objects> [; <Objects involved in the attack>`. The Output format: `{<Attack success parameter S> [: <Output message>]; {<Attack success parameter F> [: <Output message>]}`. The `Attack Success Parameter` is determined by the success probability of the attack that is associated with the host (attack target) depending on the implemented attack type. The values of attack success parameter are `Success (S)`, and `Failure (F)`. The part of `output message` shown in the < > is taken from the corresponding field of the host (target) parameters. The part of output message shown in quotation marks " " is displayed as a constant line. The *Post-Condition* format: $\{p_1 = P_1, p_2 = P_2, :, p_n = P_n\}$, where p_i — i^{th} parameter of the host (for instance, *SP, SR, TH,* etc.) which value has changed, P_i — the value of ith parameter.

Examples of the host reaction rules:

SFB: Scanning ``FTP Bounce'': Target host; Intermediate host (FTP-server) → {S: <Active ports (services) of a host>; F: ``It was not possible to determine Active ports (services)''};

`IF: ICMP message quoting: Target host → { S: <The type of operating system>; F: ''It was not possible to determine the type of operating system''}.`

6 Implementation of Attack Simulator

The software prototype of the attack simulator has been implemented. Now it is used for validation of the accepted formal framework. It consists of three components: the model of attacker, the model of the attacked computer network and the background traffic generator. Background traffic is formed taking into account the model of the attacked computer network as a set of sessions between hosts of the network. The common traffic generated by integration of streams of data from these components can be an input for IDSs evaluation and learning.

Each of the components of the attack simulator was built as an agent of multi-agent system (MAS). The design and implementation of the attack simulator is being carried out on the basis of MASDK — "Multi-Agent System Development Kit" [16]. All MAS agents generated by MASDK have the same architecture. Differences are reflected in the content of particular agent's data and knowledge bases. Each agent interacts with other agents, environment which is perceived, and, possibly, modified by agents, and user communicating with agents through his interface. *Receiver of input* and *Sender of output messages* perform the respective functions. Messages received are recorded in *Input message buffer*. The order of its processing is managed by *Input message processor*. In addition, this component performs syntax analysis and messages interpretation and extracts the message contents. The component *Database of agent's dialogs* stores for each input message its attributes like identifiers, type of message and its source. If a message supposes to be replied it is mapped the respective output message when it is sent.

Meta-state machine manages the semantic processing of input messages directing it for processing by the respective *State machines*. The basic agent's computations are executed by a set of *State machines*. The selection of scenario and therefore the output result depend on the input message content and inner state of the *State Machine*. In turn, inner state of this Machine depends on prehistory of its performance; in particular, this prehistory is reflected in the state of agent's *Knowledge base* and *Database*. One more state machine called "*Self-activated behaviour machine*" is changing its inner state depending on the state of the data and knowledge bases. In some combinations of these states it can activate functionality independently on input messages or state of the environment. Each agent class is provided with a set of particular message templates according to its functionalities. The developer carries out the specialization procedure with *Editor of message templates*, which, in turn, is a component of MASDK. Communication component of each agent includes also data regarding potential addressees of messages of given template.

The screen indicating generation of the intention "Gaining Access to Resources" *(GAR)* is depicted in Fig. 4. In this screen the attack generation at

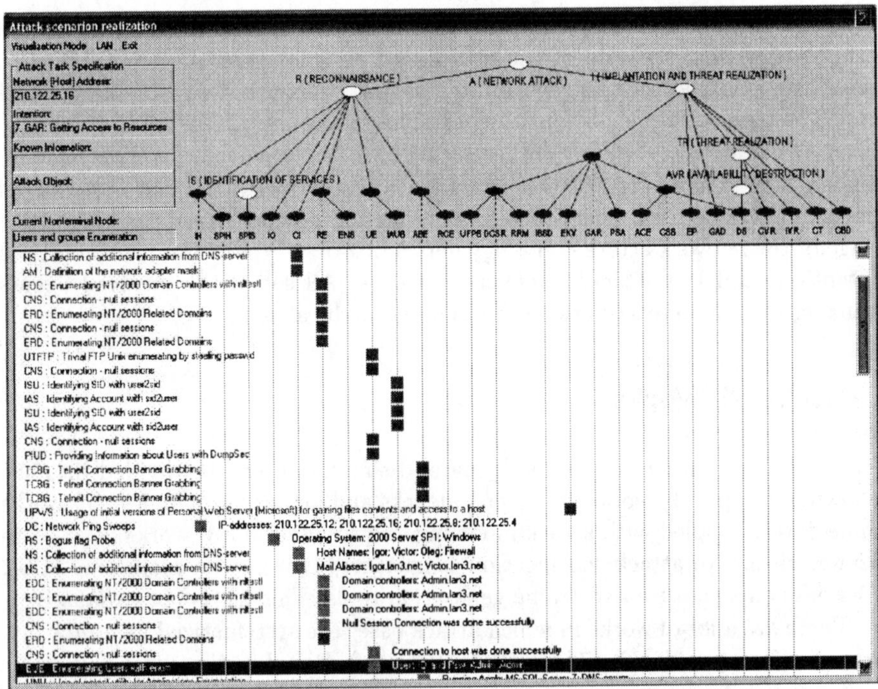

Fig. 4. Visualization of the attack development: the reconnaissance stage of the attack "Gaining Access to Resources" after execution of the action *EUE*

the reconnaissance stage after execution of the action "Enumerating Users with *enum*" *(EUE)* is fixed. In the figure the information is divided on four groups: (1) the attack task specification units are mapped in the left top of the screen; (2) to the right of them the attack generation tree is visualized; (3) the strings of the malefactor' s actions are placed in the left part of the screen below the attack task specification; (4) on the right of each malefactor's action a tag of success (failure) as green (black) quadrate and data obtained from an attacked host (a host response) are depicted.

From implementation issue, a computer network attack can be considered as a sequence of coordinated actions of the spatially distributed malefactors. Each malefactor is mapped as an intelligent agent of the same architecture possessing the similar functionality. While developing an attack, these agents interact via message exchange informing each other about current state and results of the attack in order to coordinate their further activity. These messages are represented in KQML that is standard of DARPA (for message "wrapper"), and XML (for message content).

We are developing[2] a teamwork interpretation of the malefactors' activity performing distributed attacks on the basis of joint intention theory [37]. When

[2] Currently this part of work is in progress.

implementing the complex coordinated attacks, the special meta-agent should form the common scenario of the attack and assigns areas of responsibility to other agents based on the general attack goal constituted by simulation tool user. The agents, responsible for the particular fragments (steps) of the common scenario, can in turn "employ" other agents or realize particular operations independently. For this purpose the special scenarios of operations and protocols of messaging will be used. The concrete scenario and protocol are determined with usage of the network attacks ontology depending on a type of the realizable goal (intention) and the attacked network response. All set of the involved agents realizing the concrete scenario compose a hierarchical structure.

7 Related Works

The works relevant to attack modeling and simulation can be divided into the following groups: (1) works describing attacks and attack taxonomies, (2) works immediately coupled with attack modeling and simulation, (3) works devoted to the description of attack specification languages; (4) works on evaluating IDSs; (5) works on signature and traffic generation tools. This list is not exhaustive.

There are a lot of works in which attack cases are systematized as *attack taxonomies* (for example, [2], [7], [18], [24], [25], etc.). Based on these taxonomies *we built our own taxonomy as an ontology comprising a hierarchy of intentions and actions of malefactors directed to implementation of attacks of various classes split into macro- and micro-levels.*

In different works on *attack modeling and simulation*, as a rule, attack is considered as temporal orderings of actions ([6], [8], [22], [23], [36], etc.). In [22] the state transition analysis technique was developed to model host-based intrusions. A description of an attack has a "safe" starting state, zero or more intermediate states, and (at least) one "compromised" ending state. States are characterized by means of assertions describing different aspects of the security state. The work [6] presents an approach to simulate intrusions in sequential and parallelized forms. The paper [23] suggests formal models of both network and attacks and extends the state transition analysis technique to network-based intrusion detection in order to represent attack scenarios in networks. In [8] a simple network security model "Cause-Effect Model of Information System Attacks and Defenses" was proposed. It is composed of network model represented by node and link, cause-effect model, characteristic functions, and pseudo-random number generator. In ([39], [40]) the descriptive models of the network and the attacker's capabilities, intentions, and courses-of-action are described. These models are used to identify the devices most likely to be compromised. Principles from economics are used to predict the attacker's behavior. Conceptual models of computer penetration were presented in [36]. The paper compares the traditional and "new" attack paradigms. Traditional attack paradigm includes phases of "information gathering", "exploitation", and "metastasis". The metastasis phase of the attack can be logically divided into sub-phases of "consolidation" and "continuation". The core of the new distributed metastasis methodology is a desire to utilize the distributed nature of network environment, and to perform an automation of the metastasis phase using a distributed agent-based approach. *We used in our formal model the temporal orderings of actions and proposed multi-agent teamwork-based approach for modeling of coordinated distributed attacks.*

In ([19], [32] and some other) attacks are modeled in a structured and reusable *"tree"-based form.* In [19] a high-level conceptual model of attack based on the intruder's intention (attack strategy) is presented. The paper determines intrusion intention as the goal-tree: the root node corresponds to ultimate goal of attack, and lower level nodes represent alternatives or ordered sub-goals in achieving the upper node goal. In [32] means for documenting attacks in a form of attack trees are described. Each attack tree enumerates and elaborates the ways that an attacker could cause the event to occur. Two structures are used for attack representation: an attack pattern (characterizing an individual type of attack), and an attack profile (organizing attack patterns to make it easier to apply them). As in [19] and [32] *we apply intension- and tree-based attack strategy representation, but "go further" using for node decomposition a formal framework based on context-free grammars implemented in terms of state machines.*

A model to evaluate survivability of networked systems after network incidents was developed in [31]. The model consists of three sub-models: the first one simulates the occurrence of incidents, the second one evaluates the impact of an attack on the system, and the third one assesses the survivability of the system. The model of incidents is determined as a marked stochastic process, where the incidents are the events that occur at random points in time, and the event type is the mark associated with an incident. *Besides attack generation model, our approach includes also the model of attacked computer network that evaluates the impact of an attack on the network hosts and generates reaction of the network. The attacked network is considered as environment that reacts on the malefactors' actions. The variance of attacks is ensured by the random choice of the grammar productions (or, what is the same, the state machine transition rules). The peculiarity of any attack is that the malefactor's strategy depends on the results of the intermediate actions.*

The paper [5] describes the cyber attack modeling and simulation methodology based on SES/MB (system entity structure and model base) framework and Discrete Event Simulation (DEVS) formalism. This simulation methodology allows classifying threats, specifying attack mechanisms, verifying protection mechanisms, and evaluating consequences. *Our approach has resembling purposes, but it uses stochastic formal-grammar-based specification of the malefactor's intentions and scenarios of network attacks.*

For attack fixing, reproduction, analysis, recognition, response, documenting, special *attack languages* are used [38]: event languages ([3], etc.), exploit languages ([4], etc.), reporting languages ([10], [13], etc.), detection languages ([12], [29], [34], etc.), correlation languages, response languages, and integrated languages ([9], [30], etc.). *Our formal language is related mostly to the exploit and event languages, because it is used to describe attack stages and the for-*

mat of events generated. Our attack representation language includes parts used for description of attack preconditions, attack intentions and actions, formats of actions of terminal level, and postconditions (states of the attacked hosts).

In all works devoted to the *evaluation of IDSs* the attack simulation issues are considered. In [35] a methodology and software tools for testing IDSs using scripts to generate both background traffic and intrusions are described. In evaluations performed by the Lincoln Laboratory at MIT ([26], etc.), investigators were given sniffed network traffic, audit data, and file-system snapshots. The report [11] discusses issues associated with the generation of suitable background traffic; attacks are obtained from a vulnerability database. In ([2], [27]) it was marked that benchmarking IDSs is not generic and systematic enough for evaluation needs. In [2] another approach is investigated. It consists of comparing and evaluating IDSs at the level of their specification rather than at the level of their implementation. *Our approach also presumes that IDSs can be evaluated and verified at different phases of their development and implementation. The more detailed level of attack representation is used in the attack model the more advanced level of IDS is evaluated.*

Now there are a lot of signature and traffic generation tools: FlameThrower, Fragrouter, Hailstorm, IDS Informer, MS WCAT, nidsbench, SmartBits, Stick, etc. But the majority of these tools are doing only simulated pseudorandom malicious packets. As Marcus Ranum noticed in discussions on focus-idssecurity-focus.com "Make sure that you're not only generating "signatures" but that they are within the context of apparently valid sessions - otherwise you're actually benchmarking an IDS' ability to detect false positives, not real attacks." On our sight Hailstorm (Cenzic) [17] and IDS Informer (BLADE Software) [21] have the most interesting properties. *Hailstorm* generates traffic based on patterns specifying how a packet is to be generated over the network. *IDS Informer* has been designed to allow launch S.A.F.E. (Simulated Attacks for Evaluation) attacks. The S.A.F.E. process builds the attacks based on previously recorded real attacks. *In our approach the malicious and background traffic on the terminal levels is formed within the context of valid sessions.*

8 Conclusion

In the paper, attack is considered as competition of malefactor(s) and computer network security system, i.e. attack-modeling task is considered as an adversary domain. We proposed a formal grammar-based framework for attack modeling and considered the basic issues of the attack simulator development. Formal framework is built as a hierarchy of attribute stochastic context-free grammars interconnected via the "grammar substitution" operation. The framework makes it possible to specify and to simulate a wide spectrum of attacks at various levels of detail. Software prototype of the attack simulator is developed. The attack simulator allows simulating a wide spectrum of real-life attacks. The respective software code is written using Visual C++ 6.0, Java 2 version 1.3.1, KQML and

XML languages. The basic ideas of the modeling and simulation of coordinated distributed attacks are developed.

Acknowledgments. This research is being funded by European Office of Aerospace R&D (Projects #1994 P) and grant #01-01-108 of Russian Foundation of Basic Research.

References

1. Aho, A.V., Ullman, J.D.: The Theory of Parsing, Translation, and Compiling, Vol. 1, 2, Prentice-Hall, Inc. (1972)
2. Alessandri, D., Cachin, C., Dacier, M., Deak, O., Julisch, K., Randell, B., Riordan, J., Tscharner, A., Wespi, A., Wuest, C.: Towards a Taxonomy of Intrusion Detection Systems and Attacks. MAFTIA deliverable D3. Version 1.01. Project IST-1999-11583. Malicious- and Accidental-Fault Tolerance for Internet Applications (2001)
3. Bishop, M.: A standard audit trail format. Technical report, Department of Computer Science, University of California at Davis (1995)
4. Custom Attack Simulation Language (CASL). Secure Networks (1998)
5. Chi, S.-D., Park, J.S., Jung, K.-C., Lee, J.-S.: Network Security Modeling and Cyber Attack Simulation Methodology. Lecture Notes in Computer Science, Vol.2119 (2001)
6. Chung, M., Mukherjee, B., Olsson, R.A., Puketza, N.: Simulating Concurrent Intrusions for Testing Intrusion Detection Systems: Parallelizing Intrusions. Proceedings of the 18th NISSC (1995)
7. Cohen, F.B.: Information System Attacks: A Preliminary Classification Scheme. Computers and Security, Vol.16, No.1 (1997)
8. Cohen, F.: Simulating Cyber Attacks, Defenses, and Consequences. IEEE Symposium on Security and Privacy, Berkeley, CA (1999)
9. Cuppens, F., Ortalo, R.: Lambda: A language to model a database for detection of attacks. RAID'2000, Lecture Notes in Computer Science, Vol.1907 (2000)
10. Curry, D.: Intrusion detection message exchange format, extensible markup language (xml) document type definition. draft-ietf-idwg-idmef-xml-02.txt (2000)
11. Debar, H., Dacier, M., Wespi, A., Lampart, S.: An experimentation workbench for intrusion detection systems. Research Report RZ-2998 (# 93044). IBM Research Division, Zurich Research Laboratory (1998)
12. Eckmann, S.T., Vigna, G., Kemmerer, R.A.: STATL: An Attack Language for State-based Intrusion Detection. Proceedings of the ACM Workshop on Intrusion Detection, Athens, Greece (2000)
13. Feiertag, R., Kahn, C., Porras, P., Schnackenberg, D., Staniford-Chen, S., Tung, B.: A common intrusion specification language (cisl). specification draft (1999)
14. Fu, K.S.: Syntactic Methods in Pattern Recognition, Academic Press, New York (1974)
15. Glushkov, V., Tseitlin, G., Yustchenko, E.: Algebra, Languages, Programming. Naukova Dumka Publishers, Kiev (1978) (In Russian).
16. Gorodetski, V., Karsayev, O., Kotenko, I., Khabalov, A.: Software Development Kit for Multi-agent Systems Design and Implementation. Lecture Notes in Artificial Intelligence, Vol. 2296, Springer Verlag (2002)

17. Hailstorm. Users Manual, 1.0 (2000) http://www.clicktosecure.com/
18. Howard, J.D., Longstaff, T.A.: A Common Language for Computer Security Incidents, SANDIA REPORT, SAND98-8667 (1998)
19. Huang, M.-Y., Wicks, T.M.: A Large-scale Distributed Intrusion Detection Framework Based on Attack Strategy Analysis. RAID'98, Louvain-la-Neuve (1998)
20. Icove, D., Seger K., VonStorch, W.: Computer Crime: A Crimefighter's Handbook, O'Reilly & Associates, Inc., Sebastopol, CA (1995)
21. IDS Informer 3.0. User Guide. BLADE Software (2001)
22. Iglun, K., Kemmerer, R.A., Porras, P.A.: State Transition Analysis: A Rule-Based Intrusion Detection System. IEEE Transactions on Software Engineering, Vol. 21, No.3 (1995)
23. Kemmerer, R.A., Vigna, G.: NetSTAT: A network-based intrusion detection approach. Proceedings of the 14th ACSAC, Scottsdale, Arizona (1998)
24. Krsul, I.V.: Software Vulnerability Analysis, Ph.D. Dissertation, Computer Sciences Department, Purdue University, Lafayette, IN (1998)
25. Lindqvist, U., Jonsson, E.: How to Systematically Classify Computer Security Intrusions. Proceedings of the 1997 IEEE Symposium on Security and Privacy, IEEE Computer Society Press, Los Alamitos, CA (1997)
26. Lippmann, R., Haines, J.W., Fried, D.J., Korba, J., Das, K.: The 1999 DARPA off-line intrusion detection evaluation. RAID'2000, Lecture Notes in Computer Science, Vol.1907 (2000)
27. McHugh, J.: The 1998 Lincoln Laboratory IDS Evaluation: A Critique. RAID'2000, Lecture Notes in Computer Science, Vol.1907 (2000)
28. McHugh, J.: Intrusion and intrusion detection. International Journal of Information Security, No.1 (2001)
29. Me, L.: Gassata, a genetic algorithm as an alternative tool for security audit trails analysis. RAID'98, Louvain-la-Neuve (1998)
30. Michel, C., Me, L.: ADeLe: an Attack Description Language for Knowledge-based Intrusion Detection. Proceedings of the 16th International Conference on Information Security, Kluwer (2001)
31. Moitra, S.D., Konda S.L.: A Simulation Model for Managing Survivability of Networked Information Systems, Technical Report CMU/SEI-2000-TR-020 ESC-TR-2000-020 (2000)
32. Moore, A.P., Ellison, R.J., Linger, R.C.: Attack Modeling for Information Security and Survivability. Technical Note CMU/SEI-2001-TN-001 (2001)
33. http://www.ontology.org/
34. Paxson, V.: Bro: A system for detecting network intruders in real-time. Proceedings of the 7th Usenix Security Symposium (1998)
35. Puketza, N., Chung, M., Olsson, R.A., Mukherjee, B.: A Software Platform for Testing Intrusion Detection Systems. IEEE Software, Vol.14, No.5 (1997)
36. Stewart, A.J.: Distributed Metastasis: A Computer Network Penetration Methodology. The Packet Factory (1999) (Phrack Magazine, Vol. 9, Issue 55)
37. Tambe, M.: Towards Flexible Teamwork. Journal of Artificial Intelligence Research, No.7 (1997)
38. Vigna, G., Eckmann, S.T., Kemmerer, R.A.: Attack Languages. Proceedings of the IEEE Information Survivability Workshop, Boston (2000)
39. Yuill, J., Wu, F., Settle, J., Gong, F., Huang, M.: Intrusion Detection for an On-Going Attack. RAID'99, West Lafayette, Indiana, USA (1999)
40. Yuill, J., Wu, F., Settle, J., Gong, F., Forno, R., Huang, M., Asbery, J.: Intrusion-detection for incident-response, using a military battlefield-intelligence process. Computer Networks, No. 34 (2000)

Capacity Verification for High Speed Network Intrusion Detection Systems

Mike Hall and Kevin Wiley

Cisco Systems, Inc.
12515 Research Blvd., Austin TX 78759 USA
[mlhall, klwiley]@cisco.com

Abstract. Commercially available Network Intrusion Detection Systems (NIDS) came onto the market over six years ago. These systems have gained acceptance as a viable means of monitoring the security of consumer networks, yet no commercial standards exist to help consumers understand the capacity characteristics of these devices. Existing NIDS tests are flawed. These tests resemble the same tests used with other networking equipment, such as switches and routers. However, switches and routers do not conduct the same level of deep packet inspection, nor require the higher-level protocol awareness that a NIDS demands. Therefore, the current testing does not allow consumers to infer any expected performance in their environment. Designing a new set of tests that is specific to the weak areas, or bottlenecks, of a NIDS is the key to discovering metrics meaningful to the consumers. Any consumer of NIDS technology can then examine the metrics used in the tests and profile his network traffic to these same metrics. The consumer can use standard test results to accurately predict performance on his network. This paper proposes a test methodology for standardized capacity benchmarking of NIDS. The test methodology starts with examining the bottlenecks in a NIDS, mapping these bottlenecks to metrics that can be tested, and then exploring some results from tests conducted.

1 Introduction and Scope

There are currently no industry standards for testing any aspect of Network Intrusion Detection Systems (NIDS). The NIDS industry is maturing along the same lines as the routers, switches, and firewalls that came before it, and has now reached the point where standardization of testing and benchmarking is possible. Attempting to define a testing standard is beyond the scope of this paper. Instead, the metrics and methodology used to properly verify the capacity of high speed NIDS are explored. Performance of NIDS is usually defined by false positive and false negative ratios, and speed or capacity. This paper addresses the issue of benchmarking the *capacity* of a NIDS. For the purposes of this paper we use capacity to refer to the ability of a NIDS *to capture, process and perform at the same level of accuracy under a given network load as it does on a quiescent network.*

Gauging the capacity of a NIDS is difficult. There are several variables in the characteristics of the network traffic that affect the performance of a NIDS. In the

A. Wespi, G. Vigna, and L. Deri (Eds.): RAID 2002, LNCS 2516, pp. 239-251, 2002.

last year there have been claims of NIDS performing at or near gigabit speeds. In every case, further investigation by reasonably sophisticated NIDS practitioners revealed critical flaws in the testing methodology.

The variety of technology used to perform network-based intrusion detection further complicates finding the proper metrics. The following technology is used for NIDS:

- Stateless inspection of the packets or packet headers
- Protocol decode and analysis
- Regular expression matching of packet data
- Anomaly detection

Most NIDS employ a mix of all of these methods. Some of the metrics discussed in this paper do not apply to all of the technologies. Choosing metrics and test methods valid for all NIDS in existence is impossible. Choosing a broad set of metrics that is generally applicable to most NIDS is possible. What are the proper metrics for performance testing? What testing methodology best evaluates these metrics? This paper focuses on these two questions.

The testing metrics and methodology described are intended for use on a NIDS located at the edge of an internal network functioning near the firewall or border router. The focus is further refined by looking at how these metrics apply to a NIDS using a combination of the technologies listed above. However, many of the metrics and methods included also apply to the performance of a NIDS inside the core of an enterprise network and to a NIDS employing other methods of detecting intrusions such as pure anomaly-based systems.

2 History

The majority of NIDS capacity benchmarks to date have been run by independent third parties either for publication in a trade magazine or at the request of the vendor for inclusion in marketing material. The test methodologies were developed based on experiences in the router- and switch-testing arenas.

These tests are generally not adequate for the purposes of developing a NIDS performance profile because the benchmark tests for switch and router capacity often forward packets of various sizes without regard for any protocol above IP or even the validity of the packets used. While routers and switches are typically not concerned with layer four and above, NIDS may discard packets that are not interesting. A NIDS also needs to look much deeper in a packet than a switch or a router to follow layer four and above. For example, a NIDS may discard TCP streams that are not opened using a valid three-way handshake. If a switch or router test is used the majority of the traffic might be ignored. The NIDS then performs very little deep packet inspection.

Since the results of a NIDS performance test based on these types of test methodologies are often skewed in the favor of the vendor, a consumer may believe these results are valid for his deployment and encounter strikingly different performance characteristics once the NIDS is fielded on his network.

For example, all the NIDS tests to date from Mier Communications are flawed [1]. Mier Labs concluded that two different NIDS could perform at gigabit lines rates. While the lab report is technically accurate, there is no mention anywhere in the lab report that a test using TCP port 0 packets was not going to be representative of the performance most consumers would experience. Using this type of testing methodology for NIDS is flawed. Marcus Ranum also mentions a few other flawed testing methodologies in "Experiences Benchmarking Intrusion Detection Systems" [2]. Mr. Ranum does an excellent job explaining why benchmarking NIDS is difficult.

3 Defining the Metrics

Defining metrics for any type of testing is a difficult task. For example, the materials used in bridge construction need to be tested to ensure the integrity of the bridge. A common approach to defining metrics for these types of tests involves asking the bridge engineer to identify the weak spots in the bridge design. Where is the most stress concentrated? What has the highest potential for failure when the load exceeds design specifications? The same methodology can be used for defining the metrics used for testing the performance of NIDS. The stress points for most protocol decode and pattern match in a NIDS are the same.

3.1 Fixed Resource Limitations

All computing devices have a similar list of fixed resources. There are only so many cycles available on the CPU and there are only so many bits to store all the program code, state information, and runtime conditions. There also are only so many transmission cycles available on the various buses of the computing architecture. The upper limit for system performance is approached as one or more of these resources approach its upper limit. Therefore, the test methods should include metrics that apply to these resources. The following resources are the most important to the performance of NIDS:

- Memory size
- Memory bus bandwidth
- Memory latency
- Bus bandwidth for the network interfaces (ethernet card)
- Persistent storage bandwidth (hard drive, flash, etc.)
- CPU speed and bandwidth

3.2 Packet Capture Architecture

NIDS products monitor network traffic and NIDS packet capture architectures impose physical limits on what type of traffic can be observed. For example, a NIDS built with a standard gigabit ethernet card cannot observe all minimum-sized ethernet frames sent at line rate. This equates to approximately 1.4 million packets per second, and no currently shipping standard gigabit network adapter can handle many more than 700 to 800 thousand packets per second. However, if the NIDS uses dedicated hardware or some network processing units (NPU), then it is quite possible to handle more than 1.4 million packets. Also, the host software platform for the NIDS can have significant impact on the ability of the NIDS to capture packets. Many NIDS running on a host operating system do not bind the network interface to an operating system's IP stack, and the architectures include custom network interface drivers.

Many of the more recently published NIDS performance tests actually tested only the interface bandwidth of the NIDS. This type of testing has limited use because it only shows the upper limit of how the NIDS performs if no other fixed resources are used. Typically this type of test lets the consumer understand how quickly a NIDS can ignore traffic that is uninteresting. Knowing the performance of only the packet capture architecture is useful, but it does not provide the information needed to quantify the performance of the entire system.

The metrics that affect the performance of the packet capture architecture include packets per second, bytes (or bits) per second, and physical network interface.

3.3 Packet Flow Architecture

Packet flow architecture is the overall architecture for data flow within the NIDS and includes the packet capture architecture. The metrics used in the packet capture architecture section are also valid for the packet flow architecture, assuming the packets used are of interest to the NIDS, cause deep packet inspection, and make proper use of protocols of interest to the NIDS. Using HTTP traffic to test the packet flow architecture is generally a good choice. For a NIDS that employs some method of protocol decode and state aware reassembly, HTTP traffic flows through a major portion of the packet flow architecture.

In addition, not all packets take the same amount of time to process. Buffering in the packet flow architecture allows a NIDS to recover from the packets that take a long time to inspect. Packet buffering is an important feature for reducing the number of dropped packets. Therefore, when testing a NIDS with buffering in the packet flow architecture, it is important to test with sustained rates for a length of time to ensure that the buffer is not inflating performance.

3.4 State Tracking

Any NIDS that performs TCP state tracking, IP fragment reassembly, and detection of sweeps and floods must keep track of the state of the traffic on the network. Many of the signatures used in this type of NIDS are based on a certain threshold of events occurring within a specified period of time. The only way to assess event thresholds is to keep a count until the time has expired or the threshold has been exceeded. A NIDS must track information about both the source and destination of the packets, the state of TCP connections, and the type and number of packets going to or from the hosts. All of this information must be stored somewhere within the NIDS. The storage medium for this information is the *state database*.

Database benchmarking is very mature. The database industry understands the weak points in database-like applications. The largest metrics include the time needed to insert, delete, and search through data and how those transaction times scale with the size of the data set and frequency of transactions. How do we correlate those database metrics to NIDS metrics?

State information must be inserted into the state database as new network hosts and unique connections are observed. The state information is typically removed from the database after either an alarm event has occurred, or some time has elapsed. State database searches are conducted any time the incoming packet may need to refer to prior state information. Depending on the types of signatures used, searching the database may need to be done for each packet.

The size of the state database derives from the unique hosts and connections that the NIDS considers interesting and maintains prior information about. The following metrics directly affect the performance of the state database:

- Number of unique hosts
- Number of new connections over time (i.e., TCP connections per second)
- Number of concurrent connections at any given time
- Efficiency of expiring data in the database

The connection duration in number of packets or time can be used as an indirect metric for testing the performance of the state database because the duration of a session is related to the number of connections over time as well as the number of concurrent connections. Therefore, to accurately measure the capacity of a NIDS, one must vary the number of new connections per second, the number of simultaneous open connections, and the total number of unique hosts that the NIDS must track. The ability of the NIDS to handle network loads varies as these variables are adjusted.

3.5 Packet Analysis

Memory bandwidth and memory latency are large factors in the performance of a NIDS. Much of the memory bandwidth use and latency are caused by access to memory while inspecting the packets. Different NIDS architectures exhibit different use patterns for memory. A NIDS that relies solely on regular expression matching consumes the most bandwidth and induces the most latency in the system. Inspecting each character in the packet payload and advancing a regular expression state are expensive operations.

Protocol analysis helps reduce the number of bytes that must be inspected. Every NIDS does some type of protocol decode, even if it is limited to just the IP header. Many of the commercial NIDS decode most layer seven protocols and only perform regular expression inspection on a subset of the entire packet.

The size of the packets therefore plays an important role in determining the capacity of a NIDS. Testing performance with the smallest possible average packet size reduces the amount of time available per packet for inspections. Increasing the average packet size allows more time for inspection and increases the use of memory.

The average packet size of typical traffic on the Internet is around 450 to 550 bytes per packet [3,4]. However, some networks contain averages much larger and much smaller. The average packet size is an important metric in capacity testing.

3.6 Event or Alarm Reporting

The generation of the alarm event expends CPU cycles that would otherwise be available for analysis. Additionally, the event needs to be stored in non-volatile storage. This usually means that it must be written to disk, which is a relatively slow operation, or sent over a network connection. Under normal circumstances this does not affect the operation of a NIDS. However, as the rate of alarm production increases and/or the load on the network increases, alarm event production and log maintenance can have a significant effect on NIDS performance. The event generation component of a NIDS must be able to handle the events generated by the high rates of traffic. The ability of the NIDS to notify the user varies as the alarm event rate is adjusted.

The metric used to test this component of a NIDS is simply the number of alarms per second. Tools such as stick and nessus easily set off alarm events in NIDS products. In addition, packet generators can be used to generate single packets that cause an alarm event. Testing the alarm channel does not require the traffic causing the alarms to originate from real hosts.

3.7 The Metrics

With the major stress points of a NIDS defined, it is now possible to focus on defining the metrics that can be used to quantify the capacity of a NIDS. Table 1 defines the test metrics and how they are related to the use of the fixed resources described in section 3.1.

Table 1. NIDS test metrics and corresponding resources used

Test Metrics	Resources Used
Packets per second	CPU cycles, network interface bandwidth, and memory bus bandwidth
Bytes per second (Average packet size)	CPU cycles, network interface bandwidth, and memory bus bandwidth
Protocol Mix	CPU cycles, memory bus bandwidth
Number of unique hosts	Memory size, CPU cycles, and memory bus bandwidth
Number of new connections per second	CPU cycles, and memory bus bandwidth
Number of concurrent connections	Memory size, CPU cycles, and memory bus bandwidth
Alarms per second	Memory size, CPU cycles, and memory bus bandwidth

4 Developing the Tests

Developing the tests to quantify the metrics for the potential weak points of a NIDS is non-trivial. This section explores traffic mix selection and simplification, potential problems with a test network, and a set of tests and the intended stress points under test.

4.1 Traffic Mix

Many of the metrics defined in the previous section are directly and indirectly derived from the network traffic. What is the correct mix? The correct mix for each user is one that best matches the traffic where that user plans to deploy the NIDS. Obviously a test cannot be designed that contains all the traffic mixes for all potential consumers of NIDS technology. However, after the NIDS industry agrees on a standard methodology for testing the stress points, consumers could profile their traffic mix and get a reasonable feel for how well the various products perform.

How do we define the tests to be used? There have been studies performed that describe the mix of traffic seen on the major network trunks. "The Nature of the Beast: Recent Traffic Measurements from an Internet Backbone" [3] and "Trends in Wide Area IP Traffic" [4] are two good resources for defining a general test. The information in these papers is from 1998 and 2000. For more recent data Table 2 includes results from profiling three datasets from 2002. A major university, a major U.S. government site, and a large online retailer provided the datasets. Although this data is not necessarily representative of a traffic mix found on a corporate network, it is representative of the mix that would be seen at the edge of most large networks. The metrics in this traffic provide a good starting point for the mix of the loading traffic for the test.

Table 2. Traffic metrics from three customer sites. Site 1 is a major university. Site 2 is a major US government site. Site 3 is an online retailer. In the layer 4 OTHER field, no individual protocol grouped into this field consisted of more than 3% or the total traffic.

	Site 1	Site 2	Site 3
Average Packet Size	543	501	557
Average Bandwidth	25.0Mbps	36.2Mbps	31.4Mbps
Layer 3			
% TCP	94.8%	97.6%	98.7%
% UDP	5.0%	1.2%	1.3%
% ICMP	0.2%	0.1%	0.0%
% OTHER	0.1%	1.1%	0.0%
Layer 4			
TCP connections per second	201	277	118
% HTTP	49.4%	64.6%	61.6%
% SMTP	5.1%	9.4%	9.3%
% NNTP	5.5%	0.0%	0.0%
% OTHER	40.0%	26.0%	29.1%

Table 2 shows the same general traffic characteristics found in the CAIDA data [3,4]. Currently we are unaware of any data sets for networks running at or near gigabit per second speeds. Further research is needed in this area. Due to the limited scope of this paper, it is assumed that the traffic mix will scale evenly with the increased bandwidth.

With a general understanding of the type of traffic found at the edge of protected networks, it is now possible to explore crafting tests that quantify the metrics at a level useful for NIDS consumers.

4.2 HTTP Traffic for Testing

Typically the HTTP protocol is not blocked outbound from firewalls and is a dominant portion of the traffic on the Internet. The servers and clients that implement HTTP have garnered the attention of many crackers and security professionals. HTTP based signatures make up the majority of signatures in NIDS. Fortunately this situation allows for simplification of the testing in the general case. When HTTP traffic is used to test the capacity of a NIDS, it obviously stresses a large portion of the packet flow architecture.

The use of HTTP traffic has a few other advantages as well. HTTP traffic is relatively easy and inexpensive (in time and money) to produce. Web server testing tools, such as ZDNet's WebBench can be used to generate traffic to reproduce tests inexpensively. In addition, there are several vendors selling network test equipment that utilizes a real TCP/IP stack implementation instead of using "canned" traffic. We are most familiar with the products from Caw Networks and Antera. For the test conducted in section 5, we used Caw Networks *WebReflector* and *WebAvalanche* products.

4.3 Generating Real Traffic vs. Replaying Traffic

Most NIDS shipping today perform some level of protocol decode and state tracking. Therefore, it is very important that any load traffic exhibit the same characteristics found on a consumer's network. Most of the products that allow for high speed traffic generation have some critical flaws that make them unsuitable for testing NIDS at high speeds. Some of these issues include:

- Inability to create valid checksums for all layers at high speeds
- Inability to vary the IP addresses in a more random manner than a straight increment
- Inability to maintain state of TCP connections and issue resets if packets are dropped
- Inability to play a large mix of traffic due to the limitations in buffer size for the transmitters

These issues do not plague the replay devices at slower speeds. At high speeds, the buffer size for the replay devices prohibits large traffic samples. Using replay requires the tester to use more replay interfaces. Adding more source interfaces when testing a high aggregate rate presents problems for the test network as described in the next section. Therefore, using test devices that use real TCP/IP implementation to generate the traffic is preferred.

4.4 Inter-packet Arrival Gap on High Speed Test Networks

The typical network setup for testing a gigabit NIDS includes several traffic generators, an attack network, and a victim network. All of these devices are typically connected to a switch, and the traffic is then port-mirrored, spanned, or copy-captured to the NIDS. For high speed tests, the interface for the NIDS is gigabit ethernet. The inter-packet arrival gap on gigabit ethernet is 96-ns. Inter-packet arrival gap becomes important as more interfaces are added to the switch. Each interface, regardless of interface speed, used to generate traffic increases the chances that traffic destined for the NIDS will be dropped at the switch.

Imagine that ten fast ethernet ports each generating around 80-Mbits of traffic are used during testing. Eventually several of these ports begin to transmit packets all within a few nanoseconds of each other. Since each of these transmitted packets must be copied to the NIDS, the switch forwards each of the packets to the port where the NIDS is connected. Unfortunately the NIDS is using Ethernet, which demands the 96-ns delay between packets. Since there are several packets arriving at the port at very nearly the same time, the port buffer gets full and the switch port drops packets. This problem does not manifest itself if there is a choke point such as a firewall or router used in the test network. But, if the industry tests require a router or firewall capable of the same high traffic speeds to reproduce the tests, it raises the cost of testing by a non-trivial amount. *It is therefore better to use fewer ports for generating traffic on the switch when testing.*

4.5 Potential Test Suite

No single test provides all the information needed to quantify the capacity of a NIDS. A suite of tests is used to quantify each portion of a NIDS. Only when looking at the output from all of these tests can a consumer infer performance on his network.

Establishing the Peak. Testing the network interface bandwidth establishes the peak for packet capture for the NIDS. The NIDS is never able to perform above this absolute peak on any further tests. Testing the network interface bandwidth is simple. Choose a packet of no interest to the NIDS and resend it at a high rate until the NIDS cannot count all the packets. Repeat the tests for minimum-sized packets and for maximum-sized packets. This reveals the maximum packets per second and the maximum bytes per second respectively. A good example packet for this test is a UDP packet with ports set to 0 (assuming the NIDS does not send alarms on such a packet).

The Alarm Channel. Testing the alarm channel capacity of a NIDS can be accomplished with a similar test. Choose a packet that causes an alarm. The "Ping of Death" is a good packet for this test. Send this packet at different rates of speed and check packet and alarm counts. Some NIDS buffer alarms when under a heavy load, so a quiescent period after the packets are sent may be necessary before collecting counts.

Stressing the State Database. Inserting, searching, and deleting the state information from the state database are all potential bottlenecks for the NIDS.

Varying the IP addresses of traffic requiring state tracking adds load to the database. Opening a large number of TCP connections causes the state database to contain many records. The search performance for a database is affected by the size of the dataset. For this type of test open a large number of concurrent TCP connections and then run one of the more general tests described below. The open connections will stress both the database and the overall system architecture.

General Tests with Configurable Metrics. Establishing a baseline of traffic and then varying one of the metrics can expose a NIDS weakness in certain environments. Since we are testing each specific component of the NIDS, it is not necessary to ensure the traffic looks exactly like the traffic of the end user. If the user can extract the same metrics from his traffic, then performance on his network can be inferred from test results using simpler data. In the example tests found in section 5, the traffic mix consists of only HTTP. HTTP rides on TCP, which requires state tracking. Depending on the level of protocol decoding used, HTTP may also require state tracking. In addition, HTTP signatures make up the majority of the signatures found in a NIDS. Therefore, using an HTTP-only traffic mix still stresses the NIDS in many areas. The example tests in section 5 could have also included additional protocols such as DNS, SMTP, and NNTP. However, due to time and space constraints these protocols were omitted for this test.

Table 3 shows the characteristics of the traffic mix when using Caw Networks WebReflector and WebAvalanche products. This test equipment allows for high speed testing using only two ports on the switch for generating traffic. The Caw Networks equipment also has the ability to randomly drop packets. The dropped packets cause their systems to retransmit and the traffic looks more like real world traffic. By simply varying the HTTP transaction size, many characteristics of the traffic can be manipulated. Using HTTP transaction size is just one example. The MSS[1] for the server or client can also be varied to affect the characteristics in other ways.

Table 3. Traffic mix characteristics when using Caw Networks WebReflector and WebAvalanche to generate HTTP traffic for general stress tests.

Transaction Size	Packets in Stream	Bytes In Stream	Avg. Packet Size
1000	8	1448	181
5000	13	5640	434
10000	20	10832	542
20000	33	21280	645
40000	60	42112	702
50000	73	52560	720
100000	140	104672	748
240000	326	250624	769
360000	486	375744	773
400000	540	417472	773

[1] Maximum segment size. Used in TCP to specify the maximum amount of TCP data in a single IP datagram that the local system can accept. The MSS is typically the outgoing interface's maximum transmission unit minus 40 bytes for the IP and TCP headers.

5 Example Test Results

Using the lessons from section 4, this section explores two simple tests that evaluate some aspects of the capacity of a NIDS. Other tests are then suggested as a way to further refine the knowledge gained from the example tests.

5.1 Test Network

Figure 1 shows the test network used for the example tests. All links to the switch are using a single 1-Gbps full duplex connection.

Fig. 1. A block diagram showing the network layout for the example tests. The WebReflector acts as all the web servers. The WebAvalanche acts as all the web clients. The catalyst switch is used to copy-capture the traffic to the NIDS.

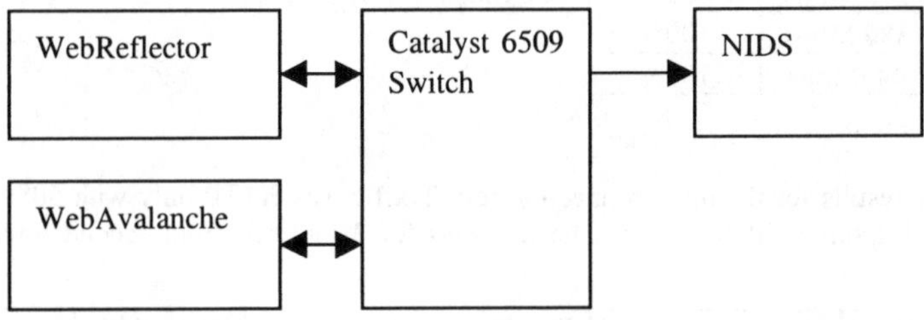

5.2 The Baseline Test

The first test baselines the capture efficiency of a NIDS in a pure HTTP environment. A full analysis load is assumed by turning on all default signatures. However, the traffic generated does not cause alarms or events to be created. The number of client hosts on the network is fixed at 5080, and the number of servers is fixed at two. In this first test, TCP sessions are allowed to run to completion as quickly as possible, therefore the number of simultaneous open sessions is fixed at less than 30 for all cases. The average packet size is varied through manipulation of the HTTP transaction size. This also results in variations in the average number of packets per TCP connection, the packets per second (KPPS), and the overall bandwidth used. The final variable manipulated is the number of new TCP connections per second. Each test is run for three minutes before capacity measurements were made. The results of the baseline test are given below in Table 4.

5.3 Adding Simultaneous Open TCP sessions

The second test introduces simultaneous open TCP sessions. A four-second delay is introduced on the server response to cause sessions to remain open. All other test variables remain constant. The bandwidth consumed, the packets per second, and the average packet size at each data point are somewhat affected by the open connections.

Without performing a correlation study, it is unclear if these factors are statistically significant. For the purposes of this paper, we assume they are not. The results from the open connection test are below in Table 5.

Table 4. The results for the baseline test. Traffic was HTTP only with 5080 client IP addresses and 2 server IP addresses. The test runs for 3 minutes. No server delay.

Avg. Packet Size	Bandwidth @ 1000 cps	Capture Efficiency	Bandwidth @ 2500 cps	Capture Efficiency	Bandwidth @ 5000 cps	Capture Efficiency
434	50 Mbps	100%	125 Mbps	100%	230 Mbps	99.95%
482	67 Mbps	100%	167 Mbps	100%	335 Mbps	97.10%
542	93 Mbps	100%	232 Mbps	100%		
645	180 Mbps	100%	446 Mbps	100%		
688	275 Mbps	100%	680 Mbps	99.30%		
702	380 Mbps	100%				
720	444Mbps	100%				

Table 5. The results for the open connection test. Traffic was HTTP only with 5080 client IP addresses and 2 server IP addresses. The test runs for 3 minutes. Four second forced server delay.

Avg. Packet Size	Bandwidth @ 1000 cps and 4000 open streams	Capture Efficiency	Bandwidth @ 2500 cps and 10000 open streams	Capture Efficiency	Bandwidth @ 5000 cps and 20000 open streams	Capture Efficiency
434	50 Mbps	100%	125 Mbps	100%	245 Mbps	68.75%
482	69 Mbps	100%	170 Mbps	100%	350 Mbps	72.26%
542	93 Mbps	100%	241 Mbps	100%		
645	181 Mbps	100%	446 Mbps	99.95%		
688	268 Mbps	100%	655 Mbps	95.01%		
702	355 Mbps	100%				
720	440Mbps	100%				

5.4 The Results

The most significant variations in capture efficiency are in the 5000 connections per second tests. Capture efficiency appears affected in the 2500 connections per second tests as well, however, this does not appear to be statistically meaningful.

The two tests only differ in the number of concurrent open TCP sessions. This implies the state database is the component under stress. With only these results it is not possible to precisely identify what operation within the database is causing the drop in capacity. The traffic may have exceeded the database's insertion rate, its time to search capacity, or the ability of the system to perform maintenance on the database and delete aging entries.

Regardless, it is apparent that a consumer whose network has an average number of open TCP sessions near 10,000, an average new TCP connections per second rate of no more than 2500 per second, and a bandwidth consumption less than approximately 400 Mbps, could field the tested NIDS with confidence that the capture efficiency would be at or near 100 percent.

5.5 Further Tests

The example tests do not provide enough information on which to base a full confidence decision. Nevertheless, by using the same methodology for developing further tests, the NIDS industry or independent labs could establish a suite of tests that could provide quantifiable results for each of the different stress points.

For example, the total number of database insertions per second for the database can be quantified by establishing a test that runs at a very low rate for all other stress points of the NIDS. The traffic must be crafted such that the database needs to start maintaining state on many different key values. One possible way of shaping the traffic is to use a packet generator and generate a valid TCP session with a full three-way handshake. The rate at which the connections are introduced must be ramped upward until the NIDS starts dropping traffic. Since these simple TCP connections consist of small packets, the bandwidth should remain low. Results need to be cross checked with the raw packet capture architecture evaluation results as described in section 4.5 to ensure that it is the database inserts and not the packets per second limit that has been reached.

6 Conclusion

As the NIDS industry matures, standardized testing will become reality. Developing these tests can be done using the same concepts of standardized testing found in other industries. Hopefully, the information found in this paper can serve as a catalyst to stimulate the development of standardized tests providing the NIDS consumer the information that is now missing. The same techniques used for capacity testing can be extended to other performance areas such as false positive ratios.

References

1. Mier Communications: Test report for ManHunt from Recourse Inc. and test report for Intrusion.com's NIDS. At: http://www.mier.com/reports/vendor.html
2. Ranum, M.: Experiences Benchmarking Intrusion Detection Systems. At: http://www.nfr.com/forum/white-papers/Benchmarking-IDS-NFR.pdf
3. Claffy, K., Miller, G., Thompson, K.: the nature of the beast: recent traffic measurements from an Internet backbone. At: http://www.caida.org/outreach/-papers/1998/Inet98/ (1998)
4. McCreary, S., Claffy, K.: Trends in Wide Area IP Traffic Patterns: A View from Ames Internet Exchange. At: http://www.caida.org/outreach/papers/2000/-AIX0005/ (2000)

Performance Adaptation in Real-Time Intrusion Detection Systems

Wenke Lee[1], João B.D. Cabrera[2], Ashley Thomas[3], Niranjan Balwalli[1], Sunmeet Saluja[1], and Yi Zhang[1]

[1] College of Computing, Georgia Institute of Technology,
801 Atlantic Drive, Atlanta, GA 30332, USA
{wenke, niranjan, sunny, yizhang}@cc.gatech.edu
http://www.cc.gatech.edu/fac/Wenke.Lee
[2] Scientific Systems Company Inc.,
500 West Cummings Park, Suite 3000, Woburn, MA 01801, USA
cabrera@ssci.com
[3] Department of Electrical and Computer Engineering,
North Carolina State University,
Raleigh, NC 27695, USA
athomas@unity.ncsu.edu

Abstract. A real-time intrusion detection system (IDS) has several performance objectives: good detection coverage, economy in resource usage, resilience to stress, and resistance to attacks upon itself. In this paper, we argue that these objectives are trade-offs that must be considered not only in IDS design and implementation, but also in deployment and in an *adaptive* manner. We show that IDS performance trade-offs can be studied as classical optimization problems. We describe an IDS architecture with multiple dynamically configured front-end and back-end detection modules and a monitor. The IDS run-time performance is measured periodically, and detection strategies and workload are configured among the detection modules according to resource constraints and cost-benefit analysis. The back-end performs scenario (or trend) analysis to recognize on-going attack sequences, so that the predictions of the likely *forthcoming* attacks can be used to pro-actively and optimally configure the IDS.

Keywords: Real-time intrusion detection, performance metrics, performance adaptation, optimization.

1 Introduction

Intrusion detection is a critical component of the defense-in-depth network security mechanisms. An intrusion detection system (IDS) monitors operating system or network activities by capturing and analyzing audit data (e.g., `BSM` [33] or `libpcap` [19] stream) to determine whether there is an attack occurring. Most systems perform *misuse detection* by pattern matching known attack behavior or effects, and some systems also employ *anomaly detection* techniques, which

A. Wespi, G. Vigna, and L. Deri (Eds.): RAID 2002, LNCS 2516, pp. 252–273, 2002.

flag unacceptably large deviation from normal profiles as (probably) the result of an attack. A real-time IDS differs from an off-line IDS in that it tries to detect and respond to an attack in real-time (i.e., when it is unfolding).

An IDS is a "mission-critical" system, one that needs to be effective and available. More specifically, its *performance objectives* include: good detection coverage, economy in resource usage, and resilience to stress [26]. Since sophisticated adversaries may try to first evade or even subvert IDSs when launching their intended attacks, another important performance objective is that an IDS must resist attacks upon itself [25,23]. These objectives can be conflicting goals. For example, for broad coverage and high detection accuracy, an IDS needs to perform stateful analysis on a lot of audit data. This requires a large amount of resources (in both memory and detection time). A resource-intensive IDS is then vulnerable to stress and overload attacks. We therefore need to carefully consider the trade-offs.

It is well known that most current IDSs, employing only misuse detection techniques, cannot detect new attacks. Worse yet, even for known attacks, the detection performance cannot be guaranteed when the IDSs are under stress (e.g., due to high traffic volume) or are targeted by evasion, overload, and crash attacks [25,23,30]. Researchers are developing attack resistance techniques. Many evasion attempts can be foiled if an IDS uses stateful analysis and employs a network traffic *normalizer* [12]. Some IDSs are carefully designed to be very "light weight" or are specially configured with high-end hardware (e.g., RealSecure with AppSwitch [34]) to cope with high-speed and high-volume traffic. However, as our analysis and experiments in this paper show, as long as an IDS is *statically* configured in run-time, an intelligent adversary can overload the IDS to a point that it will miss the intended attack with high probability.

We advocate enabling an IDS to provide *performance adaptation*, that is, the best possible performance for the given operation environment. It is extremely difficult, if not impossible, for an IDS to be 100% accurate [2]. The optimal performance of an IDS should be determined by not only its ROC (Receiver Operating Characteristics) curve of detection rate versus false alarm rate, but also its cost metrics (e.g., damage cost of intrusion) and the probability of intrusion [10]. Accordingly, performance adaptation means that an IDS should always maximize its cost-benefits for the given (current) operational conditions. For example, if an IDS is forced to miss some intrusions (that can otherwise be detected using its "signature base"), for example, due to stress or overload attacks, it should still ensure that the best value (or minimum damage) is provided according to cost-analysis on the circumstances. As a simple example, if we regard buffer-overflow as more damaging than port-scan (and for argument sake all other factors, for example., attack probability, detection probability, are equal), then missing a port-scan is better than missing a buffer-overflow. In this research, we extend work on IDS cost-analysis [10,15] to provide a framework for considering the trade-offs of IDS performance objectives, develop techniques for run-time performance measurement and monitoring, and for dynamic adaptation and reconfiguration of IDS policies and mechanisms. We currently focus our work on misuse detection systems.

The rest of the paper is organized as follows. We first analyze IDS performance issues, objectives, and discuss the need for performance adaptation. We then discuss how to enable performance adaptation in real-time IDS. We describe prototype real-time adaptive IDSs, and present experiments and results. We then compare our research with related work, and conclude the paper with a summary and a discussion on future work.

2 IDS Performance Analysis

In this section, we analyze the trade-offs in IDS performance objectives from an optimization and control perspective, and discuss the danger of static configuration and hence the need for performance adaptation. We found it necessary to introduce a more abstract formalism in order to study the problems in a general fashion. Later in the paper, we will explain how these principles can be applied in practice.

2.1 Definitions and Preliminaries

Audit Records. Audit records (or audit events, e.g., packets) are categorized according to their types. Examples of (high level) types are `telnet`, `http`, `icmp echo request`, etc. There are a total of N record types that an IDS accepts. Each audit record is either part of a normal session, or an attack of a certain label. We denote E_i as an arbitrary audit record of type i. Audit record types are characterized by their prior probabilities π_i, which denote the probability that a given record belongs to type i.

Attacks. There are a certain number of attacks associated with each audit event type. For example, a `telnet` connection (and its packets) can be part of a "port-scan", and there can be a "guess-password" or "buffer-overflow" attack in it. Denote N_i as the number of "known" attacks associated with audit event type i. That is, for type i, the IDS has analysis tasks and detection rules for only N_i attacks (other attacks are "unknown" to the IDS). We denote the attacks as A_{ij}, where $j = 1, 2, \cdots, N_i$. We say that $E_i \leftarrow A_{ij}$ when A_{ij} is present in E_i, and $E_i \leftarrow A_{i0}$ when audit event E_i is normal. There is a total of $\sum_{i=1}^{N} N_i$ known attacks to the IDS. Attacks are characterized by the following quantities:

- Prior Probability: The probability p_{ij} that an event of type i contains A_{ij}, that is,
  $p_{ij} = \mathbb{P}\mathrm{rob}(E_i \leftarrow A_{ij})$. Clearly, from the perspective of IDS, $\sum_{j=0}^{N_i} p_{ij} = 1, \quad i = 1, 2, \cdots, N, \; j = 1, 2, \cdots, N_i$, where p_{i0} is the prior probability that an audit record of type i is normal, that is, $p_{i0} = \mathbb{P}\mathrm{rob}(E_i \leftarrow A_{i0})$.
- False Alarm Cost: The cost associated with a response triggered by a false alarm that attack A_{ij} is present, denoted as C_{ij}^{α}.
- Damage Cost: The cost associated with attack A_{ij} being missed by the IDS, denoted as C_{ij}^{β}.

Analysis Tasks. Each audit record is subject to a number of analysis tasks in the IDS, including data (pre)processing, rule checking (i.e.,intrusion detection), and logging. Denote K_i be the (maximum) number of tasks for audit event type i. We denote the tasks as R_{ij}, where $j = 1, 2, \cdots, K_i$. We say that $R_{ij} \stackrel{r}{\leftarrow} A_{ij}$ when a Detection Rule R_{ij} reports the presence of attack A_{ij} in audit event E_i. We say that $R_{ij} \stackrel{r}{\leftarrow} A_{i0}$ when R_{ij} reports that event E_i is normal. The detection rules are characterized by the following quantities:

- The False Alarm Rate of R_{ij} denoted by α_{ij} is defined as $\alpha_{ij} = \mathbb{P}\text{rob}(R_{ij} \stackrel{r}{\leftarrow} A_{ij} \mid E_i \leftarrow A_{i0})$
- The False Negative Rate of R_{ij} denoted by β_{ij} is defined as $\beta_{ij} = \mathbb{P}\text{rob}(R_{ij} \stackrel{r}{\leftarrow} A_{i0} \mid E_i \leftarrow A_{ij})$

Each task R_{ij} (regardless whether it is a detection rule or not) is also characterized by its Computation Time t_{ij}.

System Configuration. The run-time configuration of an IDS is characterized by the collection (union) of its analysis tasks. That is, IDS configuration $\mathcal{P} = \bigcup_{i=1,\cdots,N} \mathcal{P}_i$, where $\mathcal{P}_i$ is the collection of tasks for event type i, that is, $\mathcal{P}_i = \bigcup_{j=1,\cdots,K_i} R_{ij}$ (note that not all tasks are detection rules). A *statically configured* IDS has a fixed set of tasks regardless of changes in run-time conditions.

2.2 Performance Metrics

Expected Value. The purpose of a real-time IDS is to detect intrusions and prevent damages. Instead of using mere statistical accuracy, we should evaluate an IDS according to its value (or cost-benefit). For each attack A_{ij}, an IDS equipped with the detection rule R_{ij} (and the necessary preprocessing and logging tasks) for A_{ij} provides the expected value:

$$\mathcal{V}_{ij} = \mathcal{C}^{\beta}_{ij}\pi_i p_{ij}(1 - \beta_{ij}) - \mathcal{C}^{\alpha}_{ij}\pi_i(1 - p_{ij})\alpha_{ij} \tag{1}$$

The first term is the loss (damage) prevented because of true detection, and the second term is the loss incurred because of false alarms. The total value of an IDS depends on its configuration, that is, its collection of analysis tasks and hence the attacks that it "covers". For the "default" configuration $\mathcal{P}$ that covers all known attacks, the value is $\mathcal{V}(\mathcal{P}) = \sum_{i=1}^{N} \sum_{j=1}^{N_i} \mathcal{V}_{ij}$.

Response Time. Figure 1 shows a generic IDS processing flow. Upon arrival in the system, audit records are placed in a (common) queue (e.g., the `libpcap` buffer). The queue has only one server, the audit data processing and intrusion analysis unit. The nature of the service performed on an audit record item depends on its type. That is, records of type i are only subject to the tasks belonging to $\mathcal{P}_i$. The processing and analysis tasks for each audit record are applied sequentially, as depicted in Figure 2. That is, each event goes through a sequence of analysis tasks. The process terminates if a detection rule R_{ij} determines that

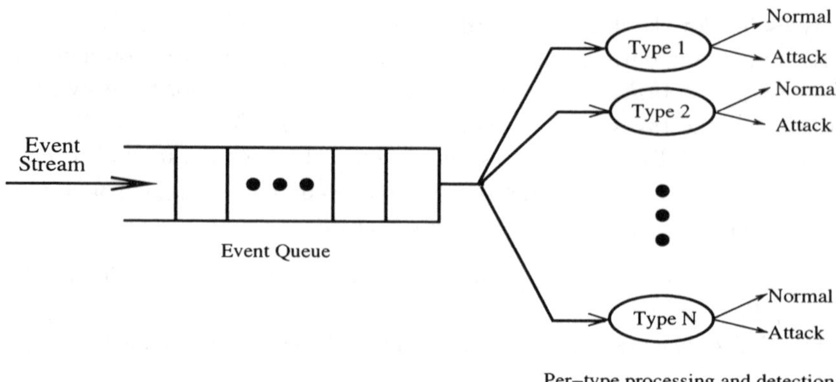

Fig. 1. The IDS Processing Flow. All events are directed to a common queue, but the nature of the service performed on each event depends on event type.

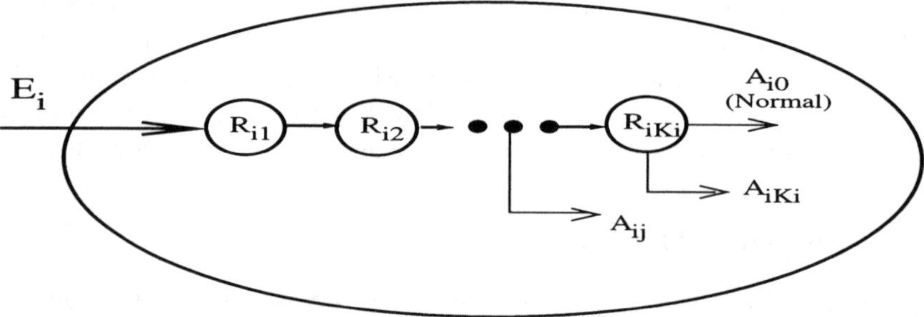

Fig. 2. Processing of events of type i. That tasks include preprocessing, rule-checking, and logging. They are applied serially.

the event is (part of) an intrusion. Or the process ends when all analysis is done and the event is deemed normal. The expected system time (queuing time plus service time - [13]) for an audit record of type i' that arrives in an IDS with configuration $\mathcal{P}$ at a time when there are m_i records of type i, $i = 1, 2, \cdots, N$ is given by:

$$T(\mathcal{P}) = (\sum_{i=1}^{N} m_i T_i) + T_{i'} \tag{2}$$

where T_i denotes the expected service time for a record of type i. The system time corresponds to the time interval elapsed between an audit record entering the system and a decision being made about the presence or absence of an attack in the event. We call it the response time of the IDS. We are interested in the "worst case" when the queue is close to full. In such a case, we have $\sum_{i=1}^{N} m_i T_i \gg T_{i'}$, that is, the queuing time dominates the service time for a typical event. Equation (2) becomes:

$$T(\mathcal{P}) = \sum_{i=1}^{N} m_i T_i, \quad \text{where } T_i = \sum_{j=0}^{N_i} p_{ij} T_{ij}, \quad \text{and } T_{ij} = \sum_{\ell=1}^{j} t'_{i\ell} \tag{3}$$

T_{ij} denotes the service time of an event of type i which is matched by detection rule R_{ij}. Here, $T_{i0} = T_{iK_i}$, since, as depicted in Figure 2, an event of type i which is labeled as normal has the same service time of an event which is matched by the last detection rule R_{iN_i}. Note that here we include ("factor in") the time of common preprocessing and logging tasks for event type i into the time of the N_i detection rules ($N_i \leq K_i$), one for each known attack associated with event type i. T_{ij} is computed as a sum of time of all previous tasks because $R_{i1}, R_{i2}, \cdots, R_{iN_i}$ are applied sequentially. Each t'_{il} is the sum of the time of detection rule R_{il} (i.e., t_{il}), and the time of common preprocessing and logging tasks for event type i. Recalling that $\sum_{j=0}^{N_i} p_{ij} = 1$, we have:

$$T(\mathcal{P}) = \sum_{i=1}^{N} \sum_{j=1}^{N_i} u_{ij} t'_{ij},$$

$$\text{where } q_{ij} = 1 - \sum_{\ell=1}^{j-1} p_{i\ell}, \text{ for } j \geq 1, \quad q_{i1} = 1, \quad \text{and } u_{ij} = m_i q_{ij} \tag{4}$$

2.3 Performance Optimization

An IDS should provide broad detection coverage to maximize its $\mathcal{V}(\mathcal{P})$. This requires that the IDS perform a thorough analysis (e.g., do stateful packet reassembly and analysis), and include many detection rules. This in turn requires a configuration $\mathcal{P}$ with many complex tasks, resulting in a large $T(\mathcal{P})$.

On the other hand, the main constraint in real-time intrusion detection is that $T(\mathcal{P})$ needs to be bounded. As audit events stream into the system queue (see Figure 1), they need to be serviced (taken off) at a rate faster than the arrival rate. Otherwise, the queue (with limited size) will be filled up, and remain so, with the not yet serviced events, thus the newly arriving events cannot be placed in the queue. This phenomenon is referred to as audit data "dropping". The consequence is that the false negative rate(s) β_{ij} of some detection rule(s) R_{ij} will increase due to missing information (evidence). The IDS value $\mathcal{V}_{ij}$ (see Equation (1)) and hence $\mathcal{V}(\mathcal{P})$ will then also decrease. Therefore, in order to provide the expected value, an IDS configuration should satisfy the constraint $T(\mathcal{P}) \leq D_{max}$, where D_{max} is the mean audit event inter-arrival time.

Our goal is then to configure an IDS to provide the best value while operating under the above constraints. That is, if an IDS cannot accommodate all desirable analysis tasks (without violating the constraints), it should just include the more valuable tasks (we also assume that additional and orthogonal optimization techniques, such as rule-set ordering, can be used). For example, an

IDS should always detect "buffer-overflow" and only analyze "slow scan" when time permits. More formally, we need to solve the following problem:

$$\max_{x_{ij}} \mathcal{V}(\mathcal{P}) = \sum_{i=1}^{N}\sum_{j=1}^{N_i} \mathcal{V}_{ij}x_{ij}$$
$$\text{subject to } \; T(\mathcal{P}) = \sum_{i=1}^{N}\sum_{j=1}^{N_i} u_{ij}t'_{ij}x_{ij} \leq D_{\max}$$
$$\text{where } \; x_{ij} = 1 \; \text{ if } R_{ij} \text{ is active in } \mathcal{P} \text{ and } \; x_{ij} = 0 \; \text{ otherwise} \tag{5}$$

The solution is the set of x_{ij} values, which specifies an IDS configuration by indicating which tasks should be included (active). This is known as the Knapsack problem (e.g., [18,22]) in the optimization literature. Note that pre-processing and logging tasks in $\mathcal{P}$ are "factored in" (included in) the detection rules in the following ways: as long as a detection R_{ij} for event type i is active, the common preprocessing and logging tasks for event type i are also included in $\mathcal{P}$; otherwise, when all detection rules for event type i are deactivated, these non-detection tasks for i can also be excluded from $\mathcal{P}$ (Alternatively, one may still want to at least log type i events after some minimum amount of processing. For simplicity sake, we omit the time and value in such a situation).

Instead of requiring exact measurements of the parameter values (e.g., p_{ij}) when solving Problem (5), a more meaningful approach is to allow a value range (with upper and lower bounds) for each parameter. For a feasible IDS configuration $\mathcal{P}$ (specified by a set of x_{ij} values), there is a range of $\mathcal{V}(\mathcal{P})$ values (while the constraint $T(\mathcal{P}) \leq D_{max}$ is always satisfied) because of the ranges of parameter values. The "worst-case" is when $\mathcal{V}(\mathcal{P})$ is the minimal. The optimization problem is then to find an IDS configuration that maximizes the minimal value. In [5], we showed that we can convert the resulting robust max-min problem into an equivalent Knapsack problem, with computational properties similar to the original problem.

2.4 Static Configuration vs. Adaptation

The above optimization problem is often implicitly studied in IDS development and (initial) deployment time. That is, developers and on-site engineers often use knowledge of threat models and assumptions on operation environments to make the appropriate design and customization decisions so that the IDS provides the best performance under the constraints. However, as examples in Section 2.4 show, current IDSs do not have the mechanisms to continuously monitor its performance and the conditions of its run-time environment. That is, they are usually statically configured in run-time. Such systems are not optimal when run-time conditions change, and are vulnerable to attacks aimed to elude IDS.

We note that a solution to Problem (5) (i.e., the optimal IDS configuration) is only valid for a given set of parameter value ranges. Among them, π_i, m_i, and p_{ij} can fluctuate with the operating conditions (e.g., network traffic). For example, when an IDS is under "stress" (i.e., high speed and/or large volume of audit data), m_i becomes much larger and so does $T(\mathcal{P})$. In overload or DoS attacks, an attacker first generates a lot of events (that may include "nuisance" attacks) to overload the IDS, and then launches the intended attack [25,23], say A_{ij}. The overloaded IDS may be "dropping" audit data, missing key evidence, and hence failing to detect attack A_{ij}. Or its detection is too late (slow) to prevent the damage of A_{ij}. In either case, $\mathcal{V}(\mathcal{P})$ will likely decrease by $\mathcal{V}_{ij}$.

In these "stress" and "overload attack" situations, while it is unavoidable that IDS performance will suffer (i.e., the intended (original) $\mathcal{V}(\mathcal{P})$ cannot be guaranteed), it is desirable that a new optimal $\mathcal{V}(\mathcal{P})$ (i.e., the best value possible under the new operating conditions) be provided. For example, instead of having a high probability of missing a more important attack A_{ij}, the IDS can decide not to include the tasks for a less important attack A_{kl} so that there will be sufficient resources (queue space and service time) available for the tasks detecting A_{ij}. Decreasing $\mathcal{V}$ by $\mathcal{V}_{kl}$ rather than $\mathcal{V}_{ij}$ is a better solution because $\mathcal{V}_{kl} < \mathcal{V}_{ij}$.

We define *performance adaptation* as the process of dynamically reconfiguring an IDS to provide the optimal value given the current run-time constraints. Note that performance adaptation cannot prevent audit data dropping or prolonged detection delay caused by stress or overload attacks, as long as the IDS has limited resources (e.g., bounded queue size) and has no control over the volume and speed of the audit data stream. The purpose of performance adaptation is essentially to manage the risk better. That is, instead of having no control over how its performance is degraded (i.e., no control over which attack will be missed) when stressed or under overload attacks, the IDS can *quickly* reconfigure to provide the best detection value under the new conditions. We consider the problem of improving the capability to deal with high-speed and high-volume audit data stream (e.g. making IDS more efficient thus reducing $(T(\mathcal{P}))$ an orthogonal issue to performance adaptation. In [8], we describe our work in designing and implementing IDS on programmable network processors. We believe that by carrying out intrusion detection tasks right at the network processor level, an IDS can keep up with very high speed and high volume traffic.

Performance adaptation relies on *performance monitoring* in run-time to detect the conditions (e.g., "stress") that cause performance degradation and to measure the parameter values needed for solving the optimization problem (5). In Section 3, we will detail our performance monitoring and adaptation techniques and their implementation in prototype real-time IDSs.

Experiments with IDSs. We conducted experiments to study how IDSs perform under stress and how a statically configured IDS can be overloaded by a determined adversary. We used Snort [27] Version 1.8.6 with a subset of its latest ruleset and a number of important preprocessors (e.g., `frag2`, `stream4`, `http-decode`, `unidecode`, `telnet`, `telnet decode`, `rpc decode`, and `portscan`)[1], as well as Bro [23] Version 0.7a90. Snort is very lightweight: it per-

[1] We used the recommended options "`snort -c snort.conf -b -A fast`'' to make it run as fast as possible.

forms packet-based analysis (its TCP stream re-assembly is incomplete). Bro, on the other hand, performs full packet re-assembly, connection stateful analysis, and even keystroke editing. In Bro, the "event engine" is responsible for processing packets and generate "events", the "policy script interpreter" is responsible for using the site-specific intrusion detection logics, coded as "event-handlers" in the C-like Bro policy language, to analyze the events. By carefully implementing the event-handlers (i.e., determining what is an "event" and how to analyze it), and setting the packet filters to decide what kind of audit data should be captured, one can customize Bro to perform only the "important" tasks and as efficiently as possible.

We ran Bro with the standard scripts that come with the distribution, plus a few additional rules to detect our test attacks. We also set the configuration file to load the `http`, `udp`, `icmp` event handlers. According to the loaded event handlers specified in the configuration file, Bro sets the packet filter at the `libpcap` layer, thereby restricting the network traffic that will be processed by its event engine. The filter can also be specified in command line when starting up Bro. For example, in our experiments, we used "(`tcp`[13]&0x7!=0) or (port `ftp`) or (port `telnet`) or (dst port 80 or dst port 8080) or (`udp` port 53) or (`icmp`)", to specify that only certain `tcp` packets (`SYN`, `FIN`, `RST` packets), or `ftp`, `telnet`, `http`, `DNS` (`udp`), or `icmp` packets are to be captured, thereby limiting the traffic by a significant amount. An adversary can still overload Bro by sending a huge amount of traffic that matches the filter. The overload situation is more severe if the packets also result in Bro events, thus overwhelming not only the packet processing and event engine level but also the policy interpreter level.

Our experiments were conducted using LARIAT [28], an IDS testbed used in the 1999 DARPA ID evaluation. LARIAT provides a configurable test environment where ID modules can be "plugged" in the testbed to capture audit data and invoke response. It provides many ways to configure background traffic and attack generation, and facilitates repeatable controlled experiments. We obtained the LARIAT software from MIT Lincoln Lab, and have built a network testbed based on LARIAT. The testbed, with several traffic generators, can produce (simulate) intranet and (both in-bound out-bound) Internet traffic. We use this testbed in our research to simulate (thousands of) virtual hosts and virtual sessions, and create high-speed and high-volume network traffic to test our algorithms and prototype IDS.

The background traffic in our experiments was generated using LARIAT scripts based on traffic profiles. The overload attack experiments performed on the IDSs were done in two stages: in the first stage, we overload the IDS by sending a huge amount of traffic, and in the second stage we launch a critical attack. This attack which would have otherwise been detected will be missed by the IDS with a non-negligible probability. There is an inherent uncertainty from the attacker's point of view whether the attack will go unnoticed by the IDS or not. But by making some intelligent guesses via trial-and-error, a determined attacker can still overload the IDS. This kind of scenario works for the attacker especially when the attack involves very few packets, for example

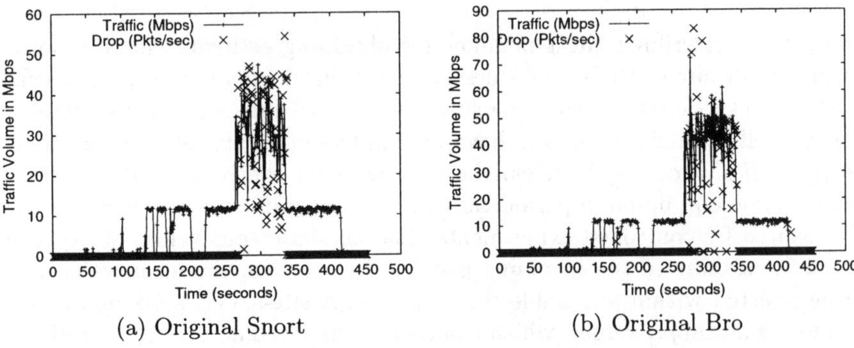

(a) Original Snort (b) Original Bro

Fig. 3. Performance of IDSs under stress: when traffic volume reaches a certain point, IDSs drop packets indiscriminately.

a buffer overflow exploit or the `WEB-IIS CMD.EXE` attack[2]. On the contrary, attacks like port-scan or guess-password that involve many more packets are more likely to be detected by the IDS even when it is dropping some packets. In our experiments, the traffic surges (spikes) is a combination of `udp` packets (to `DNS` port) and `udp` packets to a user port. Note that overloading need not always be intentional or malicious. An IDS monitoring a heavy network can be stressed at peak times. The attack packets were sent during the traffic surge. All the above traffic were captured using `tcpdump` and replayed using `tcpreplay`[3] version 1.1; thereby ensuring exactly the same traffic conditions for all experiments.

As shown in Figure 3, the results from our experiments showed that both Snort and Bro can be overloaded when the traffic volume was increased to a certain point (> 40 Mbps) and will drop packets. While Snort detected approximately 10% (2 out of 20 exploit packets sent) of the `WEB-IIS CMD.EXE` attacks that were launched during the flooding, Bro detected 20% (4 out of 20) of them. Both the IDSs were able to detect 100% of the exploit attempts when the traffic load was low (< 10 Mbps). The exact IDS configurations, traffic conditions, and attack strategies leading to these results are not very important. The point here is that an IDS can be overloaded to drop packets. Unless it can detect such conditions and quickly reconfigure appropriately, it can miss the critical attacks (that can otherwise be detected) with high probability. Thus, instead of being statically configured, a real-time IDS needs to provide performance adaptation via dynamic reconfiguration.

---

[2] This is a single packet attack. The attacker sends a malicious `GET` request to a Microsoft IIS Server. The request is as follows:
"`GET/scripts/..%5c%5c../winnt/system32/cmd.exe?/c+dir`".
See `http://www.cert.org/advisories/CA-2001-26.html` for details.

[3] See `http://tcpreplay.sourceforge.net/` for details.

2.5 Practical Considerations

We start by describing the guidelines for obtaining estimates for the various parameters in our method. For rules derived from anomaly detection schemes, β_{ij} and α_{ij} can be estimated using suitable training data sets. Misuse detection rules for well defined attacks will have well understood behavior, e.g., some may even have $\beta_{ij} = \alpha_{ij} = 0$. π_i can be estimated on the basis of typical traffic statistics, and can updated periodically on basis of traffic measurements. t_{ij} can be measured by controlled experiments. The instantaneous values of m_i reflect the traffic mixture of the incoming packets. In practice, the mean value of m_i can be selected within a suitable time window. A site-specific risk analysis can provide the initial p_{ij} values, which can then be updated according to traffic and attack history. Scenario analysis (see Section 3.3) can use information on attacks detected thus far to predict the likely forthcoming attack(s) R_{ij} along with its p'_{ij}. We can use p'_{ij} as the updated (posterior) probability in place of p_{ij}. Note that as discussed in Section 2.3, the performance optimization problem allows ranges in parameters, thus the parameters need not be measured in absolute values. This relaxation should significantly simplify the measurement tasks.

Estimating the costs C^{β}_{ij} and C^{α}_{ij} also requires site-specific risk analysis. Although it is difficult to measure exact costs, we can still learn the relative ordering of intrusions in terms of their risks (or "damage cost") [9,21,3]. One needs to first define a site-specific *attack taxonomy*. For example, attacks can be first categorized by results (e.g., root access, user access, denial-of-service (DoS), and probing), then further by techniques (e.g., DoS by resource consumption or crashing), and still further by targets (e.g., a server or a desktop) [15]. Then knowledge about the intrusions (e.g., buffer-overflow is usually more damaging than DoS) and assets (e.g., the same intrusion to a server is more damaging than to a desktop) can be used to *qualitatively* measure the damage costs in relative scales [21,15]. A false alarm cost can be the penalty if an automated response is used. For example, if a normal user session is terminated, then the cost can be the same as a DoS damage cost [15]. If an investigation is initiated, it can be the labor cost involved (wasted). Again, we can define the site-specific relative scales of false alarm costs. As discussion later in the paper will show, since the main purpose of cost-benefit analysis is to achieve performance adaptation under resource constraint, such relative scales (not exact numbers) are sufficient for determining which intrusion detection tasks should be given higher priorities.

One may argue that dealing with prior probabilities is always too hard (despite our best effort to simplify the matter), and instead of attempting to solve the optimization problem, an IDS should just use some simple reconfiguring techniques. We note that such "simple" techniques can be derived from the optimization problem by simplifying certain parameters and computations. For example, when stressed or overloaded, an IDS can cease to accept certain types of audit data (e.g., `icmp` and `udp`), and only provide full analysis for a small number of types of events (e.g., analyze `telnet`, `ftp`, and `http` only). In this example, the IDS regards the analysis tasks for certain events as more important than those for the others. Implicitly, the IDS has a ranking order of the values of the analysis tasks. This IDS reconfiguration is equivalent to solving Problem (5) by using only the values (their rankings) and setting the time required for the less important (to be excluded) events to an artificially large value (e.g., D_{max}).

If these simple techniques seem to be adequate in practice[4], what then is the value of our elaborate analysis here? We believe that it is important to provide a formal understanding of IDS performance analysis and adaptation, so that one can follow the optimization principles discussed here to improve upon various simple techniques to provide better IDS value. We also believe that measurement techniques and tools will become available and matured, and the full potential of optimization techniques will be realized.

3 Enabling Performance Adaptation in Real-Time IDS

In this section, we discuss how to enable real-time IDS to provide performance adaptation. We describe prototype systems, and present experiment results.

An adaptive IDS can include multiple intrusion detection (ID) modules, performing increasingly more complex and more time-consuming analysis, and sharing the IDS workload. For example, a front-end module performs data gathering, pre-processing (e.g., packet re-assembly), and as much of the detection work as possible in real-time. A back-end module may not have stringent real-time requirement because, for example, it uses pre-processed audit data (sent from the front-end) to analyze attack trends. Its predictions on forthcoming attacks can be used to help configure the IDS.

The front-end (real-time) module needs to provide performance adaptation. Whenever it is stressed or overloaded, it computes a new optimal IDS configuration according to the new operation conditions (see Problem (5)). The reconfiguration deactivates some (less critical) tasks (e.g., port-scan analysis) and/or cease to capture some events. These excluded tasks can be carried out by the back-end if possible, for example, if they only require pre-processed audit data and the front-end continues to capture and process the needed audit data. We call the process of moving some analysis tasks from the front-end to the back-end *load-shedding*. It essentially allocates the limited resources (i.e., buffer space and service time) to the more critical tasks and events, thus ensuring that the front-end module can provide optimal value while satisfying the constraint $T(\mathcal{P}) \leq D_{max}$. The manager needs to participate in monitoring the ID modules and initiating appropriate re-configuration because the ID modules can be under attacks (or even crashed) and thus may not be able to self-monitor and self-reconfigure. An "active filtering" module, such as a firewall, is desirable for first dropping the obvious offending packets, and thus cutting down the data volume to the ID modules. It can also be used as "admission control", for example, to slow down the data stream (e.g., hold and delay the packets) under some extreme situations to help the ID modules keep up with the traffic.

A popular approach to manage IDS workload is to have several front-end modules and use load-balancing to "split" the traffic [34]. Our research is com-

[4] We have no evidence that this is indeed the case.

plimentary to IDS load-balancing. First, performance adaptation is necessary because a front-end can still be overloaded if each traffic portion (split) is in very high volume. Second, some distributed (and network-based) attacks (e.g., port-scan) may be missed due to load-balancing because the evidence gathered at each front-end module may be below the detection threshold. We then need a correlator, which is essentially a back-end module in our architecture, to detect these attacks. Third, there are complex analysis tasks that should be performed in a back-end module rather than a front-end module because of their computational time and space requirements. These tasks include attack scenario analysis (more in Section 3.3), and alarm correlation and reduction, which are considered very important and desired IDS features [30]. In our approach, the back-end can also carry out some analysis tasks shed from the front-end.

3.1 Performance Monitoring

According to Problem (5), the IDS can provide the expected value $\mathcal{V}(\mathcal{P})$ only when the constraint $T(\mathcal{P}) \leq D_{max}$ is satisfied. The IDS thus needs to self-monitor the run-time conditions, and reconfigure itself to operate under the (new) constraints when necessary. As discussed in Section 2.3, D_{max} should be the mean audit event inter-arrival time.

There are two approaches in monitoring $T(\mathcal{P}) \leq D_{max}$. In *internal measuring*, since the front-end ID module knows the arrival time and detection time of each audit event, it can compute both D_{max} and $T(\mathcal{P})$ (and including m_i, t_{ij}, π_i, and p_{ij}) as a moving average. Alternatively, to avoid the overhead, the front-end can periodically check (e.g., via `libpcap`) whether it is dropping audit events, and if so, conclude that it needs to reconfigure. In *external testing*, the manager periodically sends out a simulated attack that contains an event marked "attack-simulation". The front-end, upon "detecting" this simulated attack, is required to reply to the manager the $T(\mathcal{P})$ value along with the sequence number of the simulated attack. The manager can detect the condition where the returned $T(\mathcal{P})$ is out of bound (according to historical data), and thus concluding that the front-end is overloaded. If the manager receives no reply, it concludes that the front-end is at a "fault" state (e.g., crashed due to crash attacks [23], or an infinite-loop due to implementation errors), and can take immediate action such as activating another (replacement) front-end module.

3.2 Dynamic Reconfiguration

As described in Section 2.1, an IDS configuration is characterized by its collection of run-time analysis tasks. Although an IDS may have a very comprehensive set of tasks that it can use, its optimal configuration, that is, the solution to Problem (5), may include only a subset of these tasks because of run-time constraints. When performance adaptation is enabled in the IDS, this subset (the active tasks) is dynamic, that is, re-computed whenever necessary, rather than static. An implementation of the dynamic task set is to equip the ID modules with a common and complete set of analysis tasks, and have non-overlapping bit masks specifying which tasks are activated at each module.

In Section 2.5, we discussed the practical considerations in measuring the parameters needed to compute the optimal IDS configuration. We can use some heuristics to improve the parameter estimations. The back-end can perform attack scenario analysis (will be discussed in Section 3.3) and supply p'_{ij}, the probability of an attack given the traffic and attack conditions seen thus far. We can use p'_{ij} as the updated (posterior) probability of intrusion in place of p_{ij} when computing $\mathcal{V}(\mathcal{P})$ and $T(\mathcal{P})$. However, we need to avoid being fooled by an intelligent attacker who tries to divert IDS resource from his intended attacks, e.g, by first launching (nuisance) attacks that seem to lead to some other possible attacks so that they will have artificially (and falsely) much higher probabilities than his intended one(s). One solution is to *always* capture audit data and perform analysis tasks for the critical services. This is equivalent to always setting the values of these tasks the highest and their required time the smallest.

3.3 Scenario Analysis

A scenario is a sequence of related attacks that together accomplish a malicious end-goal. We can use scenario analysis to predict the likely forthcoming intrusions to make better load-shedding decisions.

We can use site-specific threat models to form a base set of known scenarios. A *scenario graph* is a directed graph where an edge from node a_i to a_j labeled with p'_{ij} and condition(s) $cond_{ij}$ specifies that after a_i occurs, a_j will occur next with probability p'_{ij} if $cond_{ij}$ is true. A path specifies an attack scenario. In run-time, each reported attack is described by a set of attributes: name (type), timestamp, target_IP, target_port, etc. The back-end "attaches" each attack to a node in the network topology graph using the target attributes, and examines whether the attack is part of an existing scenario(s) there. Based on the currently recognized (partial) scenarios, the back-end reports to the front-end and the manager the possible attacks, their probabilities, and their likely targets. The attack and probability information are used to compute $\mathcal{V}(\mathcal{P})$ and $\mathcal{T}(\mathcal{P})$ for load-shedding decisions (see Section 3.2). The target information is very useful to determine (if necessary) what portion of the audit data the front-end can stop capturing. We are actively studying how to automatically update scenario information and discover new scenarios.

3.4 Prototype Systems

We next describe prototype systems, and present some experiment results. Our main goals here are to see how performance monitoring and dynamic configuration mechanisms can be built into an IDS, and how such adaptive IDS perform in overload situation.

Adaptive Bro. Our adaptive IDS comprises of a front-end IDS, a back-end IDS and a manager module. We use Bro version 0.7a90 (on OpenBSD 2.9) with our modifications as the front-end IDS. The back-end module runs on a different machine and is connected to the front-end on a private network. The manager runs on a third machine and is on the same network that Bro is monitoring.

We modified Bro in two main areas. The first is in adding bookkeeping functions for the purpose of performance monitoring. Note that a Bro "event" is different from an "audit event" (an audit record arriving at IDS event queue) described in Section 2.1. The latter is equivalent to a packet. In Bro, events are generated from the processing of packets, and intrusion analysis (i.e., rule checking and logging) is performed on events rather than on packets. We discussed the constraint $T(\mathcal{P}) \leq D_{max}$ in Section 2.3. $T(\mathcal{P})$ is equivalent to the expected packet service time in Bro, and D_{max} is equivalent to the mean inter-packet arrival time. Clearly, this constraint must still hold for Bro, otherwise there will be packet dropping (and the quality of event data will suffer) and detection performance can suffer. To accommodate the notion of event-level (versus of packet-level) analysis in Bro, we use $T'(\mathcal{P})$ to represent expected event service time, which is the interval between the arrival of the first packet of the event to the completion of analysis of the last packet of the event (also the completion of the event). We use D'_{max} to represent the mean inter-event arrival (or generation) time. For Bro events, we should use the more meaningful constraint $T'(\mathcal{P}) \leq D'_{max}$. It is easy to see that using $T'(\mathcal{P}) \leq D_{max}$ is incorrect, and $T'(\mathcal{P}) > D'_{max}$ will likely lead to $T(\mathcal{P}) > D_{max}$.

Our Adaptive Bro thus has the following measurements: number of packets received per second, number of packets dropped per second, mean inter-event arrival time, and a counter of each event. The packets received and dropped is available from the `libpcap` `pcap_stats` function and is already implemented in the Bro `HeartBeat` function. The interval between two heart beats can be configured. We used 1 second in our experiments. By recording the number of events generated within a time interval (which is 0.1 seconds), mean inter-event arrival time is computed as an average. Bro initiates reconfiguration in two cases: if it detects that there are dropped packets; or if it discovers that $T'(\mathcal{P}) > D'_{max}$.

The second area of changes to Bro is adding dynamic reconfiguration mechanisms. Recall that the process of reconfiguration is to then compute a new optimal solution to Problem (5) according to the new run-time constraints, and then deactivate some analysis tasks and/or cease to capture certain audit data types according to the newly computed configuration. We implemented a dynamic programming Knapsack algorithm. The parameters associated with the event-level analysis tasks are initially measured using benchmark experiments and stored in a system configuration file. For example, the service time for a specific event is the average time taken by Bro to process packets, generate the event, and analyze the event (e.g., match it against rules). The parameters are then loaded in an array in Bro start-time, and can be dynamically updated. For example, π_i and m_i are measured as moving averages in run-time, and p'_{ij} from the scenario analysis function in the back-end can replace p_{ij}. The computed configuration is represented as an array of flags (Bro script variables). These flags are checked before the event analysis tasks (handlers) are invoked. If all event analysis tasks for an audit type are disabled, then `libpcap` filter is also reset to cease capturing such data. Since compiling and loading a new filter at the `libpcap` layer incurs significant delay, we modified Bro to keep a set of pre-compiled filters and load them when necessary. Also, when changing packet filters, the `pcap_setfilter` function invokes the `ioctl` kernel function. It turns out that `ioctl`, while changing the filters, clears out the packets that have not been passed to the upper layer. We took out the code that clears the buffer to avoid losing those packets that might match the new filter. Finally, Bro has an option to store (remember) the "default" configuration, the start-up configuration which is considered as the 'optimal" or desirable one under normal situations, so that if it was reconfigured and has been stable (no need to reconfigure again) for several heart beats (in our experiments, we used 10), it can switch back to run the default configuration.

We briefly describe other modules in our system. The main functions of the manager are to collect statistics and intrusion reports from the Bro and the back-end, and create logs and alerts. It sends a test periodically to Bro to measure delay. Bro also sends the performance measurements (e.g., the numbers of packet received and drop) every heart beat. If the manager does not receive Bro performance measurements and or a reply from its test for a time threshold, it raises an alarm (to security staff) that Bro has probably crashed. The policies on the firewall can be dynamically configured by the front-end. For example, it can terminate an offending connection. It can also delay packets when instructed. The main functions of the back-end module include: sharing analysis load, for example, probe (scan) detection shed from the front-end, and performing attack scenario analysis. Presently, we only have very primitive scenario analysis functions.

Adaptive Snort. We also implemented an adaptive IDS using Snort version 1.8.6 with the latest rule set, and with `libpcap` version 0.6.2 and OpenBSD version 2.9. Unlike Bro, Snort applies intrusion detection rules on packet data directly rather than on "events" extracted from packet data. Snort supports "plug-ins", which can be pre-processors (e.g., de-fragmentation) or detection rules. Snort is thus more loosely coupled and easier to customize. We wanted to study how different IDS architectures influence the implementation of performance adaptation mechanisms.

In Snort, packets go through first the pre-processors then the rule trees. A RuleTreeNode determines whether a packet is a "match" and hence needs to be examined by its OptTreeNodes. In Bro, we can measure service time at the Bro event level and use event service time to include preprocessing and event analysis time because each packet contributes to a Bro event. For Snort, we need to measure the service time at the packet level. There are two cases. First, for packets that match an OptTreeNode (i.e., they match or "belong to" a particular Snort rule), the service time is preprocessing time plus rule checking time (which is the time spent traversing the rule tree up to and including the OptTreeNode). In this case, we call the service time T_R and keep a measurement for each OptTreeNode (i.e., each Snort rule). Second, for packets that do not match any of the RuleTreeNodes (i.e., they do not belong to any Snort rule), the service time is the preprocessing time plus the time traversing the RuleTreeNodes. In this case, we call the service time T_P and keep a measurement for packets of each protocol: `http`, `telnet`, `ftp`, `ssh`, `finger`, `other-tcp`, `icmp`, and `udp`. We need to include T_R and T_P measurements when computing an optimal Snort configuration. Since preprocessing is the main factor in T_P, we need to consider the "value" of preprocessing in addition to the values of the rules. We assign the highest value to preprocessing because it is always needed. If T_P is too high (e.g., when Snort is overloaded by packets that do not necessarily match rules), the Knapsack algorithm can output a configuration that does not include preprocessing. Such a configuration is not acceptable. In such a case, the following iterative process is used: use Knapsack algorithm to first determine what packet filters should be used (what protocols are allowed) in order to keep T_P low (e.g., half of the value in the previous iteration), using priorities among the protocols; then use Knapsack to compute a Snort configuration, considering both T_P and T_R; if a configuration including preprocessing is output, then terminate, otherwise, continue to iterate.

In order to efficiently enable and disable Snort rules without having to traverse the entire rule tree data structure, we implemented a direct access mechanism. It uses a two-dimensional linked list. The head nodes in one dimension are the priorities of the rules (i.e., rank orders in terms of their values), and the other dimension comprises of a list of pointers to all the rules having the same priority. This data structure is populated when parsing the rules at Snort start-up time.

Experiments. We conducted a set of experiments, similar to those described in Section 2.4, to study the performance of our prototype Adaptive Bro and Adaptive Snort. We replayed the same traffic as explained earlier using `tcpreplay`. Regarding the parameter measurements, we assigned damage costs ($\mathcal{C}_{ij}^{\beta}$) of intrusions in relative scales: 100 for root access, 50 for user access, 30 for DoS, and 2 for probing, according to analysis in [21,15]. Since we use automatic intrusion responses (using the firewall), we assign all false alarm costs ($\mathcal{C}_{ij}^{\alpha}$) the same as the DoS damage cost. Since we do not have statistics on attack distribution yet, we assign the prior probabilities (p_{ij}) of all intrusions to be the same (effectively, 1). As mentioned above, π_i and m_i are measured in run-time.

Figure 4 shows the behavior of the Adaptive Bro and Adaptive Snort when overloaded with the same background and flooding traffic used before. We describe some details as follows. For Adaptive Bro, the initial configuration is to detect all of its "known" attacks, which include more than 100 detection rules on root access (e.g., `imapd` buffer-overflow), user access (e.g., `PHF`), DoS (e.g., `smurf`, `syn-flood`), and probes (e.g., `portsweep`). The initial `libpcap` filters were set to be "(`tcp`[13]&0x7!=0) or (port `ftp`) or (port `telnet`) or (dst port

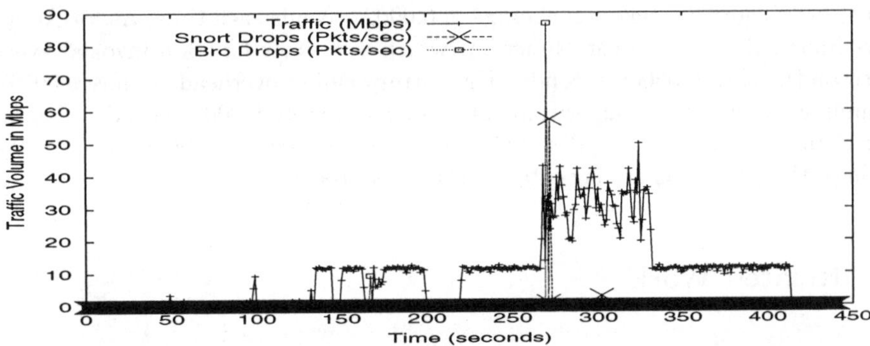

Fig. 4. Behavior of adaptive IDSs under stress: when under stress, it changes to a new configuration to minimize delay and data loss.

80 or dst port 8080) or (port `imap`) or (`udp` port 53) or (`icmp`)" as before. For Adaptive Snort, the initial configuration consisted of a subset of the default rule set (277 rules) coming with the distribution and with the important preprocessors (i.e. `frag2`, `stream4`, `portscan`, `http-decode`, `unidecode`, `rpc-decode`, `telnet-decode`) activated. Initially when the traffic is low, the inter-event arrival time is high and the systems can perform all the analysis tasks. When the traffic rises high, the inter-event arrival time drops low and Bro discovers that $T'(\mathcal{P}) > D'_{max}$. It then invokes Knapsack to compute a new optimal configuration for the current conditions. Also it can be noted that there were initial packet drops due to the heavy load, but the quick reconfiguration avoided further packet drops. This happens at time $t = 270$. This is different from the Original Bro, as shown in Figure 3, where the situation of packet drops continues. Since the flood was caused by `udp` packets, the m_i value of `udp` increased a lot and so did the "weights" (time requirements) of their analysis tasks (see Problem (5)). The reconfiguration ended up dropping all tasks for `udp` and hence the filter for picking up `udp` traffic. Similarly, for Adaptive Snort, when the system is overloaded, it performs a quick reconfiguration (disabling `udp` rules and packet filter) to avoid further packet drops.

We also experimented with the same overload attacks described in Section 2.4 against Adaptive Bro and Adaptive Snort. The systems detected the WEB-IIS CMD.EXE exploits more than 90% of time (19 out of 20) for Bro and 100% of the time for Snort (20 out of 20).

An important consideration when employing performance monitoring and dynamic reconfiguration mechanisms is their overheads. We used micro-benchmark experiments to compute these overheads. As for Adaptive Bro, at each "heart beat" (1 second in our experiment), on average 0.0002 seconds are spent on computing the numbers of packets received and drops, the mean inter-event arrival time, and the mean event service time, the overhead is thus .02%. We believe the overhead is acceptable. Changing filters (using pre-compiled filters) takes only an average of 0.00005 seconds each time, which is very fast. Running the

Knapsack algorithm takes an average of 0.0002 seconds each time. As for Adaptive Snort, the "heart beat" function (which is a timer function invoked every 1 second), takes 0.00005 seconds. The corresponding overhead is thus 0.005%. Running the Knapsack algorithm takes approximately 0.0002 seconds. The filter change (using pre-compiled filters) takes an average of 0.00005 seconds. We believe the overheads are acceptable in both systems.

4 Related Work

The problems of real-time network-based IDSs being evaded (e.g. with "ambiguous" packets), overloaded, and crashed were first discussed in [25,23]. Evasion can be prevented if an IDS uses stateful analysis and a network traffic *normalizer* [12]. Although high-end hardware platforms can be used for data capturing to ensure "no packet filter drops" [23], an IDS, with the often site-specific ID logics implemented as application-level software, can still be overloaded due to high volume filtered events. Paxson suggested that load-shedding may help a real-time IDS defend against overload attacks [23]. Several enterprise-wide and Internet-wide distributed IDSs [24,35] and agent-based architectures [4,11] have been proposed to address the issues of detection coverage and workload distribution. For example, in EMERALD [24], ID modules are deployed and configured in a hierarchical fashion according to the enterprise network topology. More recently, Kruegel et al. proposed a partitioning approach for intrusion detection in high speed network environments [14]. Compared with simple load-balancing, this approach is to partition the traffic meaningfully to a distributed set of sensors each equipped with a set of detection rules. Even with this distributed approach it is noted that the system is vulnerable if the configuration for splitting is static, resulting in some of the sensors being overloaded. Therefore some adaptive approach has to be incorporated. Our research complements these research efforts because performance monitoring and performance adaptation via dynamic reconfiguration are the necessary techniques for an IDS to adaptively resist attacks. It is often more appropriate to evaluate an IDS using the damages (costs) it has prevented [10,15]. We use cost-benefit analysis to determine the best IDS configurations given the resource constraints.

Operating systems also need load-shedding techniques (e.g., [7,20]) to support real-time and multimedia applications. Performance monitoring and Quality-of-Service adaptation approaches have been studied in other domains, e.g., network servers (e.g., [1]), and middleware (e.g., [16]). We can learn from these studies and design techniques for monitoring IDS performance and shedding workload to satisfy the real-time requirements of high priority detection tasks. Although load-balancing is typically used in a different context than load-shedding, we can still learn from the research in this classical problem. For example, in addition to measuring resource utilization, it is necessary to test the application response in order to determine the "meaningful" system load. In our approach, the manager tests the IDS response time using a simulated attack. We can also learn from the vast amount of research in dynamic real-time system scheduling (e.g., [32, 17]), and develop techniques to optimize real-time IDSs.

Attack tree analysis [29] is an off-line process. We are interested in recognizing run-time scenarios and predicting likely forthcoming attacks. Similar to the activity graphs in GrIDS [31,6], we use a scenario graph and network topology to detect scenarios.

5 Conclusion

Providing performance guarantee (assurance) should be the key requirement of IDSs, and security products in general. In this paper, we provided an analysis of IDS performance metrics and constraints, and argued that an IDS should provide the best value under operational constraints. This is essentially an optimization problem. We showed that a statically configured IDS can be overloaded by adversaries to a point that it will miss the intended attacks with high probability. We argued that an IDS should at least achieve a weaker goal: providing performance adaptation, i.e., providing the best possible performance for the given operation environment. We discussed that in order to enable performance adaptation in real-time IDS, performance monitoring and reconfiguration mechanisms are needed. We described prototype adaptive IDSs based on Bro and Snort. Results from experiments thus far validate our motivation and approach.

As for future work, we will be conducting more extensive and realistic experiments. We will refine the performance monitoring and adaptation mechanisms, focusing on lowering the overheads and making them not only customizable but also dynamically configurable.

Although we were able to add performance adaptation mechanisms to Bro and Snort, it was not without conceptual and architectural difficulties. We thus plan to follow our formal framework to design and implement an adaptive real-time IDS with built-in performance monitoring and dynamic optimization capabilities.

Acknowledgments. This research is supported in part by grants from DARPA (F30602-00-1-0603). We wish to thank Vern Paxson of ICSI/LBNL for help on Bro, and Lee Rossey and Richard Lippmann of MIT Lincoln Lab for help on LARIAT.

References

1. T. F. Abdelzaher, K. G. Shin, and N. Bhatti. Performance guarantees for web server end-systems: A control-theoretical approach. *IEEE Transactions on Parallel and Distributed Systems*, 2001. to appear.
2. S. Axelsson. The base-rate fallacy and the difficulty of intrusion detection. *ACM Transactions on Information and System Security*, 3(3), 2000.
3. R. Bace. *Intrusion Detection.* Macmillan Technical Publishing, 2000.

4. J. S. Balasubramaniyan, J. O. Garcia-Fernandez, D. Isacoff, E. Spafford, and D. Zamboni. An architecture for intrusion detection using autonomous agents. Technical report, COAST Laboratory, Department of Computer Science, Purdue University, West Lafayette, IN, 1998.
5. J.B.D. Cabrera, W. Lee, R. K. Prasanth, L. Lewis, and R. K. Mehra. Optimization and control problems in real time intrusion detection. submitted for publication, March 2002.
6. S. Cheung, R. Crawford, M. Dilger, J. Frank, J. Hoagland, K. Levitt, J. Rowe, S. Staniford-Chen, R. Yip, and D. Zerkle. The design of GrIDS: A graph-based intrusion detection system. Technical Report CSE-99-2, U.C. Davis Computer Science Department, Davis, CA, 1999.
7. C. L. Compton and D. L. Tennenhouse. Collaborative load shedding for media-based applications. In *International Conference on Multimedia Computing and Systems*, Boston, MA, May 1994.
8. D. Contis, W. Lee, D. E. Schimmel, W. Shi, A. Thomas, Y. Zhang, Jun Li, and C. Clark. A prototype programmable network processor based ids. submitted for publication, March 2002.
9. D. Denning. *Information Warfare and Security.* Addison Wesley, 1999.
10. J.E. Gaffney and J. W. Ulvila. Evaluation of intrusion detectors: A decision theory approach. In *Proceedings of the 2001 IEEE Symposium on Security and Privacy*, May 2001.
11. R. Gopalakrishna and E. H. Spafford. A framework for distributed intrusion detection using interest driven cooperating agents. In *The 4th International Symposium on Recent Advances in Intrusion Detection (RAID 2001)*, October 2001.
12. M. Handley, C. Kreibich, and V. Paxson. Network intrusion detection: Evasion, traffic normalization, and end-to-end protocol semantics. In *Proceedings of the 10th USENIX Security Symposium*, August 2001.
13. L. Kleinrock. *Queuing Systems, Vol. 1: Theory.* John Wiley & Sons, Inc., 1975.
14. C. Kruegel, F. Valeur, G. Vigna, and R. A. Kemmerer. Stateful intrusion detection for high-speed networks. In *Proceedings of 2002 IEEE Symposium on Security and Privacy*, May 2002.
15. W. Lee, W. Fan, M. Miller, S. J. Stolfo, and E. Zadok. Toward cost-sensitive modeling for intrusion detection and response. *Journal of Computer Security*, 2001. to appear.
16. L. Liu, C. Pu, K. Schwan, and J. Walpole. Infofilter: Supporting quality of service for fresh information delivery. *New Generation Computing Journal*, 18(4), August 2000.
17. C. Lu, J. A. Stankovic, T. F. Abdelzaher, G. Tao, S. H. Son, and M. Marley. Performance specifications and metrics for adaptive real-time systems. In *Proceedings of the IEEE Real-Time Systems Symposium*, December 2000.
18. S. Martello and P. Toth. *Knapsack Problems: Algorithms and Computer Implementations.* John Wiley & Sons Ltd., 1990.
19. S. McCanne, C. Leres, and V. Jacobson. `libpcap`. available via anonymous ftp to ftp.ee.lbl.gov, 1994.
20. J. Nieh and M. S. Lam. The design, implementation and evaluation of SMART: A scheduler for multimedia applications. In *Proceedings of the Sixteen ACM Symposium on Operating Systems Principles*, October 1997.
21. S. Northcutt. *Intrusion Detection: An Analyst's Handbook.* New Riders, 1999.
22. C. H. Papadimitriou and K. Steiglitz. *Combinatorial Optimization - Algorithms and Complexity.* Prentice-Hall, Inc., 1982.

23. V. Paxson. Bro: A system for detecting network intruders in real-time. *Computer Networks*, 31(23-24), December 1999.
24. P. A. Porras and P. G. Neumann. EMERALD: Event monitoring enabling responses to anomalous live disturbances. In *National Information Systems Security Conference*, Baltimore MD, October 1997.
25. T. H. Ptacek and T. N. Newsham. Insertion, evasion, and denial of service: Eluding network intrusion detection. Technical report, Secure Networks Inc., January 1998. http://www.aciri.org/vern/Ptacek-Newsham-Evasion-98.ps.
26. N. Puketza, K. Zhang, M. Chung, B. Mukherjee, and R. Olsson. A methodology for testing intrusion detection systems. *IEEE Transactions on Software Engineering*, 22(10), October 1996.
27. M. Roesch. Snort - lightweight intrusion detection for networks. In *Proceedings of the USENIX LISA Conference*, November 1999. Snort is available at http://www.snort.org.
28. L. M. Rossey, R. K. Cunningham, D. J. Fried, J. C. Rabek, R. P. Lippmann, and J. W. Haines. LARIAT: Lincoln adaptable real-time information assurance testbed. In *The 4th International Symposium on Recent Advances in Intrusion Detection (RAID 2001)*, October 2001.
29. B. Schneier. *Secrets & Lies: Digital Security in a Networked World.* John Wiley & Sons, Inc., 2000.
30. G. Shipley and P. Mueller. Dragon claws its way to the top. In *Network Computing.* TechWeb, August 2001.
31. S. Staniford-Chen, S. Cheung, R. Crawford, M. Dilger, J. Frank, J. Hoagland, K. Levitt, C. Wee, R. Yip, and D. Zerkle. GrIDS-a graph based intrusion detection system for large networks. In *Proceedings of the 19th National Information Systems Security Conference*, 1996.
32. J. A. Stankovic, C. Lu, S. H. Son, and G. Tao. The case for feedback control real-time scheduling. In *Proceedings of the EuroMicro Conference on Real-Time Systems*, June 1999.
33. SunSoft. *SunSHIELD Basic Security Module Guide.* SunSoft, Mountain View, CA, 1995.
34. Top Layer Networks and Internet Security Systems. Gigabit Ethernet intrusion detection solutions: Internet security systems RealSecure network sensors and top layer networks AS3502 gigabit AppSwitch performance test results and configuration notes. White Paper, July 2000.
35. G. Vigna, R. A. Kemmerer, and P. Blix. Designing a web of highly-configurable intrusion detection sensors. In *Proceedings of the 4th International Symposium on Recent Advances in Intrusion Detection (RAID 2001)*, October 2001.

Accurate Buffer Overflow Detection via Abstract Payload Execution

Thomas Toth and Christopher Kruegel

Distributed Systems Group
Technical University Vienna
Argentinierstrasse 8, A-1040 Vienna, Austria
{ttoth, chris}@infosys.tuwien.ac.at

Abstract. Static buffer overflow exploits belong to the most feared and frequently launched attacks on todays Internet. These exploits target vulnerabilities in daemon processes which provide important network services. Ever since the buffer overflow hacking technique has reached a broader audience due to the Morris Internet worm [21] in 1988 and the infamous paper by AlephOne in the `phrack` magazine [1], new weaknesses in many programs have been discovered and abused.
Current intrusion detection systems (IDS) address this problem in different ways. Misuse based network IDS attempt to detect the signature of known exploits in the payload of the network packets. This can be easily evaded by a skilled intruder as the attack code can be changed, reordered or even partially encrypted. Anomaly based network sensors neglect the packet payload and only analyze bursts of traffic thus missing buffer overflows altogether. Host based anomaly detectors that monitor process behavior can notice a successful exploit but only a-posteriori when it has already been successful. In addition, both anomaly variants suffer from high false positive rates.
In this paper we present an approach that accurately detects buffer overflow code in the request's payload by concentrating on the *sledge* of the attack. The sledge is used to increase the chances of a successful intrusion by providing a long code segment that simply moves the program counter towards the immediately following exploit code. Although the intruder has some freedom in shaping the sledge it has to be executable by the processor. We perform abstract execution of the payload to identify such sequences of executable code with virtually no false positives.
A prototype implementation of our sensor has been integrated into the `Apache` web server. We have evaluated the effectivity of our system on several exploits as well as the performance impact on services.

Keywords: Intrusion Detecion, Buffer Overflow Exploit, Network Security

This work has been supported by the FWF (Fonds zur Förderung der wissenschaftlichen Forschung), under contract number P13731-MAT. The views expressed in this article are those of the authors and do not necessarily reflect the opinions and positions of the FWF.

A. Wespi, G. Vigna, and L. Deri (Eds.): RAID 2002, LNCS 2516, pp. 274–291, 2002.

1 Introduction

The constant increase of attacks against networks and their resources causes a necessity to protect these valuable assets. Although well-configured firewalls provide good protection against many attacks, some services (like HTTP or DNS) have to be publicly available. In such cases a firewall has to allow incoming traffic from the Internet without restrictions. The programs implementing these services are often complex and old pieces of software. This inevitably leads to the existence of programming bugs. Skilled intruders exploit such vulnerabilities by sending packets with carefully crafted content that overflow a static buffer in the victim process. This allows the intruder to alter the execution flow of the service daemon and to execute arbitrary code that he can inject, eventually leading to a system compromise and elevating the privileges of the attacker to the ones of the service process. Such an attack is called a *buffer overflow exploit.* Recent studies [19] have indicated that these attacks contribute to a large number of system compromises as many daemons run with `root` privileges.

Intrusion detection systems (IDS) are security tools that are used to detect traces of malicious activities which are targeted against the network and its resources. IDS are traditionally classified as anomaly or signature based. Signature based systems like Snort [18] or NetSTAT [23,24] act similar to virus scanners and look for known, suspicious patterns in their input data. Anomaly based systems watch for deviations of actual from expected behavior and classify all 'abnormal' activities as malicious.

As signature based designs compare their input to known, hostile scenarios they have the advantage of raising virtually no *false alarms* (i.e. classifying an action as malicious when in fact it is not). For the same reason, they have the significant drawback of failing to detect variations of known attacks or entirely new intrusions.

Because of the ability to detect previously unknown intrusions a number of different anomaly based systems have been proposed. Depending on their source of input data, they are divided into host based and network based designs.

Host based anomaly detection systems can focus on user or program behavior. User profiles are built from login times and accessed resources [6,11,2] (e.g. files, programs) or from timing analysis of keystrokes [20]. Unfortunately, user behavior is hard to predict and can change frequently. Additionally, such systems cannot react properly when network services get compromised as no single user profile can be associated to a daemon program.

As a consequence the focus was shifted from user to program behavior. The execution of a program is modeled as a set of system call sequences [8,7] which occur during 'normal' program execution. When the observed sequences deviate from the expected behavior the program is assumed to perform something unintended, possibly because of a successful attack (e.g. buffer overflow). Other researchers use neural networks [10] and concentrate on the analysis of the input data that is passed to programs. These systems are capable of detecting buffer overflows attacks against service daemons but only *after* they have been successful and manifest themselves in abnormal behavior. This has the problem that damage might have already occurred. Another approach, followed by Stackguard [5], modifies compilers to have them insert canary words into the stack frame of vulnerable processes at runtime. Before a function is allowed to return, the canary word is checked for alteration, possibly causing a termination of an exploited process. This mechanism can prevent damage but requires the service process to be recompiled.

We present a system that analyzes the content of service requests at the network level and can a-priori prevent malicious code from being executed. This is similar to network based anomaly detection systems which do not concentrate on activities at hosts (e.g. users or programs) but focus on the packets that are sent over the network. Current network based systems [16,17,4,22] however only model the flow of packets. The source and destination IP addresses and ports are used to determine parameters like the number of total connection arrivals in a certain period of time, the inter-arrival time between packets or the number of packets to/from a certain machine. These parameters can be used to reliably detect port scans or denial-of-service (DOS) attempts. Unfortunately, the situation changes when one considers buffer overflow attacks. In this case the attacker sends one (or a few) packets including the attack code which is executed at the remote machine on behalf of the intruder to elevate his privileges. As the attacker only has to send very few packets (most of the time a single one is sufficient), it is nearly impossible for systems that use traffic models to detect such anomalies.

We propose an approach to do buffer overflow detection at the application level for important Internet services. These services usually operate in a client/server setup where a client machine sends a request to the server which returns a reply with the results. Our detection approach distinguishes normal request data from malicious content (i.e. buffer overflow code) by performing abstract execution of payload data contained in client requests. In the case of detecting long 'instruction chains' of executable code (see Section 3) a request can be dropped before the malicious effects of the exploit code are triggered within vulnerable functions (and maybe detected by another ID system afterwards).

The first section describes buffer overflow exploits in general and the possibilities of an attacker to disguise his malicious payload to evade ID systems. Then we present our approach of abstract payload execution and explain the advantages of this design. The following section introduces the results of the integration of our prototype into the `Apache` [3] web server. Finally, we briefly conclude and outline future work.

2 Buffer Overflow Exploits

Many important Internet services (e.g. `HTTP`, `FTP` or `DNS`) have to be publicly available. They operate in a client/server setup which means that clients send request data to the server, which operates on the given input and returns a reply containing the desired results or an error message. This allows virtually anyone (including people with malicious intents) to send data to a remote server daemon which has to analyze and process the presented data.

The server daemon process usually allocates memory buffers where request data received from the network is copied into. During the handling of the received data, the input is parsed, transformed and often copied several times. Problems arise when data is copied into fixed sized buffers declared in subroutines that are statically allocated on the process' stack. It is possible that the request that has been sent to the service is longer than the allocated buffer. When the length of the input is not checked, data is copied into the buffer by means of an *unsafe* `C` string function. Parts of the stack that are adjacent to the static buffer may be overwritten - a *stack overflow* occurs.

Unsafe `C` library string functions (see Table 1 for examples) are routines that are used to copy data between memory areas (buffers). Unfortunately, it is not guaranteed that the amount of data specified as the source of the copy instruction will fit into the destination buffer. While some functions (like `strncpy`) at least force the programmer to specify the number of bytes that should be moved to the destination, others (like `strcpy`) copy data until they encounter a terminating character in the source buffer itself. Nevertheless, neither functions check the size of the destination area.

Table 1. Vulnerable C Library Functions

`strcpy`	`wstrcpy`	`strncpy`	`wstrncpy`
`strcat`	`wcscat`	`strncat`	`wstrncat`
`gets`	`getws`	`fgets`	`fgetws`
`sprintf`	`swprintf`	`scanf`	`wscanf`
`memcpy`	`memmove`		

Especially functions that determine the end of the source buffer by relying on data inside that buffer carry a risk of overflowing the destination memory area. This risk is especially high when the source buffer contains unchecked data directly received from clients as it allows attackers to force a stack overflow by providing excessive input data.

The fact that `C` compilers (like `gcc` [9]) allocate both, memory for local variables (including static arrays) as well as information which is essential for the program's flow (the return address of a subroutine call) on the stack, makes static buffer overflows dangerous. Figure 1 below shows the stack layout of a function compiled by `gcc`. When an attacker can overflow a local buffer stored on the stack and thereby modify the return address of a subroutine call, this might lead to the execution of arbitrary code on behalf of the intruder.

An adversary that knows that a subroutine in the daemon process utilizes a vulnerable function (e.g. `strcpy`) can launch an attack by sending a request with a length that exceeds the size of the statically allocated buffer used as the destination by this copy instruction. When the server processes his input, a part

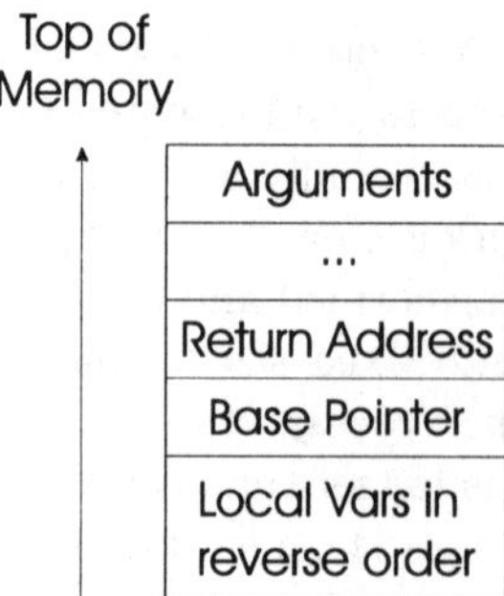

Fig. 1. Stack Layout

of the stack including the subroutine's return address is overwritten (see Figure 2 below). When the attacker simply sends garbage, a segmentation violation is very likely to occur as the program continues at a random memory address after returning from the subroutine.

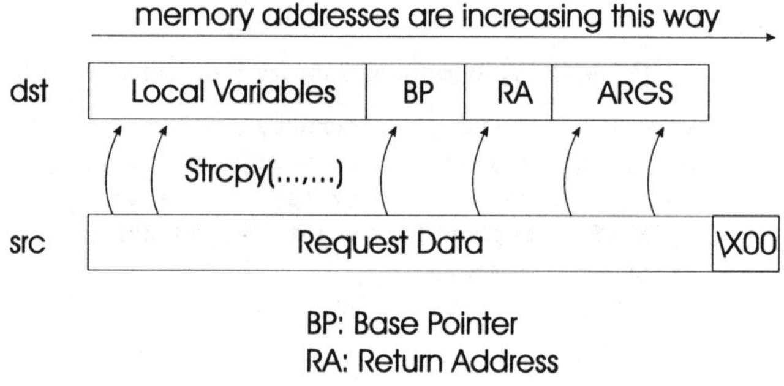

Fig. 2. Operation of `strcpy(char * dst,char * src)`

A skillful attacker however could carefully craft his request such that the return address points back to the request's payload itself which has been copied onto the stack into the destination buffer. In this case the program counter is set to the stack address somewhere in the buffer that has been overflown when the subroutine returns. The processor then resumes execution of the bytes contained in the request with the privileges of the server process (often with `root` rights).

The main problem with this technique is the fact that the attacker does not know the exact stack address where his payload will be copied to. Although the intruder can compile and analyze the service program on his machine to get a rough idea of the correct address, the exact value depends on the environment variables that the user has set. When a wrong address is selected, the processor will start to execute instructions at that position. This is likely to result in an illegal opcode exception because the random value at this memory position does not represent a valid processor instruction (thereby killing the process). Even when the processor can decode the instruction, its parameter may reference memory addresses which are outside of any previously allocated areas. This causes a segmentation violation and a termination of the process.

To circumvent this problem, the attacker can put some code in front of the exploit code itself to increase the chances of having the faked return address point into the correct stack region. This extra code is called the *sledge* of the exploit and is usually formed by many (a few hundreds are common) `NOP` (no operation) instructions. The idea is that the return address simply has to point somewhere into this long sledge that does nothing except having the processor move the program counter towards the actual exploit code. Now it is not necessary anymore to hit the exact beginning of the exploit code but merely a position somewhere in the sledge segment. After the exploit code, the guessed return address (RA) (which points into the sledge) is replicated several times to make sure that the subroutine's real return address on the stack is overwritten. A typical layout of a buffer overflow code that includes a sledge is shown in Figure 3.

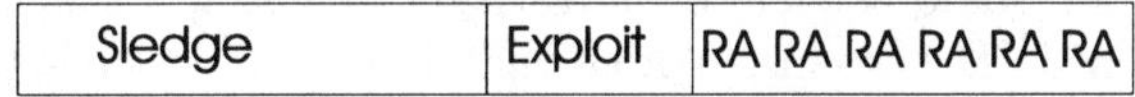

Fig. 3. Typical Structure of a Buffer Overflow Exploit

Some network based misuse IDSs (like Snort [18]) try to identify buffer overflow exploits by monitoring the network traffic and scanning packet payload for the occurrence of suspicious bytecode sequences. These sequences are drawn from actual exploits and represent strings like `/bin/sh` or operating system calls.

This suffers from the problem that there are virtually infinite variations of buffer overflow exploits that attack different vulnerabilities of the same service or express the same functionality differently. In addition, code transformation techniques like reordering or insertion of filling instructions change the signature of the exploit and render the misuse based detection useless. Some intruders have even started to encode the actual exploit with a simple routine (e.g. `ROT-13`) while placing the corresponding decode routine in front of the encrypted exploit. When the buffer overflow is executed the decode routine first decrypts the exploit segment and then executes it.

The wide variety of different exploit signatures shifted the focus of these systems to the sledge. Every attack includes a long chain of architecture specific `NOP` (no operation) instructions that precedes the actual exploit - `NOP` has a byte representation of `0x90` for the `Intel IA32` [12], for other architectures refer to [15].

Unfortunately, the sledge of a buffer overflow exploit can also use opcodes different from `NOP` causing the signature detectors to fail when these instructions are replaced by functionally equivalent ones. According to [13] there are more than 50 opcodes for the Intel IA32 architecture which are suitable for replacing NOP operations (Table 2 below enumerates a few examples).

Table 2. Single-Byte NOP Substitutes for IA32

Mnemonic	Explanation	Opcode
AAA	ASCII Adjust After Addition	0x37
AAS	ASCII Adjust After Subtraction	0x3f
CWDE	Convert Word To Doubleword	0x98
CLC	Clear Carry Flag	0xf8
CLD	Clear Direction Flag	0xfc
CLI	Clear Interrupt Flag	0xfa
CMC	Complement Carry Flag	0xf5
...	...	...

Using these operations (or any combination of those) causes the sledge to behave exactly the same as before, nevertheless its shape can be modified to evade misuse based ID systems. Notice that it is not possible to encrypt the sledge because it has to be executed before the exploit code (and any decryption routines).

The operations presented in Table 2 behave exactly like the NOP instructions on an Intel architecture in the sense that they are only a single byte long. By considering the fact that modern compilers align variables and data structures on the stack at word boundaries[1], more sledge modifications can be performed.

Instead of using single byte instructions without parameters, an intruder can even use multi-byte opcodes with arguments. One just has to make sure that executable code is present at all positions starting at word boundaries. This is necessary because the return address could point to the beginning of any word (i.e. 4 bytes) inside the sledge. This allows the attacker to choose assembler instructions like the ones depicted in Table 3 below or any similar to them. Operations that require an immediate parameter are best suited for this because there is no risk of accessing illegal memory areas (thereby creating a segmentation violation). The attacker just has to use a return address that also points to a word aligned boundary.

The only restriction that remains for the intruder when creating the exploit code and the sledge is that no NULL (0x00) characters may be present. This is due to the fact that a NULL character is interpreted as the end character by many vulnerable C functions. Because the complete attack code has to be copied, the routines may not terminate prematurely. Other than that, the intruder has virtually no limitations in designing his exploit.

[1] Word alignment means that the address of any variable allocated on the stack modulo four equals 0

Table 3. Multi-Byte NOP Substitutes for IA32

Instruction	Bytecode
`adc $0x70,%cl`	`80 d1 70`
`adc $0x70d18070, %ecx`	`81 d1 70 80 d1 70`
`and $0x55120125, %eax`	`25 01 12 44`
`jmprel 0x37`	`0xeb 0x37`
...	...

Any network based misuse system can be easily evaded when such a freedom is given in choosing the layout of the attack. Even anomaly based systems [14] that base their analysis on the payload of the packet could be fooled. Such systems operate with profiles of a 'normal' request that can be imitated with the means shown above. Network based anomaly detectors that consider only protocol information or the flow of traffic fail as well because only a few legal packets need to be transmitted.

While host based anomaly sensors can notice the effect of a completed buffer overflow and raise an appropriate alarm, this approach is undesirable because of two reasons. First, only successful attacks which cause a corresponding distortion in process behavior are reported. No indication on the number of attempts is given. Second, the system can only react to the attack after it has manifested itself as weird behavior. Potential damage that has been inflicted before the ID reacts cannot be prevented (e.g. deletion/modification of files).

We present a system that accurately detects exploits in the payload of application requests and is capable of stopping malicious activity before it effects the system even when the attacker applies all evasion mechanisms described above. The idea of our approach is to focus on the executability of the sledge (which cannot be prevented without breaking the attack) by means of abstract execution.

3 Abstract Execution

The following two properties that classify a sequence of bytes executed on behalf of a certain process are used to define *abstract execution*.

- **Correctness:** A sequence of bytes is correct, if it represents a single valid processor instruction. This implies that the processor is able to decode it. The byte sequence consists of a valid opcode and the exact number of arguments needed for this instruction (and none more). Otherwise, the sequence is incorrect.
- **Validity:** A sequence of bytes is valid if it is correct and all memory operands of the instruction reference memory addresses that the process which executes the operation is allowed to access. A memory operand of an instruction is an operand that directly references memory (i.e. specifies an address in

the memory area of the process). Validity is important because references to non-accessible memory addresses will be detected by the operating system resulting in the immediate termination of the process with a segmentation violation. A correct instruction without any (memory) operands (e.g. `NOP`) is automatically valid. We also call a valid byte sequence a valid instruction.

Definition:

A sequence of bytes is *abstract executable* if it can be represented as a sequence of consecutive valid instructions.

We partition the pool of processor instructions into two sets. One contains all instructions that alter the execution flow of a process (i.e. operations that modify the program counter) while the other set consists of the rest. The elements of the first set (e.g. `call`, `jmp`, `jne`) are called *jump* instructions. A sequence of valid instructions can be decomposed into subsequences that do not contain jump instructions. Such subsequences are called *valid instruction chains (IC)*. An instruction chain ends with a jump instruction or bytes that are not abstract executable. The length of an instruction chain is equal to the number of instructions that it consists of.

An important metrics is the **execution length** of a sequence of valid instructions. The basic idea is that the execution length combines the lengths of instruction chains that are connected by jump instructions. It can be computed for a byte sequence using the following algorithm.

Algorithm: Execution Length of a Byte Sequence

The algorithm expects two input arguments, a byte sequence `seq` and a position `pos` in this sequence. It uses an auxiliary array `visited` to mark already visited blocks whose elements are initialized with `false`. The return value is a positive integer denoting the execution length starting at position `pos`.

1. When the instruction at `pos` is invalid, return 0.
2. When the instruction at `pos` has already been visited, a loop is detected and 0 returned.
3. Find the instruction chain starting at `pos` and calculate its length L. In addition, mark all its operations as visited.
4. When the instruction chain ends with invalid bytes, return L.
5. Otherwise, the chain ends in a jump instruction. When the target of the jump is outside the byte sequence `seq` or cannot be determined statically, return $L + 1$.
6. When the jump targets an operation at position `target` that is inside the sequence, call the algorithm recursively with the position set to `target` and assign the result to L'.
7. When the jump is unconditional, return $L + L'$.

8. Otherwise, it is a conditional jump. Call the algorithm recursively for the continuation of the jump - i.e. set the position to the operation immediately following the jump instruction and assign the result to L''. Then determine the maximum of L' and L'' and assign it to L_{max}. Then return $L + L_{max}$.

Definition:

The *maximum execution length (MEL)* of a byte sequence is the maximum of all execution lengths that are calculated by starting from every possible byte position in the sequence. It is possible that a byte sequence contains several disjoint abstract execution flows and the MEL denotes the length of the longest.

4 Detecting Buffer Overflows

Following the definitions above, we expect that requests which contain buffer overflow exploits have a very high MEL. The sledge is a long chain of valid instructions that eventually leads to the execution of the exploit code. Even when the attacker inserts jumps and attempts to disguise the functionality of this segment, its execution length is high. In contrast to that, the MEL of a normal request should be comparatively low. This is due to the fact that the data exchanged between client and server is determined by the communication protocol and has a certain semantics. Although parts of that data may represent executable code, the chances that random byte sequences yield a long executable chain is very small.

The idea is that a static threshold can be established that separates malicious from normal requests by considering requests with a large MEL as malicious while those with a small execution length as regular. Because the sledge has to be executable in order to fulfill its task, a simple test is utilized to find long executable chains. Requests are analyzed immediately after they have been received by the server. This enables the system to drop potential dangerous requests before the service process can be affected and executes vulnerable functions. We have chosen to place our sensor at the application layer to circumvent the problem of encrypted network traffic faced by NIDS.

The following observation allows an improvement of the search algorithm that has to determine the MEL of requests. According to the definition of the maximum execution length, all positions in the request's byte sequence could potentially serve as a starting point for the longest execution flow. However, if the MELs of normal requests and exploits differ dramatically, it is not necessary to search for the real maximum length. It is sufficient to choose only some random sample positions within the byte sequence and calculate the execution length from these positions. Instructions that have been visited by earlier runs of the algorithm are obviously ignored. The rationale behind this improvement is the fact that it is very likely that at least one sample position is somewhere in the middle of the sledge leading to a tremendously higher MEL than encountered when checking normal requests.

5 Implementation

The algorithms to determine a single execution length and to choose reasonable sample points in the byte sequence of a request have been implemented in C. Because the recursive procedures are potentially costly, the main focus has been on an efficient realization. As every request needs to be evaluated, the additional pressure on the server must be minimized.

An important point is the decoding of byte sequences to determine the correctness and validity of instructions. As data structure we have chosen a static trie for storing all supported processor instructions together with the required operands and their types.

A trie is a hierarchical, tree like data structure that operates like a dictionary. The elements stored in the trie are the individual characters of 'words' (which are opcodes in our case). Each character (byte) of a word (opcode) is stored in the trie at a different level. The first character of a word is stored at the root node (first level) of the trie together with a pointer to a second-level trie node that stores the continuation of all the words starting with this first character. This mechanism is recursively applied to all trie levels. The number of characters of a word is equal to the levels needed to store it in a trie. A pointer to a leave node that might hold additional information marks the end of a word.

We store all supported opcodes of the processor's instruction set in the trie to enable rapid decoding of byte sequences. The leaf nodes hold information about the number of operands for each instruction together with their types (immediate value, memory location or register). This enables us to calculate the total length of the instruction at runtime by determining the necessary bytes for all operands.

It is important to notice that different instructions can be of different length, therefore a hash table is not ideally suitable. Currently, only the Pentium instruction set [12] has been stored in this trie, but no MMX and SIMD instructions are supported.

Figure 4 shows a simplified view of our trie. The opcodes for the instructions AAA (opcode 0x37), ADC (opcode 0x661140 - add with carry the ax register to the value of the register indirect address determined by eax and the one byte operand), ADC (opcode 0x80d1 - add with carry a one byte value to the cl register) and CMP (opcode 0x80fc - compare the immediate value with the register ah) have been inserted.

The algorithms used to determine an approximation of the MEL of HTTP requests have been integrated as a module into an Apache 1.3.23 web server. During the startup of the server, the trie is filled and a function to check the request is registered as a post_read_request procedure. The Apache configuration file has been adapted to make sure that our module is the first to be invoked.

Each time a request arrives at the HTTP server, our subroutine calls the URL decoding routine provided by Apache and then searches for executable instructions in the resulting byte sequence. It is necessary to decode the request first to make sure that all escaped characters are transformed into their corresponding

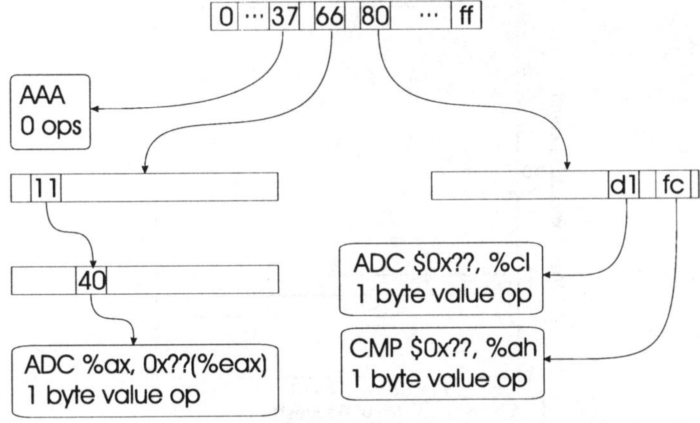

Fig. 4. Storing instruction opcodes in the trie

byte values. The module uses a definable threshold and stops the test immediately when a detected execution length exceeds this limit. We do not calculate the MEL of the request because of performance reasons. Instead we chose to calculate the execution length at equally distributed positions within the request.

6 Evaluation

6.1 Execution Length of HTTP Requests

In order to estimate the maximum execution length of regular HTTP requests, we calculated the MEL for service requests targeted at our institute's web server. Only successful requests that completed without errors have been included in our test data set and we also manually removed attack requests to avoid that buffer overflow exploits distort the data set. An additional ID system has been deployed to verify that assumption. 117228 server requests which we have been captured during a period of 7 days have been processed. The resulting MELs are shown in Figure 5 below.

Only 350 requests had a MEL value of 0 meaning that they did not contain a valid instruction at all. Most of the packets showed a maximum execution length of 3 and 4 (33211 and 31791 respectively) with the numbers decreasing for increasing lengths. The highest maximum instruction length that has been encountered was 16 which appeared for a total of 14 HTTP queries. As expected the numbers indicate that the MELs for regular requests are short.

6.2 Exection Length of DNS Requests

We performed a similar experiment as explained above on DNS data. We captured all the DNS traffic (from the inside and from the outside) to our DNS server during

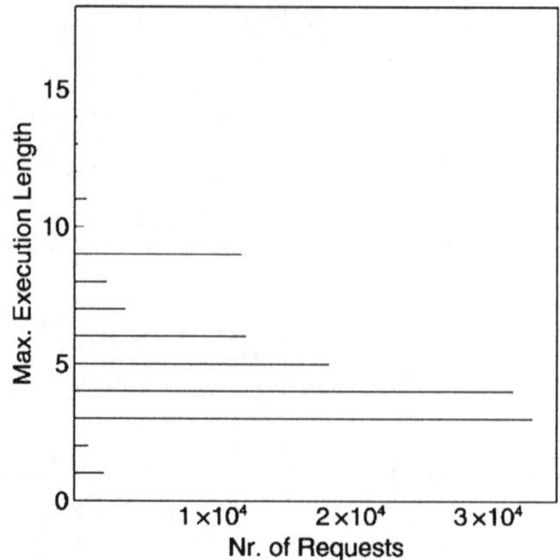

Fig. 5. Maximum Execution Length of regular HTTP Requests

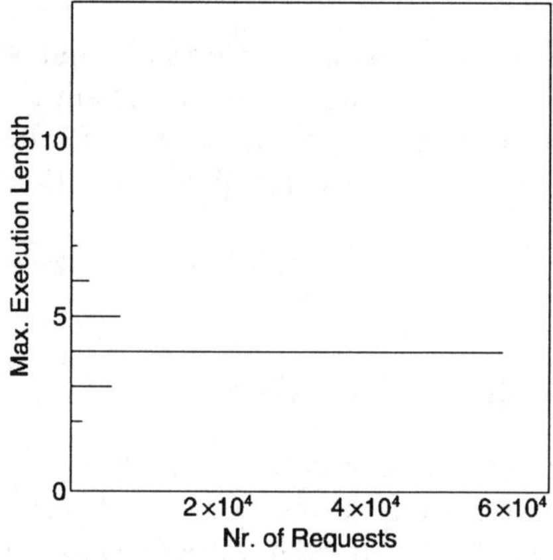

Fig. 6. Maximum Execution Length of regular DNS Requests

a period of one week. We collected 75464 requests and calculated the MEL on each of these.

As shown in Figure 6 the maximum execution length distribution has its peak at 4 with 58557 request. In descending order the MELs of 5 and 3 follow with 6531 and 5500 requests, respectively. The maximum MEL found in our sample data is 12 which has been present in only 4 requests. Therefore the maximum of all MELs of the measured DNS requests is even lower then the maximum of measured HTTP requests.

6.3 Execution Length of Exploits

In order to support our claim that buffer overflow attacks contain long valid instruction chains, a number of available exploits have been analyzed. We have chosen buffer overflow exploits against the `Internet Information Service (IIS)` (the web server from `Microsoft`), `BIND` (a `UNIX` DNS server) and `WU-FTP` (a `UNIX FTP` server, all from [19]. Although our prototype has been tested with a web server, attack code against a different service daemon has been evaluated to show the applicability of our approach to other areas as well. The results of this evaluation are listed in Table 4.

Table 4. Maximum Execution Lengths of Exploits

Exploit	Max. Execution Length (MEL)
`IIS 4 hack 307`	591
`JIM IIS Server Side Include overflow`	807
`wu-ftpd/2.6-id1387`	238
`ISC BIND 8.1, BID 1887`	216

According to the table above, the maximum execution lengths of requests that contain buffer overflow exploits is significantly higher than those of normal queries. This observation supports our assumptions presented in Section 4. For the actual detection of intrusions, a threshold has to be chosen to separate malicious from normal traffic and to raise an alert when the limit is exceeded.

For our first prototype, we simply select a 'reasonable' magic number between the maximum value gathered from the set of normal requests (16) and the minimum among the evaluated exploits (216). Because an attacker might attempt to limit the MEL by choosing a shorter sledge, the value should stay closer to the maximum of the normal requests. We decided to select 30 for the deployment of the probe. This leaves enough room for regular requests to keep the false positive rate low and forces an intruders to reduce the executable parts of his exploit to a length less than this limit to remain undetected. Such a short sledge nevertheless seriously impacts the attacker's chances to guess an address that is 'close enough' to the correct one to succeed.

An obvious shortcoming of the proposed approach is that it can't detect exploits that utilize methods to avoid executable sledges. Vulnerable services that include debug routines that output information which might be used to calculate the exact stack address can be exploited by hackers. If the attacker causes the service to execute the debug output and calculates the *exact* stack address (infoleak attack), he can create buffer overflows that don't include executable sledges.

6.4 Performance Results

To evaluate the performance impact of our module on the web server, we used the `WebSTONE` [25] benchmark provided by Mindcraft. `WebSTONE` can simulate an arbitrary number of clients that request pages of different sizes from the web server to simulate realistic load. It determines a number of interesting properties that are listed below.

- Average and maximum connection time of requests
- Average and maximum response time of requests
- Data throughput rate

The *connection time* is the time interval between the point when the client opens a TCP connection and the point when the server accepts it. The *response time* measures the time between the point when the client has established the connection and requests data and the point when the first result is received. The *data throughput rate* is a value for the amount of data that the web server is able to deliver to all clients.

The connection and response time values are relevant for the time a user has to wait after sending a request until results are delivered back. These times also characterize the number of requests a web server is able to handle under a specific load. The data throughput rate defines how fast data can be sent from the web server to the client. Because clients obviously cannot receive replies faster than the server is sending them, this number is an indication for how long a client has to wait until a request completes.

Our experimental setup consists of one machine simulating the clients that perform HTTP requests (Athlon, 1 Ghz, 256 MB RAM, Linux 2.4) and one host with the Apache server (Pentium III, 550 MHz, 512 MB RAM, Linux 2.4). Both machines are connected using a 100 Mb Fast Ethernet. WebSTONE has been configured to launch 10 to 100 clients in steps of 10, each running for 2 minutes. We did only a measurement of static pages, so no tests involved dynamic creation of results.

We measured the connection rate, the average client response time and the average client throughput for each test run with and without our installed module. The results are shown in Figures 7, 8 and 9. The dotted line represent the statistics gathered when running the unmodified Apache while the solid line represent the one with our activated module.

As can be seen above, the connection rate has dropped slightly when our sensor is activated. The biggest difference emerged when 50 clients are active and a value of 494,2 connections per second versus 500,7 connections per second with the unmodified Apache has been observed. While this maximum difference is 6.5 connections per second (yielding a decrease in the client connection rate of about 1.4 %), the average value is only 2,4 (about 0.5 %).

There has been no significant decrease in the average response time. Both lines are nearly congruent with regards to the precision of measurements.

The client throughput decreased most with 10 active clients when it dropped from 75,90 Mbit per second to 73,70 MBit per second. This is an absolute difference of 2,2 MBit per second. (about 2,9 %). On average, the client throughput only decreased by 0,8 Mbit per second (about 1,05%).

The trie consumed about 16 MB of memory during the tests. While this seems to be a large number at first glance, one has to take the usual main memory equipment of web servers into account where a Gigabyte of RAM is not

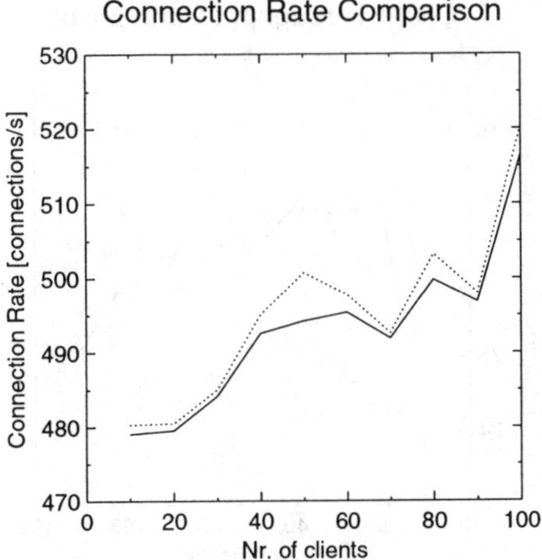

Fig. 7. Client Connection Rate

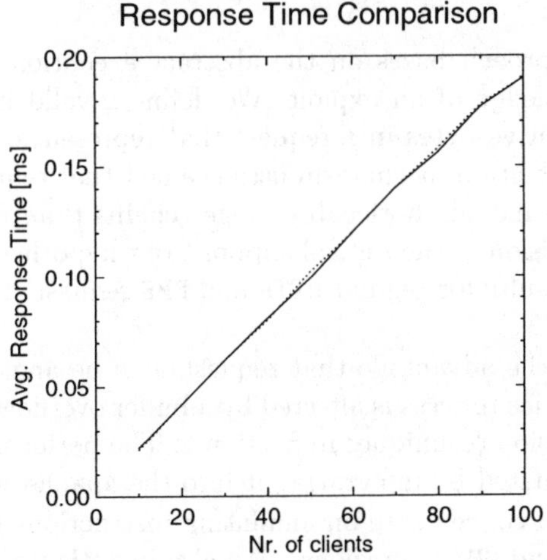

Fig. 8. Average Response Time

uncommon. In addition, this data structure makes very fast tests possible and is a classical trade-off in favor of speed.

7 Conclusion and Future Work

This paper presents an accurate way of detecting buffer overflow exploit code in Internet service requests. We explain the structure and constraints of these attacks and discuss methods used by intruders to evade common detection techniques.

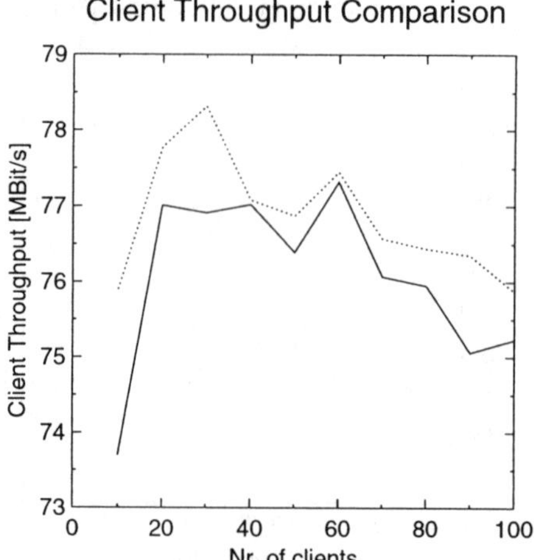

Fig. 9. Client Throughput

Our analysis approach bases on the abstract execution of the packet payload to detect the sledge of an exploit. We define a valid instruction chain as a number of consecutive bytes in a request that represent executable processor instructions. The detection mechanism uses the fact that requests which contain buffer overflow code include noticeably longer chains than regular requests. In addition to the provision of theoretical support, our hypothesis has been verified by comparing the results for regular `HTTP` and `DNS` requests to ones with exploit code.

The system has the advantage that requests can be analyzed and denied a-priori before the service process is affected by a buffer overflow. It is also resistant to the presented evasion techniques in Section 2. The performance impact of the probe has been evaluated by integrating it into the `Apache` web server.

Further work will concentrate on emulating instructions that have not been included yet (`SIMD` and `MMX` operations). We also investigate whether it is useful to perform a 'full' emulation of the effects of the instructions (not only to check the basic executability) in order to detect buffer overflow exploits with a self modifying sledge.

Additionally, we plan to collect experimental data for other protocols like `FTP` and `NFS` to validate that our proposed approach is also applicable there.

References

1. AlephOne. Smashing the stack for fun and profit. *Phrack Magazine*, 49(14), 1996.
2. Debra Anderson, Thane Frivold, Ann Tamaru, and Alfonso Valdes. *Next Generation Intrusion Detection Expert System (NIDES)*. SRI International, 1994.
3. The Apache Software Foundation. http://www.apache.org.

4. M. Bykova, S. Ostermann, and B. Tjaden. Detecting network intrusions via a statistical analysis of network packet characteristics. In *Proceedings of the 33rd Southeastern Symposium on System Theory*, 2001.
5. Crispin Cowan, Calton Pu, David Maier, Heather Hinton, Peat Bakke, Steve Beattie, Aaron Grier, Perry Wagle, and Qian Zhang. Automatic detection and prevention of buffer-overflow attacks. In *7th USENIX Security Symposium*, January 1998.
6. Dorothy Denning. An intrusion-detection model. In *IEEE Symposium on Security and Privacy*, pages 118–131, Oakland, USA, 1986.
7. Laurent Eschenauer. Imsafe. http://imsafe.sourceforge.net, 2001.
8. Stephanie Forrest, Steven A. Hofmeyr, Anil Somayaji, and Thomas A. Longstaff. A sense of self for Unix processes. In *Proceedinges of the 1996 IEEE Symposium on Research in Security and Privacy*, pages 120–128. IEEE Computer Society Press, 1996.
9. The GNU Compiler Collection. http://gcc.gnu.org.
10. A. Ghosh and A. Schwartzbard. A study in using neural networks for anomaly and misuse detection. In *USENIX Security Symposium*, 1999.
11. Judith Hochberg, Kathleen Jackson, Cathy Stallins, J. F. McClary, David DuBois, and Josephine Ford. NADIR: An automated system for detecting network intrusion and misuse. *Computer and Security*, 12(3):235–248, May 1993.
12. Intel. *IA-32 Intel Architecture Software Developer's Manual Volume 1-3*, 2002. http://developer.intel.com/design/Pentium4/manuals/.
13. Home of K2. http://www.ktwo.ca.
14. Christopher Kruegel, Thomas Toth, and Clemens Kerer. Service Specific Anomaly Detection for Network Intrusion Detection. In *Symposium on Applied Computing (SAC)*. ACM Scientific Press, March 2002.
15. Mudge. Compromised: Buffer-Overflows, from Intel to SPARC Version 8. http://www.l0pht.com, 1996.
16. Peter G. Neumann and Phillip A. Porras. Experience with EMERALD to date. In *1st USENIX Workshop on Intrusion Detection and Network Monitoring*, pages 73–80, Santa Clara, California, USA, April 1999.
17. Phillip A. Porras and Peter G. Neumann. EMERALD: Event Monitoring Enabling Responses to Anomalous Live Disturbances. In *Proceedings of the 20th NIS Security Conference*, October 1997.
18. Martin Roesch. Snort - Lightweight Intrusion Detection for Networks. In *USENIX Lisa 99*, 1999.
19. SecurityFocus Corporate Site. http://www.securityfocus.com.
20. Jude Shavlik, Mark Shavlik, and Michael Fahland. Evaluating software sensors for actively profiling Windows 2000 computer users. In *Recent Advances in Intrusion Detection (RAID)*, 2001.
21. E. Spafford. The Internet Worm Program: Analysis. *Computer Communication Review*, January 1989.
22. Stuart Staniford, James A. Hoagland, and Joseph M. McAlerney. Practical Automated Detection of Stealthy Portscans. In *Proceedings of the IDS Workshop of the 7th Computer and Communications Security Conference*, Athens, 2000.
23. Giovanni Vigna and Richard A. Kemmerer. NetSTAT: A Network-based Intrusion Detection System. In *14th Annual Computer Security Applications Conference*, December 1998.
24. Giovanni Vigna and Richard A. Kemmerer. NetSTAT: A Network-based Intrusion Detection System. *Journal of Computer Security*, 7(1):37–71, 1999.
25. WebSTONE - Mindcraft Corporate Site. http://www.mindcraft.com.

Introducing Reference Flow Control for Detecting Intrusion Symptoms at the OS Level

Jacob Zimmermann, Ludovic Mé, and Christophe Bidan

Supélec, France
{jacob.zimmermann, ludovic.me, christophe.bidan}@supelec.fr

Abstract. This paper presents a novel approach to policy-based detection of "attacks by delegation". By exploiting unpredictable behaviour such as unknown side-effects, race-conditions, buffer overflows, confused deputies etc., these attacks aim at achieving their goals (i.e. executing some illegal operation) as legal consequences of other legitimate operations. The proposed approach enforces restrictions on whether an operation can be executed as a consequence of another, in order to detect that kind of attacks. We propose a proof-of-concept application to a Unix system and show its ability to detect novel attack scenarii that seek the same intrusion goals.

1 Introduction

To enforce a given security policy, one has to address two separate problems. First, the policy has to be implemented using existing mechanisms: access control, firewall, authentication system, etc. Then, it is necessary to detect policy violations, i.e. intrusions, and eventually apply appropriate counter-measures. Current intrusion detection technology relies mostly on two approaches: *signature based detection* and *anomaly based detection*. These methods have proven to be fairly effective and are widely used. However, they suffer also from several problems. A signature-based Intrusion Detection System requires an active maintenance of its attack database. Novel attacks are generally not detected. An anomaly-based IDS may generate a high amount of false positives, even if the observed actions are perfectly legitimate [1,2]. In any case, dealing with legitimate but unplanned behaviour is problematic.

These problems can be addressed in multiple ways. Sophisticated knowledge-based systems involving multiple IDSes and advanced alarm interpretation models have been proposed [3,4,5,6,7]. Nevertheless this is a complex, expensive approach.

Another possible way to deal with these problems is *policy-based* detection [8]. A policy-based IDS detects anomalies that violate *policy rules* rather than a learned behaviour considered to be "normal". For instance, a policy could state that telnet should not be used at all, or that ftp should be used only at certain hours and on specific sites. The IDS should then rely on a firewall-like module to verify if these constraints are respected and raise an alarm when a violation occurs.

A. Wespi, G. Vigna, and L. Deri (Eds.): RAID 2002, LNCS 2516, pp. 292–306, 2002.

The problem is that in many cases, it is actually hard to tell whether a specific action violates the security policy or not. For instance, the OS access-control primitives actually provide some form of policy-based intrusion detection, in the sense that they can forbid operations that violate a given policy (in Unix, "users other than `root` cannot read `/etc/shadow`"). Still, experience shows that attacks are possible: for instance, by exploiting side-effects such as buffer overflows or coordination between multiple subject identities, one can gain supplementary privileges to perform operations that should theoretically be forbidden to him. Thus, a policy that states which operations are forbidden can be defeated by performing series of operations that are not illegal per se, but which ultimately lead to the same goals.

We believe that to overcome these problems, a security policy should be implemented in terms of what goals should *not* be achieved, no matter how. We propose to detect *intrusion symptoms*, rather than intrusions themselves. We focus on intrusions where the attacker attempts to achieve some forbidden goal, as opposed to denial-of-service attacks which are out of scope of our work. Such attacks include buffer overflow exploits, side-effects exploits, and race conditions exploits. We propose a policy-based intrusion detection model suitable for runtime detection of such attack symptoms. Our proposed approach is able to deal with novel attack scenarii and requires no empirical profile of "normal behaviour".

It is a known fact that the current weakness of intrusion detection techniques is partly due to their lack of coherent theoretical foundations [2]. Research efforts to build well-defined intrusion detection models started to appear only recently [9]. We think that this kind of model is needed, and we propose one in this paper.

The model itself is described in section 2. In section 3, we discuss an application of this model to intrusion detection in an usual Unix system. Section 4 presents our prototype implementation and practical examples.

2 Model

In this section, we describe the proposed approach to intrusion symptom detection.

To implement a given security policy in current operating systems, access control mechanisms bind precise access rights to each subject. As long as the subject executes operations specifically allowed by his rights, the security policy is enforced. However, existing security flaws may allow an attacker to modify (and, most certainly, extend) his rights in an unpredictable way. For this very reason, existing access control mechanisms alone are not able to prevent such attacks.

To detect symptoms of these, we propose to define operation domains that match a given security policy, i.e. sets of operations that can be executed and combined in any way without the security policy being harmed. Any legal operation (in the sense of the security policy) is thus permitted in at least one operation domain. For instance, the operation *"read /home/user/document.txt"* would be allowed in the domain defined by *user*'s rights, whereas *"write*

/etc/shadow" would be possible only in a domain that allows the password to be changed. However, none of these two domains allows the operation *"cp /home/user/document.txt /etc/shadow"*, executing this operation would involve more than one domain, and thus is illegal.

More precisely, a computer system is modeled as a set of objects, each object having a set of methods that allow to read or update the object's state[1]. Thus any possible system operation (for example, in the Unix system call sense) has precise semantics in terms of elementary object methods. Bounding the capability to perform an object method call to a confined domain as described above restricts in effect the possible use of system operations: if all required privileges to perform an operation exist but belong to different operation domains, such operation is by definition illegal. If attempted, it is the symptom of an intrusion.

We propose the "reference flow control" model as a practical way to implement this approach.

2.1 References

A *reference* represents the capability to execute an elementary object method call in some operation domain. Much like hidden capabilities [10], references exist on their own and can be associated to processes, but their existence is not tied to the execution of a process. Unlike a capability, a reference is not bound to a subject or an executing processes but to a *reference bag* that represents the operations allowed in a domain.

Definition: Given an object o, a method m and a reference bag S, the reference $R_S(o.m)$ is the capability to call method m on o from within the domain represented by S.

Thus, any possible system operation requires one or several references to be authorized: for example, let us consider a Unix access control analogy, where file opening is done through the file methods *openread* and *openwrite*. To be authorized in a domain associated to the reference bag S, the operation

```
open(/etc/shadow, O_RDWR)
```

requires a reference $R_S(/etc/shadow.openread)$ and a reference $R_S(/etc/shadow.openwrite)$. It also requires a reference $R_S(/etc.openread)$ in order to access the `shadow` file in the `/etc` directory, a reference $R_S(/.openread)$, and so on. All these requirements are met within the same reference bag S for the operation to be legal. However, unlike the Chinese Wall model [11], where operation callers are bound explicitly to exclusive operation domains according to their behaviour, in our approach, any reference bag S is *a priori* usable to perform the operation. This allows the caller to execute a potentially wide

[1] The notion of state is used as an abstraction here. A real system object carries actually more than one state: for example, a file object has a "contents" state, a "permissions" state, a "last write date" state and so on. In the model, these would be considered as separate objects.

range of possible operations, but restrictions are set on whether an operation is allowed to be combined with another - in other words, whether the latter operation can exploit some effect of the former.

A security policy definition consists then in the definition of reference bags. Of course, the precise reference requirements of operations are also subject to the security policy. For instance, it is natural that reading the contents of `/etc/shadow` requires appropriate references that should appear only in very specific reference bags; however, reading the screen coordinates of the mouse cursor will probably not be considered as a security-critical operation, so it will not have any particular requirement and, most certainly, references related to the mouse cursor object will not even exist at all in the system. In section 4.2, we will see that a very straightforward reference bags definition actually allows attacks to be detected; however the problem of implementing any possible security policy using reference bags remains an open issue.

In addition to requiring references, operations can create new objects or destroy existing ones, along with the corresponding references. The `open` operation above, if successful, creates a "file descriptor" object (`fd`) with (among others) *read* and *write* methods defined for it that allow the file contents to be accessed.

By definition, if an operation is executed in some domain, its consequences must be allowed to be executed in the same domain. For this reason, it is required that newly created references are inserted into the same reference bag that allowed the operation execution[2]. Thus, any operation depends on a reference bag to be legal and potentially modifies this bag, by creating references (as in the `open` example) and/or by deleting references (for example, when a file descriptor is closed). This is expressed using the following notation:

Definition: Ω being an operation and S a reference bag such that Ω is legal in S, we note Ω^S its execution allowed by references from S. We note $Ref(\Omega^S)$ the reference bag that results from Ω^S.

Reference requirements of operations and this Ref relation actually specify an operating system's model. In the example above:

$$Ref\left[\text{close(fd)}^S\right] = S \setminus \{R_S(fd.read), R_S(fd.write)\}$$

If `close(fd)` is executed using references from bag S, then its execution will remove references $R_S fd.read$ and $R_S fd.write$ from this bag.

2.2 Reference Flow

As stated above, our focus is on dealing with situations where the attacker aims at extending his rights to perform operations that raise illegal consequences. Thus, if an operation Ω_2 is a consequence of an operation Ω_1(we note $\Omega_1 \Rightarrow \Omega_2$), it must obey the same security restrictions as Ω_1. The decision whether $\Omega_1 \Rightarrow \Omega_2$ can rely on the system's internals knowledge (when such knowledge exists; it is

[2] In this paper, we further denote system operation executions simply as "operations"

known for instance that accessing a special file under the `/dev` hierarchy on a Linux system can involve loading a device driver from `/lib/modules`), a logical dependency between these operations (some information flows from Ω_1 to Ω_2, i.e. Ω_1 has set an object state that is accessed by Ω_2) or a combination of both.

In practice, this means that an operation Ω_1 transmits the reference bag that made it possible to its consequence Ω_2. If Ω_2 is executed by another process, references flow to that process, thus allowing it to execute new operations as consequences of Ω_2. This mechanism actually enforces access control rules throughout sequences of causally dependent operations, which play here the role of "subjects".

To be able to enforce security policies against attacks by delegation, one needs thus to define *Ref* rules for such causal dependencies. These rules implement the operation semantics. Let us consider an operation Ω_2^S executing `fd2=open(File,O_RDONLY)`, noted Ω_2^S`::fd2=open(File,O_RDONLY)`, S being an appropriate reference bag. If the file writing was created by some operation Ω_1^Q, then

$$Ref(\Omega_2^S) = Ref(\Omega_1^Q) \cup \{R_Q fd2.read\}$$

If `File` can be opened for reading, then its actual content must be read within the same domain it was written in (here denoted by the reference bag Q), to preserve security restrictions across causal dependency.

On the other hand, opening the file for overwriting is not the consequence of any previous operation except the opening (as it erases the contents), thus for Ω_3^S`::fd2=open(File,O_WRONLY|O_TRUNC)`, the rule is

$$Ref(\Omega_3^S) = S \cup \{R_{S'} fd2.write\}$$

In most cases, an operation involves actually more than one method call. Appending to a file (using O_APPEND) means that both reading and writing are consequences of the latest file write, while opening in a read-write mode (O_RDWR) is a special case: reading is a consequence of the operation that wrote the file (here Ω_1^Q) while writing is a consequence of the opening. This requires reference creation in two separate reference bags. Thus, in this case, `open` represents actually two independent operations (opening for reading and opening for writing) that both create references to the same object `fd2`.

Such rules must be defined for any considered operation. In most cases, they are trivial (as in the `close` case). For operations that involve object state access, they have the form above. This is also true for object creation : for instance in Unix, this means actually writing into a directory whereas accessing the created object involves reading the directory.

2.3 Implementation on Existing Systems

It is straightforward to implement discretionary access control using references. Since access permissions are not granted on a "subject/object" basis, user identities are used for authentication purposes only. As the authentication process itself is treated as a separate problem, we represent it here by an unique special `login(u)` operation method call that sets the calling process' default reference bags conforming to the user's identity u This is functionally equivalent to the real Unix model where the authentication procedure sets the new process' `uid` and `gid`. Since the `login` operation success relies on correct authentication and not on execution context, it has no reference requirements.

An Access Control List is then the set of user identities whose default reference bags include the required references to perform a given object method call. An ACL change modifies the user's default reference bags in the authentication component; furthermore this operation can be broadcast to all processes if the change is required to be immediately effective (a simple way to do this is to state that any operation is a logical consequence of an ACL change). There is no direct equivalent of the Unix "object owner" concept, as "subjects" are not really represented here, thus there are no "`chown`" and "`chgrp`" operations. If needed, object ownership can be modeled as a special usage of ACLs: upon process initialization, a reference bag is created that contains only references to objects owned by the process' identity.

3 Application to Intrusion Detection

In this section, we present a proof-of-concept intrusion detection scheme for a generic Unix-like operating system. Our goal is to provide users and applications with a known environment and some form of backward compatibility.

As we stated in the introduction, we focus specifically on intrusions where an attacker aims at gaining some unauthorized privilege. If we suppose the system implementation is such that it won't allow any user to perform an operation without having the required rights, attacks we are interested in consist in complex series of steps that involve possibly more than one subject identity to get eventually an operation performed, but by definition each step *per-se* does not violate the security policy. We call these "attacks by delegation".

We describe how a Unix-like security policy can be implemented using the proposed model. Intrusion detection issues are then discussed.

3.1 A Model of an Unix-Like Operating Environment

We consider that a security policy within the Unix operating environment obeys the following rules:

- Operation on objects are authorized according to discretionary access control;
- Subject identities require authentication. We take no hypothesis about the authentication process itself, though.

For the system to be secure, it must also conform to a third rule that is not met by current implementations, as experience shows:

– No Unix subject should be able to exploit a privilege that does not correspond to its identity.

The implementation discussed in this example follows the scheme described in section 2.3. Moreover, if we consider only the original Unix access control model, relying solely on "user", "group" and "other" permissions for three operations (`read`, `write` and `execute`) with no extended ACLs, we can further simplify the definition by considering only these three classes of references. Each user identity is then simply modeled by a reference bag that contains the corresponding `read`, `write` and `execute` references, in the same way, there is a reference bag representing each group and an additional reference bag `other`. The default reference bags for each process contain the bag corresponding to the process' `uid`, all reference bags corresponding to groups the user is member of, and the `other` bag.

For instance, considering the following file:

```
-rw-r--r-- 1 bob users 3275 mar 21 09:14 README
```

There exists a $R_{uid:bob}(README.openread)$ reference, a $R_{gid:bob}(README.openread)$ reference, a $R_{other}(README.openread)$ reference and a $R_{uid:bob}(README.openwrite)$ reference.

Such a definition provides actually a coarse view of the system but remains functionally equivalent to the atomic model. It is usable for a basic implementation of the model, because it reduces considerably the amount of references and reference bags to be considered while matching well the default Unix semantics, but on the other hand, it suffers of the same limits as the Unix security policy itself. It is not possible, for example, to devise a policy such that "files from `/tmp` should not be copied into `/etc`" using this scheme, although it can be easily defined by introducing additional reference bags.

We consider that enforcing a security policy on `root`'s actions is pointless if `root`'s privileges are unrestricted and encompass those of all other users. Thus we take into account only operations performed by regular users. Nevertheless, in Unix, users can still perform operations that do not fit in their privileges using `sudo`. By analogy, the equivalent of the `/etc/sudoers` file can be implemented using specific reference bags. For example, the operation of editing `/etc/shadow` is authorized by a bag *editshadow* that contains all references required to read and write this file.

3.2 Detecting Intrusion Symptoms

Considering the model defined in the previous section, the reference bags created by `login(u)` represent the operations the user u is authorized to make use of. An attacker's goal is then to execute for his own purpose operations that do not belong to these bags. The security policy's purpose is to make such goals impossible.

For example, if no reference bag S containing $R_S(/etc/shadow.openread)$ is created by `login(bob)`, then on the one hand, no consequence of `login(bob)` will involve a reference to open `/etc/shadow` for reading. On the other hand, since the reference $R_S(/etc/shadow.openread)$ will flow to any consequence of `open(/etc/shadow,O_RDONLY)`, we can say that these two operations are not authorized to have a common consequence, which would be precisely the symptom of an attempt by `bob` to read `/etc/shadow`.

In its simplest form, the intrusion symptom detector can raise an alarm upon such an intrusion attempt. This provides a particular form of an *access control* system. Such an alarm indicates an operation violating the security policy, that can be dealt with in different ways:

- the operation can simply be prohibited, as is done in the default Unix access control model
- the operation can be purposely authorized in order to catch the attacker in obvious offense
- the intrusion attempt can be reported along with an execution log, for forensic analysis purposes

Another option is an *early warning* approach: an alarm is raised as soon as an illegal combination of references appears (i.e. as soon as different reference bags owned by the same process contain references allowing an operation that could not be executed otherwise). Such an alarm indicates an operation that *cannot* be executed and could be correlated to further actual policy violation alarms, which would provide valuable forensic information.

In any case, the security policy is expressed using reference bags definitions. Intrusions appear to violate this policy, but how an illegal operation is eventually achieved is not taken into account, thus there is no need for a known attack scenarii database, neither a learned authorized usage profile.

4 Experimental Implementation

This section describes a prototype we use to implement the proposed approach on a Linux operating system. We show on two examples how real intrusion symptoms can be detected. A discussion of some strong points and weaknesses of this approach follows.

4.1 Prototype

By now, our implementation perform non-runtime security policy checking by relying on execution logs, like most other intrusion detection systems do. Note however that this is a pure convenience approach we use to ease development and testing. These logs represent system execution scenarii that can easily be replayed by a simulator. Our further goal is to replace the simulator by an in-core implementation, thus the need for logs will be eliminated to allow runtime security policy enforcement.

The logs are generated by a specific Linux kernel module. This module hooks on the kernel API and a report is transmitted to the system logger each time a system call is executed. In its purpose, this is similar to the `strace` utility, but has a system-wide effect in that the logs contain *all* system calls observed during the period the module is active, with the following properties:

- Local process order of operations is respected;
- Local order of object access operations is respected.

As Linux follows, with a few exceptions, the Unix "everything is a file" paradigm, the set of system calls to observe is actually quite restricted:

- Process and thread creation/termination;
- User and/or group modifying;
- Input/output descriptor handling;
- System V IPC access control.

Logs generated by this module are then parsed by a reference propagation simulator. For each process that existed at the log beginning, a default set of reference bags is provided, based on the process' identity and group membership as described in section 3.1. The simulator itself implements reference propagation rules that match closely information flow in an Unix system, i.e. reference flowing from a cause to its effect:

- Opening a file for reading is a consequence of the last write to that file, thus references flow from a write operation to the next read operation, as described in section 2.2;
- By analogy and to conform with standard Unix semantics, the `accept` operation creates a reference to the newly created socket. This reference is defined to belong to the same reference bag as the references used to execute the corresponding `connect` operation. At the moment, only local socket connections are treated.

System V shared buffers apart, there are no memory buffer access semantics defined in Unix. The only reference we can define to handle them are thus simple memory pointers. The only general rule that we can state is that reading from a memory buffer is causally dependent on writing to that buffer. Because this is obviously a weak rule that can be easily bypassed by a malicious program, a more accurate solution consists in overloading the `malloc` and `free` library calls so that memory buffer allocation is emulated through a `mmap` operation. The corresponding temporary file access can then be monitored.

As stated above, we suppose that authentication is performed through a supposed `login` operation. To mimic at best the Unix model, we consider the operations `setuid` and `setgid`, whose effect is to construct reference bags according to the new user's identity. More precisely, current `user` and `group` bags are suppressed and replaced by new ones; the other bag is identical for all users, thus needs no change. In addition, references to objects that are tied to the program's execution (memory objects, pipes etc...) and thus are not explicitly taken into account in the default reference bag definition are transferred to the new user bag, to remain coherent with Unix semantics (only the "owner" process of these objects can access them).

For authentication to be enforced, `setuid` and `setgid` are authorized only as a consequence of `login`. More precisely:

$$Ref\,[\text{login(u)}] = \{R_{setuid:u}(setuid.u)\}$$

A `setuid(u)` operation simply wraps the *setuid.u* method call and thus requires a $R_S setuid.u$ reference. The login operation creates these references in the *setuid* : *u* bag which never contains any other reference: this isolates these references from other reference flows.

4.2 Examples of Attack Detections

In this section, we describe two examples of classes of attacks that can be detected using our proposed approach. These are presented as a proof of concept, a throughout evaluation of the approach and its practical performance will be subject to a further publication.

Race-conditions using symbolic links. As a first simple example, we propose to discuss an old classical attack using "lpr," that has the advantage of being well-known and very simple, yet sufficiently illustrative. Although this particular problem was solved, we can argue that it was not solved by updating or refining the security policy definition, but by modifying "lpr" itself to include specific, ad-hoc inode number checking. This "patch-and-pray" approach proved to be ineffective, since other attacks such as the `/bin/mail` vulnerability [12] rely on an identical principle. The "lpr" attack consists in the following steps:

1. `disconnect the printer`
2. Ω_1`::lpr -s /home/bob/mydoc.ps`
3. Ω_2`::rm /home/bob/mydoc.ps`
4. Ω_3`::ln -s /etc/shadow /home/bob/mydoc.ps`
5. `connect back the printer`
6. Ω_4`::/etc/shadow is printed by the lpr daemon`

Here the user `user` exploits a side-effect of a standard, legitimate system feature (in this case, symbolic links) in a way that doesn't break any access control rules by itself, yet leads to illegal behaviour because it allows him to print the contents of a file (`/etc/shadow`) even if he does not have a read access permission.

It is important to note that in this example, the initial request to print `/home/user/mydoc.ps` could actually have been submitted by any user - even one that is allowed to read `/etc/shadow`. Moreover, nothing prevents the user `user` from creating links to `/etc/shadow` as long as they are under the control of the same access rules as the file they link to. So this attack involves no illegal operation in the access control sense.

However, the proposed model allows to detect Ω_4 as a security policy violation. We suppose that the lpr daemon itself is permitted to read `/etc/shadow` (in practice, this is case), i.e. Ω_4 will be executed using any reference bag rs (as "read shadow") where reading /etc/shadow is possible. Therefore, Ω_4 is not forbidden by itself. We consider also that the user bob operates with reference bags that do not contain references to read /etc/shadow.

When creating the symbolic link, operation Ω_3 generated references to the symlink, since otherwise, the it could not be accessed. By default, a symlink by itself is considered to be writable, readable and executable, depending on the type of the file it points to. Thus:

$$ref3 = Ref(\Omega_3^{uid:bob}) = uid : bob \cup symrefs$$

where $symrefs$ denotes the read, write and execute references to file /home/bob/mydoc.ps in the bag uid:bob.

By creating the link, Ω_3 writes to the /home/bob directory. Ω_4 performs actually two steps. The first step, denoted $\Omega_{4.1}$, reads the /home/bob directory and searches the mydoc.ps file. Therefore, $\Omega_3 \Rightarrow \Omega_{4.1}$ by definition. The reference propagation rule states then that:

$$ref4.1 = Ref(\Omega_{4.1}^{rs}) = ref3 \cup \{R_{uid:bob}(fd.read)\}$$

fd being a file descriptor to the /home/bob directory.

The second step of Ω_4, denoted $\Omega_{4.2}$, actually opens the file. Trivially, $\Omega_{4.1} \Rightarrow \Omega_{4.2}$. Thus $\Omega_{4.2}^{ref4.1}$ should be executed. However, the semantics of a symbolic link as defined in Unix, state that a read operation is permitted on the symlink if it is permitted on the file the link points to. That is, opening /home/bob/mydoc.ps for reading requires both $R_S(/home/bob/mydoc.ps.openread)$ and $R_S(/etc/shadow.openread)$.

Given this, we see that $\Omega_{4.2}^{ref4.1}$ is not possible, since $R_S(/etc/shadow.openread) \notin ref4.1$. Neither it is possible to consider $\Omega_{4.2}$ as an independent concurrent operation, i.e. as $\Omega_{4.2}^{rs}$, since rs itself does not contain required references to read /home/bob/mydoc.ps. In fact, the operation would be possible as $\Omega_{4.2}^{ref4.1 \cup rs}$, but this is by definition an intrusion symptom.

OpenSSH vulnerability. This vulnerability was published in November 2001 [13]. It is interesting to note that this is a well-known attack that originally affected the telnet daemon, which was patched. It was discovered only recently that a similar technique could be used against OpenSSH to yield exactly the same symptom (a regular user getting a root shell).

The OpenSSH daemon has a controversial feature that allows users to define their own environment variables to be set on login. If so configured, this can be exploited to introduce a Trojan horse that will prevent the user's login shell from being started with uid correctly set. In practice, the LD_PRELOAD variable is used to require the OpenSSH daemon to load an attacker-supplied shared library (usually called libroot.so) that overrides the setuid system call, effectively forcing the the attacker's session to be always started with uid set to 0. Thus, this is a means for the attacker to start a `root` shell without going through appropriate authentication.

The attack includes then the following major steps:

1. Ω_1::Create and install `libroot.so`
2. Log in as a regular user
3. Ω_2::`libroot.so` is loaded
4. Ω_3::The session's `uid` is set to 0
5. A `root` shell is started

It is evident that $\Omega_1 \Rightarrow \Omega_2$. Similarly, the `libroot.so` file requires to be loaded a buffer which is written by operation Ω_2 and read by Ω_3 when calling the executable code, however, memory buffer access monitoring as described above is needed to effectively establish that Ω_3 requires a read reference to this buffer. The only reference to allow reading this buffer is $R_{uid:bob}libroot.so.read$, which results from Ω_2. In addition, Ω_3 requires by definition also $R_{setuid:root}setuid.root$ to perform the actual `setuid` call. As it is not possible to meet these requirements within the same reference bag, Ω_3 is an intrusion symptom.

4.3 Discussion

We described how two different, realistic attacks can be detected using the proposed model. We can observe, however, that in the first example there is nothing really specific to lpr, any race-condition attack involving symbolic links may be detected in an identical way. It is interesting to observe that as seen in this example, the reference propagation rules actually restrict the usage of symbolic links. To be usable at all, a symbolic link must be created such that it points to an object that is accessible form within the same domain. By analogy, in Unix terms this means that if some user is not allowed to access a particular object, neither would he be able to create symlinks to that object. This behaviour, if implemented in Unix instead of the default one, would effectively solve the problem and such attacks would not be possible.

In the OpenSSH example, reference propagation forbids using a shared library to overload operations in another operation domain. Again, this is in no way specific to OpenSSH, the same arguments hold for other attacks relying on `libroot.so`, such as the original telnet attack. However, while the model effectively prevents the attack from succeeding, such a restriction on the use of shared libraries is probably needlessly strong for practical use. Weaker reference propagation rules for the specific case of dynamic library function calls should be experimented.

The most promising aspect is that the model as presented appears to be able to detect various "intrusions by delegation" given a security policy specification. In this sense, the proposed approach provides a policy-based intrusion detector. No knowledge of particular program behaviour nor particular attack scenarii was required for the policy to be defined.

Obviously, authentication is the delicate part. The model makes no hypothesis about the authentication process, it simply considers it as a virtual operation that is assumed to behave correctly according to the security policy. While it is possible to detect attacks that circumvent authentication (as shown in the OpenSSH example), the model itself cannot be used to detect attacks that rely on authentication errors or cheating.

The examples show also that the security policy definition depends on a granularity choice, which reflects the application context. In the present case, our focus was on elementary system operations. A coarser approach, for instance considering the receiving of a connection and opening a session as a single operation in OpenSSH would provide a much more elegant and straightforward view of the problem and, if needed, would allow for an easy definition of special reference flow rules. However this means making hypotheses on the OpenSSH daemon's internals (in this case, the fact that it simply and immediately executes what it reads on the socket) that may or may not be true. It is also possible to defeat the operation-based scheme by introducing hidden channels between memory objects. For example:

```
1. read(fd1, a, data_len);
2. for(i=0 ; i<data_len ; i++) b[i]=a[i];
3. write(fd2, b, data_len);
```

If the effect of step 2 is known, it is clear that the a reference in step 1 actually flows to step 3 as the b reference. Unfortunately, this does not appear if we consider only system operations, information flow control is needed in such a case.

For this reason, the proposed approach is suitable to handle attacks that exploit system-level features or problems; it falls short to detect attacks exploiting such hidden channels.

5 Related Work

By definition, the proposed approach belongs to the EM class of security policy enforcement mechanisms, as defined by F. B. Schneider in [14]. In its pure form, EM is implemented using security automata to detect illegal steps sequence prefixes. On the one hand, in our case sequences of steps are implicit, resulting from the causal dependency relation. On the other hand, the definition of an illegal step is trivial. For these reasons, the proposed EM implementation seems suitable for use in cases where illegal steps (i.e. intrusion goals) are well-defined, but the possible ways to achieve them are either unknown or even hardly definable.

Certain race-condition attacks can be detected by the noninterference-based intrusion detection model recently proposed by C. Ko and T. Redmond in [9]. In this model, intrusion symptoms appear as non-commuting sequences of privileged and unprivileged operations. We can observe that by considering a strict "privileged vs. unprivileged" distinction, then the reference propagation rule along with the requirement that any operation should require only one reference bag actually enforces operation commutability. However, the model proposed by the authors requires knowledge of system call commutability, thus using it to detect for instance Trojan horses that implement backdoor system calls (which can be easily done on Linux) seems to be non-trivial. In contrast, our proposed approach relies on the weaker assumption of knowledge of system calls semantics in terms of reference creation and deletion, with the default behaviour that references are preserved unless specified otherwise (this results from the reference propagation rule). In addition, our proposed model is more fine-grained, as by definition, it does not consider system calls to be atomic operations.

Research is very active in the information flow control field. Robust and very mature formalisms exist [15,16] and may be useful for static security policy validation in our case. Applications of information flow for access control purposes have been proposed [17,18,19]. We will examine how reference flow control could be used to implement such policies on a general-purpose operating system. However, real information flow control is not applicable in this case without static program analysis, however, hidden channels exploits cannot be detected using the proposed approach.

6 Conclusion

We have presented an approach to detect intrusion symptoms by controlling the flow of references. As shown in this paper, the proposed approach has the ability to detect the same symptom achieved in a variety of different scenarii, even scenarii unknown at the time the security policy is defined. It is easy to control reference flow by observing executed system operations.

We are currently experimenting our proposed approach using a reference flow simulator with actual execution traces. In the short term, we will develop a runtime implementation. An important part of our work, along with intrusion detection accuracy and policy specification effectiveness, will be the performance impact of such a module on the system.

The distributed intrusion symptom problems are yet to be addressed. This would require dealing with remote objects through "distant references", which in turn involves distant reference authentication. Existing digital signature techniques may provide an usable framework for such a system.

References

1. J. Allen, A. Christie, W. Fithen, J. McHugh, J. Pickel, and E. Stoner. State of the practice of intrusion detection technologies. Technical Report SEI-99TR-028, CMU/SEI, 2000.
2. John McHugh. Intrusion and intrusion detection. *International Journal of Information Security*, July 2001.
3. D. Schnackenberg, K. Djahandari, and D. Sterne. Infrastructure for intrusion detection and response. In *Proceedings of the DARPA Information Survivability Conference and Exposition (DISCEX'00)*, 2000.
4. Frédéric Cuppens. Managing alerts in a multi-intrusion detection environment. In *Proceedings of the 17th Annual Computer Security Applications Conference (ACSAC 2001)*, December 2001.

5. R. P. Goldman, W. Heimerdinger, S. A. Harp, C. W. Geib, V. Thomas, and R. L. Carter. Information modeling for intrusion report aggregation. In *Proceedings of the DARPA Information Survivability Conference and Exposition*, June 2001.
6. Frédéric Cuppens and Alexandre Miège. Alert correlation in a cooperative intrusion detection framework. In *Proccedings of the IEEE Symposium on Security and Privacy*, 2002.
7. Benjamin Morin, Ludovic Mé, Hervé Debar, and Mireille Ducassé. M2D2: A formal data model for IDS alert correlation. In *Proceedings of the Fifth International Symposium on the Recent Advances in Intrusion Detection (RAID'2002)*, 2002.
8. Prem Uppuluri and R. Sekar. Experiences with specification-based intrusion detection. In W. Lee, L. Mé, and A. Wespi, editors, *Proceedings of the Fourth International Symposium on the Recent Advances in Intrusion Detection (RAID'2001)*, number 2212 in LNCS, pages 172–189, October 2001.
9. Calvin Ko and Timothy Redmond. Noninterference and intrusion detection. In *Proccedings of the IEEE Symposium on Security and Privacy*, 2002.
10. Daniel Hagimont, Jacques Mossiere, Xavier Rousset de Pina, and F. Saunier. Hidden software capabilities. In *International Conference on Distributed Computing Systems*, pages 282–289, 1996.
11. David F.C. Brewer and Michael J. Nash. The chinese wall security policy. In *Proceedings of the IEEE Symposium on Research in Security and Privacy*, pages 206–214. IEEE Computer Society Press, May 1989.
12. CMU CERT/CC. Ca-1995-02: Vulnerabilities in /bin/mail. http://www.cert.org/advisories/CA-1995-02.html, January 26 1995.
13. CMU CERT/CC. Vu#40327: Openssh uselogin option allows remote execution of commands as root. http://www.kb.cert.org/vuls/id/40327, November 2001.
14. Fred B. Schneider. Enforceable security policies. *Information and System Security*, 3(1):30–50, 2000.
15. John Rushby. Noninterference, transitivity, and channel-control security policies. Technical Report CSL-92-02, SRI, dec 1992.
16. J. McLean. A general theory of composition for trace sets closed under selective interleaving functions. In *Proceedings of the IEEE Symposium on Research in Security and Privacy*, May 1994.
17. E. Ferrari, P. Samarati, E. Bertino, and S. Jajodia. Providing flexibility in information flow control for object-oriented systems. In *Proceedings of the IEEE Symposium on Security and Privacy*, pages 130–140, 1997.
18. H. Mantel and A. Sabelfeld. A generic approach to the security of multi-threaded programs. In *Proceedings of the 13th ProIEEE Computer Security Foundations Workshop*, pages 200–214, June 2001.
19. Steve Zdancewic, Lantian Zheng, Nathaniel Nystrom, and Andrew C. Myers. Untrusted hosts and confidentiality: Secure program partitioning. In *Proceedings of the 18th ACM Symposium on Operating Systems Principles*, 2001.

The Effect of Identifying Vulnerabilities and Patching Software on the Utility of Network Intrusion Detection*

Richard Lippmann, Seth Webster, and Douglas Stetson

Lincoln Laboratory MIT, 244 Wood Street, Lexington, MA 02173-9108
lippmann@ll.mit.edu

Abstract. Vulnerability scanning and installing software patches for known vulnerabilities greatly affects the utility of network-based intrusion detection systems that use signatures to detect system compromises. A detailed timeline analysis of important remote-to-local vulnerabilities demonstrates (1) Vulnerabilities in widely-used server software are discovered infrequently (at most 6 times a year) and (2) Software patches to prevent vulnerabilities from being exploited are available before or simultaneously with signatures. Signature-based intrusion detection systems will thus never detect successful system compromises on small secure sites when patches are installed as soon as they are available. Network intrusion detection systems may detect successful system compromises on large sites where it is impractical to eliminate all known vulnerabilities. On such sites, information from vulnerability scanning can be used to prioritize the large numbers of extraneous alerts caused by failed attacks and normal background traffic. On one class B network with roughly 10 web servers, this approach successfully filtered out 95% of all remote-to-local alerts.

Keywords: Intrusion detection, vulnerability, attack, network, signature, false alarm, scan, probe, DoS, worm, exploit, Internet, patch

1 Introduction

Intrusion detection systems are not used in isolation. They are almost always used in conjunction with defensive mechanisms including frequent installation of software updates or patches to eliminate known vulnerabilities, firewalls, and vulnerability scanning to find known vulnerabilities on protected machines. In recent years, the cost and difficulty of using these defenses has decreased dramatically and they have become more widespread. Low-cost personal firewalls are available to protect

* This work was sponsored by the Federal Aviation Administration under Air Force Contract F19628-00-C-0002. Opinions, interpretations, conclusions, and recommendations are those of the authors and are not necessarily endorsed by the United States Government.

A. Wespi, G. Vigna, and L. Deri (Eds.): RAID 2002, LNCS 2516, pp. 307–326, 2002.

individual hosts. Scanners that catalog vulnerabilities in large networks have become more capable and comprehensive with almost daily updates in rules used to detect vulnerabilities. In addition, software patches have become easier to install and many modern operating systems now include mechanisms to automatically apply security patches and other updates. The purpose of this paper is to assess the impact these defensive techniques have on the utility of network intrusion detection systems.

This paper focuses on network-based intrusion detection systems that detect known attacks using separate signatures for each attack and on the use of these systems to detect successful attacks where remote attackers compromise monitored machines (remote-to-local attacks). It was motivated by anecdotal evidence suggesting that these important attack types are rarely detected by network-based intrusion detection systems. On well-protected sites, network-based intrusion detection systems detect scans, some denial of service (DoS) attacks, and failed remote-to-local attacks, but rarely do they detect successful remote-to-local system compromises. This occurs even though there are many signatures for such attacks, there are 100's to 1000's of unsuccessful remote-to-local attacks per day, and analysts spend hours each day examining alerts.

This paper has two goals. The first is to analyze how defensive techniques interact with network intrusion detection to determine when these systems will detect successful remote-to-local attacks. The second is to demonstrate that knowledge of the defensive posture of a site can enhance the utility of network intrusion detection by reducing the number of alerts that analysts must examine to detect system compromises.

The remainder of this paper first reviews important characteristics of the most common type of network-based intrusion detection system. It also describes the use of firewalls to protect internal machines and to isolate, in a Demilitarized Zone (DMZ), systems that offer network services to the public. Following this, the importance of detecting and responding to remote-to-local attacks is discussed in relationship to the utility of detecting and responding to other types of attacks that are commonly found by network intrusion detection systems, including DoS and reconnaissance or scan attacks.

The concepts of "windows of vulnerability" and "windows of visibility" are then introduced. These represent time intervals when protected hosts can be compromised by exploiting recently discovered vulnerabilities and when these compromises will be detected by network intrusion detection systems. A detailed timeline analysis follows for eight important vulnerabilities. Each timeline indicates when vulnerabilities were announced, when software patches, intrusion detection signatures, and vulnerability scanner rules were available, and when vulnerabilities enabled major Internet worm incidents. The implications of these timelines for well-protected hosts in a DMZ where no hosts have known vulnerabilities are then discussed and statistics on the frequency of discovery of serious vulnerabilities are presented. These show it is feasible to update server software whenever serious vulnerabilities are discovered for a small number of hosts. The implications of these analyses for poorly protected sites and for normal sites with many protected hosts are then discussed. The paper then focuses on an analysis of intrusion detection alerts at one site to demonstrate that knowledge of known vulnerabilities for protected hosts can be used to prioritize alerts and focus on those that may represent successful systems compromises. This is followed by a summary and discussion of further issues.

2 Network-Based Intrusion Detection Systems

Network-based intrusion detection systems that use signatures to detect known attacks are pervasive because many hosts can be monitored by one intrusion detection system and no changes are required on monitored hosts. Descriptions and evaluations of research, commercial, and open-software network-based systems are available in [13,15,18,25]. These systems rely on signatures or rules to detect known attacks by comparing the contents of captured packets to signatures or rules. For example, to determine if an attacker is attempting to access a backdoor file left behind by the code-red worm [2], the string "scripts/root.exe" could be searched for in incoming traffic on TCP port 80. Signature-based systems normally do not detect new attacks. In some instances a new attack may be detected using an old signature [13,15], but this appears to be infrequent. It usually occurs only for attacks such as buffer overflows that exploit similar vulnerabilities. A new attack is typically not detected unless a new signature is developed for that attack and this signature is downloaded and installed in the intrusion detection system. Network-based intrusion detection systems also often do not distinguish between successful attacks and failed attempted attacks. For example, for the above code-red backdoor signature, the signature may create an alert whenever an external user attempts to execute the backdoor "root.exe" file whether or not the executable exists or whether it runs successfully.

The evaluations of intrusion detection systems referenced above noted many practical limitations of these systems including failing to match signatures when the network traffic load is too high, producing many false alarms, and being vulnerable to the insertion and evasion attacks described in [20]. The initial analysis in this paper assumes an ideal intrusion detection system without these practical limitations. It will be assumed that a network-based intrusion detection system issues an alert whenever an attacker attempts to exploit a known vulnerability, that there are no alerts for attacks that exploit novel new vulnerabilities, that there are no false alarms, and that after a new attack signature is installed, all instances of that attack are detected.

The initial analysis of this paper also focuses on secure networks where internal machines are behind a well-configured firewall, where only a small number of machines in a DMZ provide public networked services, and where a network intrusion detection system monitors traffic between the protected network and the Internet. This configuration is common at many government, educational, and commercial sites. It also influences system administrators to be concerned most about security for the small number of vulnerable machines in the DMZ, instead of the many more machines located behind a firewall.

Table 1. Three categories of remote attacks.

Attack Type	Description
Remote-to-Local	Remote attacker obtains privileges on a protected local machine that are normally reserved for local users and inaccessible to remote users
DoS	Deny access to a network service
Scan	Obtain information about hosts and network services

3 The Importance of Successful Remote-to-Local Attacks

The three categories of attacks shown in Table 1 are generally considered the most important types that can be detected using network-based intrusion detection. Remote-to-local or privilege-gaining attacks permit remote users to obtain privileges they are normally denied. The most serious remote-to-local attacks provide attackers with root-level privileges on Unix hosts and Administrator privileges on Microsoft Windows hosts. Denial of service (DoS) attacks deny access to a network service. Episodic DoS attacks send a small number of packets that induce a software or protocol fault while continuous DoS attacks consume a resource by sending a continuous stream of packets. Scan or reconnaissance attacks provide an attacker with information about hosts and network services. Other important attack categories including "abuse of Internet privileges" and "poor security practices" (e.g. connecting to restricted web sites, downloading restricted material, using telnet instead of ssh) are not included in Table 1 because these behaviors are site-specific and can be detected using traffic monitoring.

Table 2 shows the importance of these three attack types by considering the potential damage, the most common local site-specific response, the response cost, and the effect of the response on other, future attackers, who launch an identical attack against the same victim. This table applies only to known old attacks where software patches are available to prevent exploitation of the known vulnerability. The first row of this table analyzes successful remote-to-local and episodic DoS remote-to-local attacks. Episodic DoS attacks are included in this row because they have similar characteristics to remote-to-local attacks.

This paper focuses on detection of successful remote-to-local attacks in the first row of Table 2 because these are the most damaging and have enabled recent worldwide Internet security incidents including many worms and DDoS attacks [2,3,6,7,8,12,23]. As indicated, detecting these attacks as they occur allows system administrators to react by shutting down and cleaning up the compromised systems and protecting against future attacks by installing software upgrades and patches. Such rapid action can prevent theft of corporate data, monetary loss from theft of financial information, and much longer down times caused by further compromises

Table 2. The effectiveness of detecting and responding to known attacks.

Attack	Attack Damage	Immediate Local Response	Response Cost	Effect on Future Attackers
Successful Remote-to-Local and Episodic DoS	High	Shutdown, Analyze, Cleanup, Patch	High	Block This Attack Type
Failed Remote-to-Local and Episodic DoS (Probe)	None	Record/Block Source	Low	None
Scan	None	Record/Block Source	Low	None
Continuous DoS	Mod-High	Block Attacker(s)	Moderate	None

launched from the initial compromised machine. The dollar amount of losses due to remote-to-local attacks reported in [19] is roughly 5 to 35 times greater than that of DoS attacks depending on whether the cost of "Theft of proprietary information" is added to the cost of "System penetration by outsider" when assessing the cost of remote-to-local attacks. While these attacks can cause the most severe damage, they are also the easiest to prevent from reoccurring. Intrusion detection systems can detect the initial remote connection for these attacks or post-compromise actions including communications from the attacker, attempts to compromise other hosts, scanning from the compromised host, and packet streams sent from compromised hosts participating in DoS attacks. Detecting the initial remote connection simplifies the response and lessens the damage.

The next two rows in Table 2 contain failed remote-to-local and failed episodic DoS attacks (often called probes) and scan or reconnaissance attacks. Both types of attacks provide an attacker with information about remote hosts and network services but do no damage to the victim. Responses are also often ineffective in preventing future scans or probes. These attacks are ubiquitous and are enabled by many automated tools. Our experience on three separate class B address spaces, is that it is common to observe more than 40 separate scans per day from as many different sources with from hundreds to thousands of packets sent from each source to non-existent IP addresses. The Snort intrusion detection system [21] also often detects thousands of remote-to-local attacks each day. On one well-protected DMZ, careful hand verification of these alerts demonstrated that all were either caused by failed attacks or normal background traffic. Detecting scans and failed remote-to-local attacks provides situation awareness and sometimes indirectly detects previously successful remote-to-local attacks by identifying internal compromised machines that are scanning external hosts. The often-suggested response of blocking packets from external scanning or probing hosts may or may not be effective. It will block a novice attacker who scans and then attacks from the same source address but not more capable attackers who scan from compromised machines. It will also not block an attacker who uses public lists of IP addresses to find target machines or an attacker who falsifies the source IP address for probes. A non-local response not shown in Table 2 is to contact the system administrator of the external scanning machine. This can be extremely effective when possible, but requires identifying the location of the scanning machine and gaining the cooperation of an unknown administrator. Detecting probes and scans thus provides some situation awareness and protection from simple attacks from the same source, but it does not usually protect against future attacks of the same type from different sources.

Continuous DoS attacks are the third category in Table 2. Distributed DoS (DDoS) attacks where the attacker first compromises many hosts using remote-to-local attacks and installs agents that simultaneously send packets to one victim have become a major concern in the past few years [6]. Although they don't typically damage the target site, they prevent others from completing online purchases and obtaining information and can cause loss of revenue [19].

Intrusion detection systems are useful to detect continuous DoS attacks, however such attacks are relatively easy to detect at the victim by traffic monitoring tools including many tools used for network management. Blocking packets from the attack source(s) stops these attacks. Local blocking at a victim site will stop the current attack but not necessarily prevent future attacks of the same type. More global responses that involve modifying the Internet infrastructure can limit the severity of these types of attacks, but this requires cooperation across many sites. Intrusion detection systems are thus not necessary to detect DoS attacks and stopping an ongoing DoS attack provides little protection against future DoS attacks. Alternatively, detecting the original remote-to-local attacks used to create a group of agents can prevent these attacks.

In summary, remote-to-local attacks are the single most important category of known attacks that a network-based intrusion detection system can detect. They have the potential to inflict the most damage and installing software patches prevents other attackers from exploiting the same vulnerability. Detecting probes and scans provides situation awareness, but these attacks do no damage. Detecting continuous DoS attacks is important, but often possible without intrusion detection systems and does not lead to as great a financial loss as successful host compromises [19].

4 Windows of Vulnerability and Visibility

The ability to detect and respond to successful remote-to-local attacks depends on details concerning when vulnerabilities are discovered and publicized, when software patches or other fixes are made available and installed, and when signatures are made available and installed in intrusion detection systems. Figure 1 shows the primary events that determine if successful remote-to-local attacks will be detected and two of the many possible orderings of events. In both time lines, a new vulnerability is discovered and an exploit is developed that makes use of this vulnerability. After this,

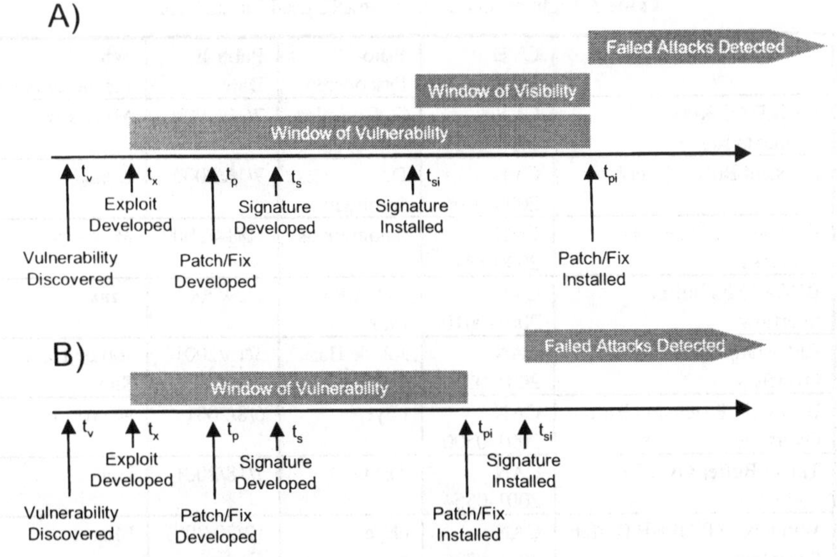

Fig. 1. Hosts are vulnerable to new exploits and there always is a window of vulnerability from the time a new exploit is developed until vulnerable systems are patched. There is a window of visibility, as shown in (A), only if a signature is installed in an intrusion detection system before a patch is installed and there is no window of visibility, as shown in (B), if the signature is installed after the patch.

a fix or software patch is made available that makes exploits using the new vulnerability impossible and a signature is developed for intrusion detection systems to detect attempted attacks that exploit the vulnerability. It is up to a system administrator to install the published signature and/or the patch and these can be installed in any order or never.

Independent of the order of other events, systems are always vulnerable to attacks from the time an exploit is developed until the time software patches or other fixes are installed. This time interval always exists and it will be called the "window of vulnerability" as in [1]. In addition, there may be a time interval when successful known attacks are detected by intrusion detection systems. This interval, if it exists, will be called the "window of visibility". The upper time line of Figure 1 labeled (A) shows a situation where a new signature is installed before the software patch and there is a window of visibility when new attacks are detected. The lower time line shows a situation when the signature is installed after software patches and there is no window of visibility when successful attacks are detected. There will only be a window of visibility when the time signatures are installed (t_{si}) is before the time patches are installed (t_{pi}). It is difficult to predict if there will be a window of visibility in practice because this depends on how long it takes to develop patches and signatures and on the strategy used by system administrators. There will be a window of visibility if patches are developed after signatures. There will not be a window of

Table 3. Eight remote-to-local attacks used for analysis.

Name	CVE Number	Who-Discovered	Publish Date	Who-Patched/Fixed
IIS MDAC RDS Vulnerability	CVE-1999-1011	G. Gonzalez	7/19/1999	Microsoft
rpc.statd Buffer Overflow	CVE-2000-0666	D. Jacobwitz	7/16/2000	Linux
IIS Unicode Directory Traversal	CVE-2000-884	Anonymous	10/14/2000	Microsoft
BIND TSIG buffer overflow	CVE-2001-0010	COVERT Labs	1/29/2001	Unix
RPC snmpXdmid Buffer Overflow	CAN-2001-0236	Job de Haas	3/15/2001	Job de Hass, Sun
IIS ISAPI Extension Buffer Overflow	CAN-2001-0500	eEye	6/8/2001	Microsoft
Telnet Buffer Overflow	CAN-2001-0554	TESO	7/18/2001	Unix
Windows XP UPNP Buffer Overflow	CAN-2001-0876	eEye	12/20/2001	Microsoft

visibility if patches are developed before signatures and administrators place a priority on minimizing the window of vulnerability by installing patches before signatures.

5 A Detailed Analysis of Eight Important Remote-to-Local Attacks

A detailed analysis was made of the eight remote-to-local vulnerabilities shown in Table 3 to determine likely sequences for those events shown in Figure 1. The first six of these vulnerabilities were selected from among the 20 most important security incidents cataloged in [22]. The last two are serious recently-discovered vulnerabilities. Information on these attacks is available from the NIST ICAT meta database [14] and in a recent security advisory [10]. Table 3 contains the attack name, the Common Vulnerabilities and Exposures (CVE) number [4], the person or organization that discovered the vulnerability, the date the vulnerability used by the attack was published, and the identity of the organization that published the patch. Dates of attacks shown range from mid 1999 to late 2001, both individuals and security companies discovered vulnerabilities, and the affected software developers provided patches. In this table, the terms "Unix" and "Linux" in the last column indicate that many different software developers of these operating systems developed patches.

Figure 2 shows time lines for all eight attacks. Each separate horizontal region corrresponds to one attack, the horizontal axis shows dates from October 1998 to

February 2002, and symbols show the dates of different events. Open squares indicate when vulnerabilities were published and plus signs indicate when fixes or patches were available. These dates were obtained by going back to the original mailing lists or web postings where the vulnerabilities were announced. The patch/fix dates correspond to the dates patches were released for all but the RPC snmpXdmid Buffer Overflow vulnerability. Here, the date is that of the fix suggested by the vulnerability discoverer and Sun released a patch more than five months after the vulnerability was discovered. The availability date for signatures was determined by finding two dates. One was when a signature was available for a widely used open-source intrusion detection system named Snort [21] with performance comparable to that of many commercial systems [18]. The second date was when a security test was available in an open-source vulnerability scanner named Nessus [17] that also performs comparably to commercial vulnerability scanners [9].

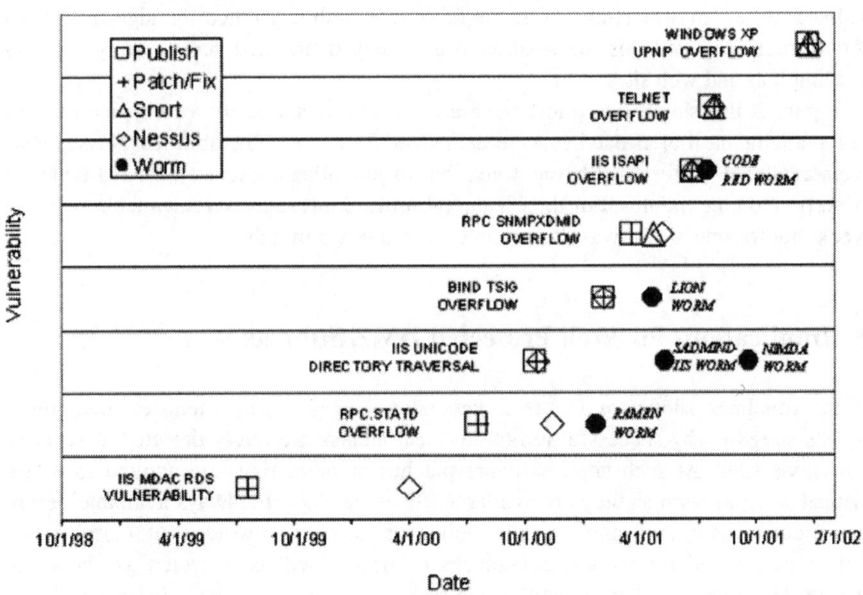

Fig. 2. Dates eight important vulnerabilities were publicly released, software patches or fixes were made available, a Snort signature was available, a Nessus vulnerability-testing rule was available, and (if applicable) when major Internet worms exploited the vulnerability.

The Snort and Nessus dates represent those that might be provided by a user group that carefully tracks publication of new vulnerabilities. Dates were found by searching postings of new security test listings for Nessus and both software repositories and new signature postings for Snort. It was found that Snort signatures were not being updated frequently until roughly April 2001. Times for Snort signatures were thus left out if they occurred before this date while dates for all Nessus security tests are listed. If an attack in Figure 2 was part of a major Internet worm incident, then the date of the worm release is shown by a solid dot. Information on the worms shown is available in [2,3,7,8,12,23].

Figure 2 shows that patches or software fixes were almost always available soon after vulnerabilities were publicly released. This agrees with the analysis of three older attacks performed in [1]. For many vulnerabilities, the discoverer notified software developers and held back public release until a software patch was available. In others the discoverer presented a fix to prevent the vulnerability from being exploited or software developers rapidly produced patches following the vulnerability announcement. Patches were available on the day the vulnerability was announced or the next day for five of the eight vulnerabilities and within ten days for all vulnerabilities. In almost all cases, the discoverer publicly noted the significance of the vulnerability and this information was widely distributed across many security mailing lists and web sites.

Figure 2 also shows that Snort signatures and Nessus security tests typically were not available until after patches were developed. For five vulnerabilities, Nessus tests were available within roughly one week, but for the others, tests weren't available for from two to nine months. For three vulnerabilities Snort rules were available within a week, but for one, a rule wasn't available for roughly a month.

6 Implications for Well Protected DMZ Subnets

If the timelines shown in Figure 2 generalize to other vulnerabilities, then these results suggest why successful remote-to-local attacks are rarely detected at security conscious sites. At such sites, software patches or other fixes are applied to a few critical hosts as soon as they are available. Since patches are always available before or simultaneous with signatures, this implies that there is no window of visibility and only unsuccessful remote-to-local attacks or probes will be detected as shown in Figure 1B. Signature-based network intrusion detection system will never detect successful remote-to-local attacks at such sites. They will simply verify that failed attempts occur.

The strategy of rapidly installing software updates on critical servers would be too costly for widespread adaptation if new vulnerabilities were discovered too frequently. Figure 3 shows that this is not the case. It shows the dates of high severity remote-to-local vulnerabilities over six years as recorded in the NIST ICAT meta-database [14]. These are vulnerabilities that can be exploited remotely and have been rated "severe" by NIST. They include remote-to-local attacks that gain root or administrator privileges, episodic DoS attacks, and other attacks that a security-aware system administrator would want to prevent. These dates are taken from the ICAT database and not from the original vulnerability announcements so they differ slightly

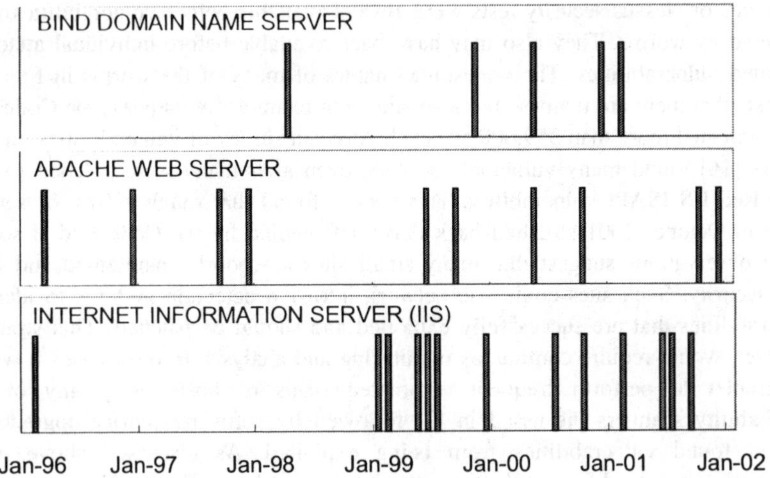

Fig. 3. Dates for high severity remote-to-local attacks recorded in the ICAT database for bind DNS server, Apache web server, and Internet Information Server (IIS) Software.

from the dates in Figure 2. Vulnerabilities that were discovered within one day are counted as single vulnerabilities because they are typically patched simultaneously.

This figure shows the number of severe vulnerabilities for the two most popular web servers (Apache and IIS) that together account for roughly 85% of all web servers [16] and for the BIND software package that is widely used as a domain name server [11]. These three software packages account for half of the vulnerabilities shown in Figure 2. All packages have had 5 or more serious vulnerabilities over six years with at most 15 serious vulnerabilities over six years for any one package, and at most 6 in any one year. Rates of high-severity remote vulnerabilities for other servers (e.g. WU-FTPD and Sendmail) are similar. Figure 3 shows that installing software patches on a small number of machines in a DMZ with limited services (e.g. web, mail, FTP, and DNS servers) requires daily monitoring of security alerts and substantial effort only every few weeks to months. It is thus a practical strategy for securing a small number of machines. A signature-based network intrusion detection system on a well-maintained DMZ provides backup protection, but should never detect successful remote-to-local attacks.

7 Implications for Small Poorly Protected Sites

At poorly protected sites with few hosts, where patches are not installed rapidly, Figure 2 suggests that signatures would have been available to detect attacks that were part of major Internet worm incidents and there would have been a window of visibility where successful attacks were detected, as shown in Figure 1A. Snort signatures or Nessus security tests were always available before vulnerabilities were exploited by worms. They also may have been available before individual attackers exploited vulnerabilities. The widespread nature of many of the worms in Figure 2 suggests that there are many vulnerable sites. For example, at its peak, the Code Red worm infected more than 359,000 sites [2]. A recent survey of vulnerabilities on web servers [16] found many vulnerable servers, even after many were patched for the Code Red IIS ISAPI vulnerability. This survey found that roughly 10% of servers tested in October 2001 still had back doors left behind by the Code Red II worm. These observations suggest that many small sites are poorly maintained and have little security. Such sites could use network intrusion detection systems to identify vulnerabilities that are successfully exploited and should be patched. This strategy, however, would require continuous monitoring and analysis. In most cases it would be simpler to perform frequent automated scans of hosts using any of the vulnerability scanners discussed in [9] followed by software/security upgrades to prevent found vulnerabilities from being exploited. As discussed above, such upgrades are practical for small sites without too many hosts. For the Nessus scanner, a strategy of automated scanning and patching performed every two weeks would have prevented hosts from being compromised by the Internet worms shown.

8 Implications for Normal Large Sites

Most sites are not as completely protected as the above secure DMZ or as poorly protected as sites used by worms that exploited year-old vulnerabilities. Most of us probably work at sites with known, but recent, vulnerabilities. This is supported by our own vulnerability testing and that of others (e.g. [16]) that finds substantial percentages of vulnerable hosts. It is also supported by a recent survey [19] where 40% of sites responding reported that one or more systems were compromised from a remote attacker over the preceding year. Vulnerabilities exist because it is difficult to eliminate all known vulnerabilities. At sites with 100's to 1000's of hosts, servers, routers, switches, and other equipment that is protected by firewalls, it is practically impossible to maintain software patches on all systems and also enforce firewall filtering and proxy rules required to prevent new vulnerabilities from being exploited. Although software management tools are being developed to simplify this task, they are not in widespread use yet and patches are frequently not installed until they are fully tested because they sometimes do not work or they disable important capabilities. It is also difficult to coordinate security concerns across many system administrators, to keep track of existing hardware, and to make sure that installing new equipment does not create new vulnerabilities. Software upgrades or fixes for known vulnerabilities also can sometimes not be installed because they disable an essential network capability or make software incompatible. There will thus almost always be machines with known vulnerabilities at any large site and intrusion detection systems will issue alerts when these machines are successfully compromised. A major problem, however, is that there are often so many alerts due to failed remote-to-local attacks and normal background traffic that alerts for successful attacks are missed.

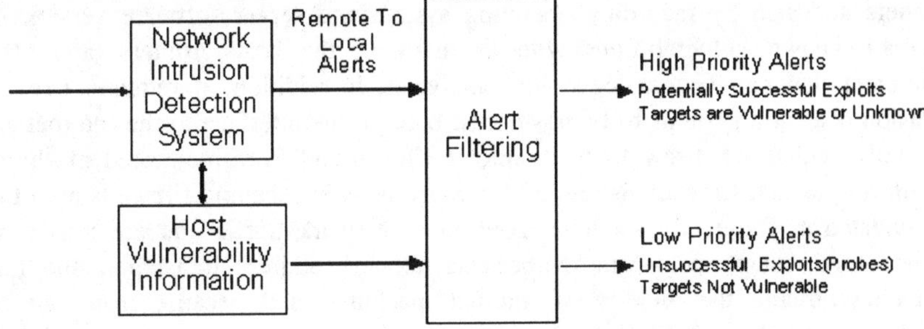

Fig. 4. Alert filtering based on vulnerabilities of protected hosts separates remote-to-local alerts into high and low priority bins

9 Using Vulnerability Information to Prioritize Alerts

Our experience is that network intrusion detection systems often produce 100's to 1000's of remote-to-local alerts per day on a class B network. Many of these correspond to failed attacks, some are false alarms caused by normal network traffic, and a very small number are caused by successful attacks. It is essential to find the few successful attacks.

It would be trivial to prioritize attacks if intrusion detection systems reliably indicated whether attacks succeeded or failed. Unfortunately, most don't. It is possible to determine the success of well-known scripted attacks, such as the code-red worm, by detecting actions performed after a successful compromise. It is also sometimes possible to monitor the response from web and other servers for evidence of a compromise. In general, however, it is difficult for a network monitor to determine when attacks succeed. In addition to the issues described in [20], many other characteristics of network monitoring make this difficult. First, some intrusion detection systems analyze only single packets or the packet stream from outside to inside addresses and either do not analyze the response from the attacked machine or do not correlate the response with the incoming request. Second, even if the response is analyzed, sometimes incoming and outgoing packets travel over separate paths and either only incoming or only outgoing packets are visible to the intrusion detection system. Third, many attacks produce no visible immediate response (e.g. episodic DoS attacks or attacks that install backdoors) or communicate back to the attacker using a different transport mechanism than was used to launch the attack. Communications back to the attacker can be sent hours or days after the attack, encrypted or "tunneled" through a common TCP service, and sent back to a different address than was used to launch the attack (e.g. see the HTTP tunnel attack in [13]).

One approach to determine which intrusion detection alerts correspond to successful remote-to-local attacks is to use information concerning known vulnerabilities and hosts to filter alerts into high and low priority bins as shown in Figure 4. This requires site-specific data that can be obtained from vulnerability scanners and also by recording operating system and server software versions to associate known vulnerabilities with as many of the hosts, routers, and other equipment that can be cataloged and analyzed. In addition, it requires network intrusion detection systems to be positioned both on the interface to the Internet and internally behind any firewalls to monitor traffic to and from monitored machines. Monitoring behind firewalls is required because machines behind firewalls are often not updated as frequently as those exposed to network traffic and are thus often vulnerable to many more attacks, because network address translation and load balancing obscure the identity of internal machines, and because many attacks originate from behind firewalls [19].

Figure 4 illustrates how alerts for remote-to-local attacks can be filtered into high and low priority bins. The high-priority bin shown at the top right of this figure contains alerts for vulnerable hosts. These are alerts corresponding to vulnerabilities that are known to exist and alerts for hosts where no information is available concerning a particular vulnerability such as alerts from recently installed hosts. The second low-priority bin shown at the bottom right of Figure 4 contains presumably failed remote-to-local attacks against hosts not vulnerable to detected exploits. To find successful attacks, an analyst would first examine alerts in the high-priority bin and then examine alerts in the low-priority bin. When the time available for analyzing alerts is limited, and all alerts cannot be hand examined, this will result in more detections of successful attacks by focusing on more important alerts. In practice, alerts in the low-priority bin should be sampled (especially previously unseen alert types for existing hosts) because new software releases sometimes inadvertently re-enable old vulnerabilities, software tools that may contain known vulnerabilities are often installed without notifying system administrators, and vulnerability data may be incorrectly recorded or transferred. This approach will be successful if few alerts are left in the high-priority bin and the vulnerability analysis is correct.

An analysis of remote-to-local alerts from the Snort intrusion detection system at one class B site containing roughly 10 Microsoft IIS web servers was performed to determine the percentage of alerts that are left in the high-priority bin with this approach. This analysis suggests that this approach can be extremely effective in reducing the number of alerts that require immediate attention by a system administrator or security analyst. There were roughly 845 alerts per day over roughly two weeks that could have indicated successful remote-to-local attacks. The top 27 types of alerts generated more than 5 alerts per day each and together were responsible for generating roughly 830 remote-to-local alerts a day. Of these 830, roughly 95% could be placed in the lower-priority bin based on knowledge of the software patches, operating systems, services, and software versions running on the web servers. This left only 5% or 42 alerts per day from one alert type in the high priority bin. This alert was a generic signature used to detect web traversals by scanning for "..\" and "../" in web requests. It detected primarily failed attacks or probes for a variety of web traversal attacks and false alarms for relative path addressing used in web pages. A detailed analysis of these alerts involving full-time intrusion detection analysts and site system administrators indicated that the low-priority alerts were all failed attacks or false alarms due to normal traffic.

At this site, this approach successfully reduced the number of high-priority remote-to-local alerts by roughly a factor of twenty (from roughly 800 to 40 per day). It also didn't require extensive information concerning software on protected hosts. Table 4 shows the information required to assign a low priority to the most common alerts. Information was required concerning the host operating system, the existence of web server extensions, and the installation of specific IIS patches.

It was not simple to determine specific questions that could be used to prioritize alerts. First, alert types corresponding to potential remote-to-local attacks had to be identified. Then, a detailed analysis of alerts and associated vulnerabilities was required to identify software components necessary for the success of attacks. In many cases, this was made overly complex by poor descriptions of alerts that required a detailed analysis of alert signatures themselves, poor cross referencing from alerts to information describing vulnerabilities, and non-standardized and distributing documentation of vulnerabilities. This analysis was made somewhat easier by the use of the CVE vulnerability numbering system [4] to associate alerts to vulnerabilities. Some alerts that could not be assigned to a low priority were false alarms caused by normal background traffic. Categorizing these alerts required a detailed analysis of signatures and background traffic. Ideally, an analysis to identify background traffic that can cause false alarms and also to specify conditions that must be met for an alert to stay at a high priority could be performed by intrusion detection developers and users. The result would be a shared list of conditions that must be met for an exploit that triggers an alert to succeed and also a list of normal traffic that might cause false alarms.

Table 4. Reasons for assigning a low priority to the most frequent remote-to-local alerts. Generic "http directory traversal" alerts could not be assigned a low priority.

Alert Name	Ave Alerts per Day	If Low Priority, Why
Web cgi redirect	140	Not running cold fusion
Sun RPC high port access	110	No Sun Servers
front page shtml.exe	85	No FrontPage
iis-vti_inf	67	IIS Patched
front page shtml.dll	63	No Frontpage
http directory traversal	54	
Shaft to client	30	No UNIX Servers
DDoS Mstream client to handler	25	No UNIX Servers
ISAPI Overflow ida	23	IIS Patched

The analysis presented in Figure 3 also requires accurate and timely information concerning IP addresses for hosts exposed to the Internet and knowledge of their operating systems, software, and patches. Vulnerability and network scanners simplify the task of gathering this information for large numbers of workstations. As

Figure 2 shows, rules for at least one scanner are now being rapidly updated to detect new vulnerabilities. Scanners are being used at many sites to gather information on operating system and software versions as well as information on large numbers of known vulnerabilities. One limitation of scanners is that it is difficult to monitor all hosts and discover new hosts continuously. This limitation can partially be overcome by continuously performing passive analysis of traffic within network-based intrusion detection systems. The double arrow between intrusion detection and host vulnerability components in Figure 4 indicates this type of analysis. Passive analysis can be used to detect new hosts recently connected to the Internet that should be analyzed for vulnerabilities. It also can potentially identify the operating systems of monitored hosts. For example, it might be possible to determine the operating system type of protected hosts using a combination of "passive operating system fingerprinting" of TCP packet headers as described in [24], analysis of banners produced by server software, and analysis of the services offered by hosts. Although this approach has not been carefully explored, a study reported in [5] used packet header information to identify operating systems and filter remote-to-local alerts. Experiments on a few networks demonstrated that often roughly 30% of the remote-to-local alerts could be sent to the low-priority bin by simple filtering based on operating system type. This approach should only be used to detect recently connected hosts and perform a preliminary analysis of those hosts. The analysis should preferably be extended and confirmed by active scanning and consultation with system administrators.

10 Summary and Discussion

It is has recently become easier to scan for known vulnerabilities in hosts and to obtain software patches designed to eliminate these vulnerabilities. This capability has had two seemingly contradictory effects on the usefulness of network intrusion detection systems. On small sites, such as small DMZ networks, network intrusion detection systems might never detect successful systems compromises because vulnerabilities can be patched before intrusion detection systems have been updated to detect associated attacks. On small sites, vulnerability scanning and patch installation delegates network intrusion detection into a backup role. On large sites, it is too expensive to install patches and network intrusion detection systems may detect system compromises. These important detections, however, are often hidden among thousands of unimportant alerts caused by failed attacks and normal background traffic. On such sites, information on vulnerabilities and protected hosts can be used to prioritize alerts and focus on those that might represent successful exploitation of known vulnerabilities. On large sites, vulnerability scanning and information on protected hosts can thus make network intrusion detection useful and practical.

The analyses that led to these conclusions focused on dangerous remote-to-local attacks where a remote attacker achieves restricted privileges on protected hosts and compromises those hosts. It also focused on the common approach to network intrusion detection where signatures are used to detect known attacks. The first part of this paper explored the time sequence for availability of software patches, intrusion detection signatures, and vulnerability scanner rules following announcements of new vulnerabilities. It was found that software patches are almost always available before new intrusion detection signatures. "Windows of visibility" where intrusion detection systems detect successful system compromises will never occur if software patches are installed as soon as they are available. In addition, it was found that vulnerability-scanning rules are available soon after new vulnerabilities are announced and, so far, they have been available well before a vulnerability was used as part of a widespread worm attack. Timelines from eight recent important vulnerabilities support these sequences of events. Signature release times were for the Snort network intrusion detection system and vulnerability scanner rule times were for the Nessus security scanner. It is expected that this order of events will continue and that intrusion detection signatures and rules to scan for vulnerabilities will be distributed simultaneous with the public disclosure of new vulnerabilities or only slightly after. This makes it possible to protect against exploits that use new vulnerabilities by scanning a site to detect vulnerable hosts and installing software patches whenever a new vulnerability is announced. Such a strategy would minimize the "window of vulnerability" where a host is susceptible to exploitation using a new known vulnerability. As noted above, it would also mean that there would never be a "window of visibility" where a network intrusion detection system using signatures detects successful remote-to-local attacks. One limitation of this analysis is that few attacks were examined and non-malicious security groups or individuals discovered all attacks. Even if vulnerabilities were discovered and exploited by a malicious group, the analysis suggests that software developers would develop and release patches soon after these "in-the-wild" exploits were discovered and that intrusion detection signatures would be provided simultaneously with patches. This work also only analyzed signature and rule development dates for one network intrusion detection system and one vulnerability scanner. Since highest priority is normally placed on developing patches, it is likely that similar results would be obtained for other intrusion detection systems and scanners.

The strategy of installing patches as soon as they are available was demonstrated to be practical for small numbers of hosts in DMZ's because the discovery rate for major vulnerabilities of Internet server software packages was shown to range from zero to seven per year. With three or four server software packages in a DMZ, serious security software patches may be required at most a few times a month. On such well-protected sites, network intrusion detection systems should never detect successful remote-to-local attacks. They could, however, serve as a backup and provide some "defense in depth." They can verify that patches are installed correctly by detecting failed remote-to-local attacks (probes), detect DoS attacks, and provide situation awareness by detecting scans.

Intrusion detection systems are more useful on large sites with known vulnerabilities. It is often necessary to use hosts with known vulnerabilities because it may be too costly or time consuming to update all the required software, because using vulnerable software may be the only solution to satisfy operational or compatibility requirements, or because no centralized authority is available to scan and enforce a security policy across all vulnerable hosts. On sites containing systems with known vulnerabilities, network intrusion detection systems can detect successful system compromises, but this requires constant monitoring and labor-intensive analysis of thousands of alerts per day caused by failed attacks and normal background traffic. An analysis of alerts from the Snort intrusion detection demonstrated that knowledge of the operating system, software packages, and vulnerabilities of monitored hosts provides information that can be used to prioritize these alerts. Of roughly 830 remote-to-local alerts that occurred each day, 95% corresponded to vulnerabilities that did not exist on a site with roughly 10 web servers and could be categorized as low priority alerts. This left only 5% or roughly 40 high priority alerts per day that could have been caused by successful system compromises. Prioritization didn't require knowing all details such as patch levels of all software running on monitored hosts. It often required only being able to prove that the host isn't susceptible to a vulnerability by knowing that is is not using a specific operating system (e.g. it is not UNIX) or that it is not running a particular server extension (e.g. it is not running FrontPage). Useful prioritization does not necessarily require maintenance of a large database including detailed patch level information. It requires lists of the identities of monitored hosts, of their operating systems, and of server software. Even knowledge of operating systems and major server types can help prioritize alerts.

In summary, vulnerability scanning and applying software patches should always be a component of site security. On small sites, such as small DMZ networks, these tactics can make network intrusion detection systems serve a secondary role because vulnerabilities can be patched before intrusion detection systems have new signatures required to detect attacks. On large sites, it is too expensive to patch all systems and network intrusion detection systems can serve a useful function by detecting successful compromises. On such sites, information on vulnerabilities and protected hosts is required to prioritize alerts and focus on those that might represent successful exploitation of known vulnerabilities.

11 Limitations and Future Work

Further analyses can be performed to explore the interrelationships between vulnerability scanning, software patching, gathering information on protected hosts and intrusion detection. This paper was limited to known attacks, primarily to remote-to-local attacks, and to network intrusion detection systems that use signatures to detect known attacks. Conclusions do not apply to network intrusion detection systems that detect novel new attacks without signatures. Such systems could provide an early warning function by guiding forensic analysis of new attacks that could lead to new patches and signatures to block and detect future exploitation of newly detected vulnerabilities. As noted above, further remote-to-local attacks could be analyzed and timeline information could be obtained for other intrusion detection systems and vulnerability scanners. Such new results are unlikely to change the major conclusions except when patches are so difficult to develop that they are distributed after signatures are available.

Some of our current work focuses on passive analysis of network traffic to determine host information required to filter alerts. Passive analysis can provide a list of protected hosts and alert analysts to newly installed unauthorized hosts that may be running software with known vulnerabilities. It might also be able to provide information concerning operating systems and server software by examining software banners and packet headers.

Acknowledgements. We would like to thank Mike Wilson, Hermann Segmuller, and Gene Solloway for helping analyze Snort intrusion detection alerts and both Marc Zissman and Peter Heldt for advice and administrative expertise.

References

1. Arbaugh, W.A., W.L. Fithen, and J. McHugh, Windows of Vulnerability: A Case Study Analysis, IEEE Computer, 2000. 33,(12), 52-59, http://www.cs.umd.edu/~waa/pubs/Windows_of_Vulnerability.pdf.
2. CAIDA, Code-Red Worms: A Global Threat, Cooperative Association for Internet Data Analysis (CAIDA), 28 November 2001, http://www.caida.org/analysis/security/code-red/.
3. Chien, E., W32.Nimda.A@mm Worm, Symantec Corporation, 18 September 2001, http://www.symantec.com/avcenter/venc/data/w32.nimda.a@mm.html.
4. CVE, Common Vulnerabilities and Exposures, The MITRE Corporation, 2002, http://www.cve.mitre.org/.
5. Dayioglu, B. and A. Ozgit, Use of Passive Network Mapping to Enhance Signature Quality of Misuse Network Intrusion Detection Systems, in Proceedings of the Sixteenth International Symposium on Computer and Information Sciences, 2001, http://www.dayioglu.net/publications/iscis2001.pdf.
6. Dittrich, D., Distributed Denial of Service (DDoS) Attacks/tools, University of Washington, Seattle, 2001, http://staff.washington.edu/dittrich/misc/ddos/.
7. Dougherty, C., S. Hernan, J. Havrilla, J. Carpenter, A. Manion, I. Finlay, and J. Shaffer, CERT Advisory CA-2001-11 sadmind/IIS Worm, CERT Coordination Center, 8 May 2001, http://www.cert.org/advisories/CA-2001-11.html.
8. Fearnow, M. and W. Stearns, Lion Worm, SANS Institute, 29 March 2001, http://www.incidents.org/react/lion.php.
9. Forristal, J. and G. Shipley, Vulnerability Assessment Scanners, Network Computing, 8 January 2001, http://www.networkcomputing.com/1201/1201f1b1.html.
10. Hassell, R., R. Permeh, and M. Maiffret, UPNP - Multiple Remote Windows XP/ME/98 Vulnerabilities, eEye Digital Security, 20 December 2001, http://www.eeye.com/html/Research/Advisories/AD20011220.html.
11. Internet Software Consortium, ISC Berkeley Internet Name Domain (BIND) Domain Name System (DNS), January 2002, http://www.isc.org/products/BIND/.

12. Lestat, M., The Ramen Worm and its use of rpc.statd, wu-ftpd and LPRng Vulnerabilities in Red Hat Linux, SANS Institute, 7 February 2001, http://rr.sans.org/malicious/ramen.php.
13. 13. Lippmann, R.P., J.W. Haines, D.J. Fried, J. Korba, and K. Das, The 1999 DARPA off-line intrusion detection evaluation. Computer Networks, 2000. 32: pp. 579-595.
14. Mell, P. and T. Grance, The ICAT Metabase CVE Vulnerability Search Engine, National Institute of Standards and Technology, January 2002, http://icat.nist.gov.
15. Mueller, P. and G. Shipley, To Catch a Thief, Network Computing, 2001, http://www.networkcomputing.com/1217/1217f1.html.
16. Netcraft Web Server Survey, Netcraft Ltd., Bath England, October 2001, http://www.netcraft.com/survey/index-200110.html.
17. Nessus, The Nessus Security Scanner, 2002, http://www.nessus.org.
18. NSS Group, Intrusion Detection Systems Group Test (Edition 2), Oakwood House, Wennington, Cambridgeshire, England, December 2001, http://www.nss.co.uk/ids/.
19. Power, R., 2001 CSI/FBI Computer Crime and Security Survey, Computer Security Institute, Spring 2000, http://www.gocsi.com/forms/fbi/pdf.html.
20. Ptacek, T.H. and T.N. Newsham, Insertion, Evasion, and Denial of Service: Eluding Network Intrusion Detection. Secure Networks, Inc., 1998, http://secinf.net/info/ids/idspaper/idspaper.html.
21. Roesch, M. Snort - Lightweight Intrusion Detection for Networks, in USENIX 13th Systems Administration Conference - LISA '99. Seattle, Washington, 1999, http://www.snort.org.
22. SANS, The Twenty Most Critical Internet Security Vulnerabilities (Updated). Bethesda, MD, System Administration, Networking, and Security (SANS) Institute, 2001, http://www.sans.org/top20.htm.
23. SANS, NIMDA Worm/Virus Report – Final, System Administration, Networking, and Security (SANS) Institute, October 2001, http://www.incidents.org/react/nimda.pdf.
24. Spitzner, L., Know Your Enemy: Passive Fingerprinting, Honeynet Project, January 2002, http://project.honeynet.org/papers/finger/.
25. Yocom, B., K. Brown, and D.V. DerVeer, Review: Intrusion-Detection Products Grow Up, Network World Fusion, 2001, http://www.nwfusion.com/reviews/2001/1008rev.html.

Author Index

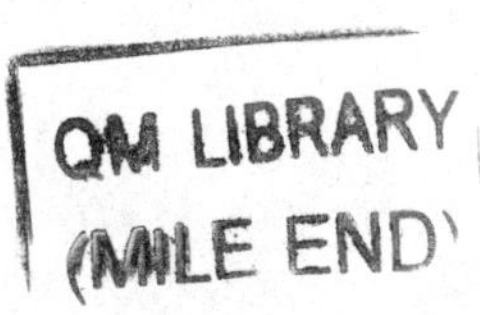

Lecture Notes in Computer Science

For information about Vols. 1–2420
please contact your bookseller or Springer-Verlag

Vol. 2343: I. Chong (Ed.), Information Networking. Proceedings, Part I, 2002. XIX, 729 pages. 2002.

Vol. 2344: I. Chong (Ed.), Information Networking. Proceedings, Part II, 2002. XX, 825 pages. 2002.

Vol. 2421: L. Brim, P. Jančar, M. Křetínský, A. Kučera (Eds.), CONCUR 2002 – Concurrency Theory. Proceedings, 2002. XII, 611 pages. 2002.

Vol. 2422: H. Kirchner, Ch. Ringeissen (Eds.), Algebraic Methodology and Software Technology. Proceedings, 2002. XI, 503 pages. 2002.

Vol. 2423: D. Lopresti, J. Hu, R. Kashi (Eds.), Document Analysis Systems V. Proceedings, 2002. XIII, 570 pages. 2002.

Vol. 2424: S. Flesca, G. Ianni (Eds.), Logics in Artificial Intelligence. Proceedings, 2002. XIII, 572 pages. 2002. (Subseries LNAI).

Vol. 2425: Z. Bellahsène, D. Patel, C. Rolland (Eds.), Object-Oriented Information Systems. Proceedings, 2002. XIII, 550 pages. 2002.

Vol. 2426: J.-M. Bruel, Z. Bellahsène (Eds.), Advances in Object-Oriented Information Systems.Procedings, 2002. IX, 314 pages. 2002.

Vol. 2427: M. Hannebauer, Autonomous Dynamic Reconfiguration in Multi-Agent Systems. XXI, 284 pages. 2002. (Subseries LNAI).

Vol. 2428: H. Hermanns, Interactive Markov Chains. XII, 217 pages. 2002.

Vol. 2429: P. Druschel, F. Kaashoek, A. Rowstron (Eds.), Peer-to-Peer Systems. Proceedings, 2002. IX, 339 pages. 2002.

Vol. 2430: T. Elomaa, H. Mannila, H. Toivonen (Eds.), Machine Learning: ECML 2002. Proceedings, 2002. XIII, 532 pages. 2002. (Subseries LNAI).

Vol. 2431: T. Elomaa, H. Mannila, H. Toivonen (Eds.), Principles of Data Mining and Knowledge Discovery. Proceedings, 2002. XIV, 514 pages. 2002. (Subseries LNAI).

Vol. 2432: R. Bergmann, Experience Management. XXI, 393 pages. 2002. (Subseries LNAI).

Vol. 2433: A.H. Chan, V. Gligor (Eds.), Information Security. Proceedings, 2002. XII, 502 pages. 2002.

Vol. 2434: S. Anderson, S. Bologna, M. Felici (Eds.), Computer Safety, Reliability and Security. Proceedings, 2002. XX, 347 pages. 2002.

Vol. 2435: Y. Manolopoulos, P. Návrat (Eds.), Advances in Databases and Information Systems. Proceedings, 2002. XIII, 415 pages. 2002.

Vol. 2436: J. Fong, C.T. Cheung, H.V. Leong, Q. Li (Eds.), Advances in Web-Based Learning. Proceedings, 2002. XIII, 434 pages. 2002.

Vol. 2437: G. Davida, Y. Frankel, O. Rees (Eds.), Infrastructure Security. Proceedings, 2002. XI, 339 pages. 2002.

Vol. 2438: M. Glesner, P. Zipf, M. Renovell (Eds.), Field-Programmable Logic and Applications. Proceedings, 2002. XXII, 1187 pages. 2002.

Vol. 2439: J.J. Merelo Guervós, P. Adamidis, H.-G. Beyer, J.-L. Fernández-Villacañas, H.-P. Schwefel (Eds.), Parallel Problem Solving from Nature – PPSN VII. Proceedings, 2002. XXII, 947 pages. 2002.

Vol. 2440: J.M. Haake, J.A. Pino (Eds.), Groupware: Design, Implementation and Use. Proceedings, 2002. XII, 285 pages. 2002.

Vol. 2441: Z. Hu, M. Rodríguez-Artalejo (Eds.), Functional and Logic Programming. Proceedings, 2002. X, 305 pages. 2002.

Vol. 2442: M. Yung (Ed.), Advances in Cryptology – CRYPTO 2002. Proceedings, 2002. XIV, 627 pages. 2002.

Vol. 2443: D. Scott (Ed.), Artificial Intelligence: Methodology, Systems, and Applications. Proceedings, 2002. X, 279 pages. 2002. (Subseries LNAI).

Vol. 2444: A. Buchmann, F. Casati, L. Fiege, M.-C. Hsu, M.-C. Shan (Eds.), Technologies for E-Services. Proceedings, 2002. X, 171 pages. 2002.

Vol. 2445: C. Anagnostopoulou, M. Ferrand, A. Smaill (Eds.), Music and Artificial Intelligence. Proceedings, 2002. VIII, 207 pages. 2002. (Subseries LNAI).

Vol. 2446: M. Klusch, S. Ossowski, O. Shehory (Eds.), Cooperative Information Agents VI. Proceedings, 2002. XI, 321 pages. 2002. (Subseries LNAI).

Vol. 2447: D.J. Hand, N.M. Adams, R.J. Bolton (Eds.), Pattern Detection and Discovery. Proceedings, 2002. XII, 227 pages. 2002. (Subseries LNAI).

Vol. 2448: P. Sojka, I. Kopeček, K. Pala (Eds.), Text, Speech and Dialogue. Proceedings, 2002. XII, 481 pages. 2002. (Subseries LNAI).

Vol. 2449: L. Van Gool (Ed.), Pattern Recognotion. Proceedings, 2002. XVI, 628 pages. 2002.

Vol. 2451: B. Hochet, A.J. Acosta, M.J. Bellido (Eds.), Integrated Circuit Design. Proceedings, 2002. XVI, 496 pages. 2002.

Vol. 2452: R. Guigó, D. Gusfield (Eds.), Algorithms in Bioinformatics. Proceedings, 2002. X, 554 pages. 2002.

Vol. 2453: A. Hameurlain, R. Cicchetti, R. Traunmüller (Eds.), Database and Expert Systems Applications. Proceedings, 2002. XVIII, 951 pages. 2002.

Vol. 2454: Y. Kambayashi, W. Winiwarter, M. Arikawa (Eds.), Data Warehousing and Knowledge Discovery. Proceedings, 2002. XIII, 339 pages. 2002.

Vol. 2455: K. Bauknecht, A M. Tjoa, G. Quirchmayr (Eds.), E-Commerce and Web Technologies. Proceedings, 2002. XIV, 414 pages. 2002.

Vol. 2456: R. Traunmüller, K. Lenk (Eds.), Electronic Government. Proceedings, 2002. XIII, 486 pages. 2002.

Vol. 2458: M. Agosti, C. Thanos (Eds.), Research and Advanced Technology for Digital Libraries. Proceedings, 2002. XVI, 664 pages. 2002.

Vol. 2459: M.C. Calzarossa, S. Tucci (Eds.), Performance Evaluation of Complex Systems: Techniques and Tools. Proceedings, 2002. VIII, 501 pages. 2002.

Vol. 2460: J.-M. Jézéquel, H. Hussmann, S. Cook (Eds.), «UML» 2002 – The Unified Modeling Language. Proceedings, 2002. XII, 449 pages. 2002.

Vol. 2461: R. Möhring, R. Raman (Eds.), Algorithms – ESA 2002. Proceedings, 2002. XIV, 917 pages. 2002.

Vol. 2462: K. Jansen, S. Leonardi, V. Vazirani (Eds.), Approximation Algorithms for Combinatorial Optimization. Proceedings, 2002. VIII, 271 pages. 2002.

Vol. 2463: M. Dorigo, G. Di Caro, M. Sampels (Eds.), Ant Algorithms. Proceedings, 2002. XIII, 305 pages. 2002.

Vol. 2464: M. O'Neill, R.F.E. Sutcliffe, C. Ryan, M. Eaton, N. Griffith (Eds.), Artificial Intelligence and Cognitive Science. Proceedings, 2002. XI, 247 pages. 2002. (Subseries LNAI).

Vol. 2465: H. Arisawa, Y. Kambayashi, V. Kumar, H.C. Mayr, I. Hunt (Eds.), Conceptual Modeling for New Information Systems Technologies. Proceedings, 2001. XVII, 500 pages. 2002.

Vol. 2467: B. Christianson, B. Crispo, J.A. Malcolm, M. Roe (Eds.), Security Protocols. Proceedings, 2001. IX, 241 pages. 2002.

Vol. 2469: W. Damm, E.-R. Olderog (Eds.), Formal Techniques in Real-Time and Fault-Tolerant Systems. Proceedings, 2002. X, 455 pages. 2002.

Vol. 2470: P. Van Hentenryck (Ed.), Principles and Practice of Constraint Programming – CP 2002. Proceedings, 2002. XVI, 794 pages. 2002.

Vol. 2471: J. Bradfield (Ed.), Computer Science Logic. Proceedings, 2002. XII, 613 pages. 2002.

Vol. 2473: A. Gomez-Perez, V.R. Benjamins, Knowledge Engineering and Knowledge Management. Proceedings, 2002. XI, 402 pages. 2002. (Subseries LNAI).

Vol. 2474: D. Kranzlmüller, P. Kacsuk, J. Dongarra, J. Volkert (Eds.), Recent Advances in Parallel Virtual Machine and Message Passing Interface. Proceedings, 2002. XVI, 462 pages. 2002.

Vol. 2475: J.J. Alpigini, J.F. Peters, A. Skowron, N. Zhong (Eds.), Rough Sets and Current Trends in Computing. Proceedings, 2002. XV, 640 pages. 2002. (Subseries LNAI).

Vol. 2476: A.H.F. Laender, A.L. Oliveira (Eds.), String Processing and Information Retrieval. Proceedings, 2002. XI, 337 pages. 2002.

Vol. 2477: M.V. Hermenegildo, G. Puebla (Eds.), Static Analysis. Proceedings, 2002. XI, 527 pages. 2002.

Vol. 2478: M.J. Egenhofer, D.M. Mark (Eds.), Geographic Information Science. Proceedings, 2002. X, 363 pages. 2002.

Vol. 2479: M. Jarke, J. Koehler, G. Lakemeyer (Eds.), KI 2002: Advances in Artificial Intelligence. Proceedings, 2002. XIII, 327 pages. (Subseries LNAI).

Vol. 2480: Y. Han, S. Tai, D. Wikarski (Eds.), Engineering and Deployment of Cooperative Information Systems. Proceedings, 2002. XIII, 564 pages. 2002.

Vol. 2483: J.D.P. Rolim, S. Vadhan (Eds.), Randomization and Approximation Techniques in Computer Science. Proceedings, 2002. VIII, 275 pages. 2002.

Vol. 2484: P. Adriaans, H. Fernau, M. van Zaanen (Eds.), Grammatical Inference: Algorithms and Applications. Proceedings, 2002. IX, 315 pages. 2002. (Subseries LNAI).

Vol. 2486: M. Marinaro, R. Tagliaferri (Eds.), Neural Nets. Proceedings, 2002. IX, 253 pages. 2002.

Vol. 2487: D. Batory, C. Consel, W. Taha (Eds.), Generative Programming and Component Engineering. Proceedings, 2002. VIII, 335 pages. 2002.

Vol. 2488: T. Dohi, R. Kikinis (Eds), Medical Image Computing and Computer-Assisted Intervention – MICCAI 2002. Proceedings, Part I. XXIX, 807 pages. 2002.

Vol. 2489: T. Dohi, R. Kikinis (Eds), Medical Image Computing and Computer-Assisted Intervention – MICCAI 2002. Proceedings, Part II. XXIX, 693 pages. 2002.

Vol. 2491: A. Sangiovanni-Vincentelli, J. Sifakis (Eds.), Embedded Software. Proceedings, 2002. IX, 423 pages. 2002.

Vol. 2493: S. Bandini, B. Chopard, M. Tomassini (Eds.), Cellular Automata. Proceedings, 2002. XI, 369 pages. 2002.

Vol. 2496: K.C. Almeroth, M. Hasan (Eds.), Management of Multimedia in the Internet. Proceedings, 2002. XI, 355 pages. 2002.

Vol. 2498: G. Borriello, L.E. Holmquist (Eds.), UbiComp 2002: Ubiquitous Computing. Proceedings, 2002. XV, 380 pages. 2002.

Vol. 2499: S.D. Richardson (Ed.), Machine Translation: From Research to Real Users. Proceedings, 2002. XXI, 254 pages. 2002. (Subseries LNAI).

Vol. 2502: D. Gollmann, G. Karjoth, M. Waidner (Eds.), Computer Security – ESORICS 2002. Proceedings, 2002. X, 281 pages. 2002.

Vol. 2503: S. Spaccapietra, S.T. March, Y. Kambayashi (Eds.), Conceptual Modeling – ER 2002. Proceedings, 2002. XX, 480 pages. 2002.

Vol. 2505: A. Corradini, H. Ehrig, H.-J. Kreowski, G. Rozenberg (Eds.), Graph Transformations. Proceedings, 2002. IX, 459 pages. 2002.

Vol. 2509: C.S. Calude, M.J. Dinneen, F. Peper (Eds.), Unconventional Models in Computation. Proceedings, 2002. VIII, 331 pages. 2002.

Vol. 2511: B. Stiller, M. Smirnow, M. Karsten, P. Reichl (Eds.), From QoS Provisioning to QoS Charging. Proceedings, 2002. XIV, 348 pages. 2002.

Vol. 2516: A. Wespi, G. Vigna, L. Deri (Eds.), Recent Advances in Intrusion Detection. Proceedings, 2002. X, 327 pages. 2002.